THE ABDUS SALAM MEMORIAL MEETING

the
abdus salam
international centre for theoretical physics

united nations educational, scientific and cultural organization

international atomic energy agency

THE ABDUS SALAM MEMORIAL MEETING

Trieste, Italy 19–22 November 1997

Editors

J. Ellis *(CERN)*
F. Hussain *(ICTP)*
T. Kibble *(Imperial College, London)*
G. Thompson *(ICTP)*
M. Virasoro *(ICTP)*

World Scientific
Singapore • New Jersey • London • Hong Kong

Published by

World Scientific Publishing Co. Pte. Ltd.

P O Box 128, Farrer Road, Singapore 912805

USA office: Suite 1B, 1060 Main Street, River Edge, NJ 07661

UK office: 57 Shelton Street, Covent Garden, London WC2H 9HE

Library of Congress Cataloging-in-Publication Data

Abdus Salam Memorial Meeting (1997 : Trieste, Italy)

The Abdus Salam Memorial Meeting : Trieste, Italy, 19–22 November 1997 / editors, J. Ellis . . . [et al.].

p. cm.

Includes bibliographical references (p. –).

ISBN 9810236190

1. Particles (Nuclear physics) -- Congresses. 2. Standard model (Nuclear physics) -- Congresses. I. Ellis, John, 1946– . II. Title.

QC793.A285 1997

539.7'2--dc21 98-49297

CIP

British Library Cataloguing-in-Publication Data

A catalogue record for this book is available from the British Library.

This book is printed on acid-free paper.

Printed in Singapore by Uto-Print

FOREWORD

The Abdus Salam Memorial Meeting was held from the 19th to the 22nd of November, 1997 on the first anniversary of the death of Prof. Abdus Salam, Nobel Laureate and the Founder-Director of the International Centre for Theoretical Physics. It was an opportunity for many of his colleagues and students to pay homage to his memory. The participants included Salam's family. Former students of Salam arrived from the farthest corners of the globe to pay tribute to his memory.

The scientific papers presented at the Meeting by some of the top experts in the field reflected the long lasting passion of Salam for the theory of the fundamental forces. Many of the contributions include personal reminiscences of Salam. The Meeting was opened with Tom Kibble's recollections of Salam at Imperial College. John Ellis reviewed the current status of the Standard Model of particle physics, "Salam's legacy". Most of the other contributions were concerned with developments beyond the Standard Model based on ideas of supersymmetry and superstring theory including recent developments like duality, D-branes and related topics. The contributions give a good picture of the field as it stood at the time of the meeting. In the last years of his life Salam was very interested in the developments that were taking place in string theory and we are sure that this volume would have been looked on by him favourably.

On the 21st of November there was a special ceremony to celebrate the renaming of the ICTP as the Abdus Salam International Centre for Theoretical Physics, with some very moving reminiscences paying homage to Salam as a great human being and physicist. Apart from many of Salam's students, colleagues and collaborators, representatives of the Italian Government, the International Atomic Energy Agency, UNESCO and the local authorities were all present at this ceremony. These reminiscences are being published separately.

The Organisers of the Meeting would like to thank all the speakers and participants. Special thanks go to all the staff of the Centre for their unstinting help in the organisation and smooth running of the Meeting. In particular we would like to thank Anne Gatti and Katrina Danforth of the Director's office for their contributions.

J. Ellis
T. Kibble
F. Hussain
G. Thompson
M. Virasoro

CONTENTS

RECOLLECTIONS OF ABDUS SALAM AT IMPERIAL COLLEGE

TOM KIBBLE
Blackett Laboratory, Imperial College,
London SW7 2BZ, United Kingdom

I shall talk about the work of Professor Abdus Salam and his Group at Imperial College from the perspective of a member of the Group, concentrating mainly but not exclusively on the period leading up to the discovery of the unified electro-weak theory.

It is a great honour for me to have been asked to give the opening talk at the Abdus Salam Memorial Meeting here at the Centre that is to bear his name. Like many other speakers, I owe an enormous debt of gratitude to Abdus Salam. He had a profound influence on my career; I have always felt that it was a stroke of great good fortune that I joined the Imperial College Theoretical Physics Group in 1959, less than three years after it was founded by Abdus Salam.

1 The Imperial College Theoretical Physics Group

Salam came to Imperial College in January 1957, as Professor of Applied Mathematics — later Theoretical Physics — a post he held until his retirement in 1993, though latterly on a part-time basis. He came at the invitation of Patrick Blackett, then Head of the Physics Department at Imperial. Blackett had been looking for a top-rank young theoretical physicist, and had consulted Bethe, who recommended Salam. Salam accepted, despite Neville Mott's warning against joining that "plumbers' college"! It was arranged that his friend and collaborator Paul Matthews should go too, as a Reader. Over the next few years, Salam and Matthews together built up at Imperial College one of the finest theoretical physics groups in Britain, and indeed the world.

It was a very exciting place to be. We had numerous visitors: Steven Weinberg, Murray Gell-Mann, Ken Johnson, Lowell Brown, Stanley Mandelstam, John Ward, to name but a few. The year I arrived, 1959, was also the year that Salam became the youngest Fellow of the Royal Society at the age of 33. Besides Salam and Matthews there was then one other faculty member, Salam's former student John C. Taylor. I became a Lecturer in 1961.

Initially we were part of the Mathematics Department, housed in the old Huxley Building, the site of the Royal College of Science in Thomas Huxley's time, now the Henry Cole wing of the Victoria & Albert Museum. Our quarters were very cramped and unsuitable, having been carved out of much larger spaces. Salam had the only decent office. I shared one that had just half a window; we could hear every word of the lecture in the neighbouring lecture theatre, and no doubt they could hear our dicussions too. The students, of whom there were many, occupied one big office, which is also where we met for tea and coffee.

So in 1960 when the fine new Physics Building opened — now the Blackett Laboratory — it was not hard for Blackett to persuade Salam to move into Physics, though he did so over the strong opposition of the Head of Mathematics, Harry Jones. In Physics, we had much better rooms and facilities, and, even more important, the proximity of experimenters, in particular the High Energy Nuclear Physics Group under Clifford Butler. We had, I recall, a fine new seminar room, but no chairs. Paul Matthews solved that problem, however, by organizing the students for a dawn raid to capture the chairs that had been intended for the Optics seminar room; I'm not sure whether they ever discovered the culprits!

What about the physics we were doing? There were, I believe, two main themes of our research: symmetries and unification. Let me deal first with symmetries.

2 Symmetries

This was of course the era when we were faced with an avalanche of new particles: first the pions and muons, then the strange particles, K, Λ, Σ and Ξ, and finally the many hadronic resonances. It was natural to try to group these particles into multiplets.

This was by no means a new idea. In 1932 Heisenberg [1] had suggested treating the proton and neutron as two states of a single entity, the *nucleon*, with what we would now call *isospin* operators acting on the doublet

$$N = \begin{pmatrix} N^+ \\ N^0 \end{pmatrix} = \begin{pmatrix} p \\ n \end{pmatrix}, \tag{1}$$

by analogy with the Pauli spin operators acting on the up and down states of an electron spin. The idea of an approximate SU(2) isospin *symmetry*, respected by the strong nuclear forces but broken by electromagnetic ones, was hypothesized by Cassen & Condon in 1936.[2] In 1937, Kemmer [3] showed that it could be extended to cover Yukawa's mesons, the supposed carriers of the strong nuclear forces. He pointed out that if the mesons formed an isospin triplet,

$$\underline{\pi} = \begin{pmatrix} \pi^+ \\ \pi^0 \\ \pi^- \end{pmatrix}, \tag{2}$$

then one could write down an isospin-invariant interaction, of the form

$$\bar{N}\gamma_5 \underline{\tau} \cdot \underline{\pi} N. \tag{3}$$

With all the new particles being discovered, it was very natural to try to extend the concept of isospin symmetry. The pioneers were Japanese physicists, beginning with the Sakata model of 1956. Sakata [4] suggested that all hadrons be regarded as bound states of three basic ones, (p, n, Λ). Then in 1959, Ikeda, Ogawa & Ohnuki [5] proposed an extended version of isospin, an even more approximate SU(3) symmetry, acting on the fundamental triplet

$$\begin{pmatrix} p \\ n \\ \Lambda \end{pmatrix}. \tag{4}$$

Salam & Ward took up this idea and in 1961 [6] proposed a gauge theory of strong interactions with an SU(3) octet of gauge bosons. The full "eightfold way" SU(3) symmetry theory, with the baryons in an octet and decuplet rather than a triplet, was proposed independently by Murray Gell-Mann [7] and by Yuval Ne'eman, [8] then a student of Salam's at Imperial College.

Salam himself made many contributions then and later to unravelling the hadronic symmetries. But this was not, I believe, his first love. His real ambition had always been to find a fundamental theory of all the particle interactions, to do for the strong and weak interactions, and perhaps even gravity, what QED had done for the electromagnetic. He saw the discovery of symmetries not as an end in itself but rather as a means to an end, as clues to the nature of the underlying fundamental theory. This brings me to the second of the major themes of our work.

3 Unification

Salam was convinced from an early date — certainly well before he came to Imperial College — that the fundamental theory would be a gauge theory. This was not a popular view. Many people at the time thought that field theory had had its day, particularly for the strong interactions, where it was to be superseded by S-matrix theory. This was an appealingly economical idea: everything was supposed to follow from fundamental principles of covariance, causality and unitarity. Salam, however, always looked for an underlying field theory.

The origin of gauge theory goes back to the pioneering work of Hermann Weyl.[9] But the key paper that really put the idea on the map was that of Yang & Mills [10] in 1954. The same model was in fact written down independently by Ronald Shaw, a student of Salam's in Cambridge.[11]

Yang & Mills formulated for the first time what we now call the *gauge principle*. They started by considering a theory of a nucleon doublet,

$$\psi = \begin{pmatrix} \psi_p \\ \psi_n \end{pmatrix}, \tag{5}$$

with interactions invariant under the infinitesimal isospin transformations $\psi(x) \to \psi(x) + \delta\psi(x)$, with

$$\delta\psi(x) = \tfrac{1}{2} i \underline{\delta\alpha} \cdot \underline{\tau}\psi(x). \tag{6}$$

So long as we neglect electromagnetic interactions, this symmetry should be exact. Then which direction in the two-dimensional isospin space represents the proton and which the neutron is a matter of arbitrary convention, a choice of axes. Yang & Mills argued that it is inconsistent with the spirit of relativity to suppose that making this choice here and now constrains it everywhere and for all times; one should expect to be able to make independent choices in different regions of space-time. That would require that the symmetry be extended from a global to a local symmetry, of the form

$$\delta\psi(x) = \tfrac{1}{2} i \underline{\delta\alpha}(x) \cdot \underline{\tau}\psi(x). \tag{7}$$

It is of course well known that this extension is impossible unless we introduce an extra *gauge* field $A_\mu(x)$, and replace the derivatives $\partial_\mu\psi$ of ψ by *covariant* derivatives

$$D_\mu\psi = \partial_\mu\psi + \tfrac{1}{2} i \underline{A}_\mu \cdot \underline{\tau}\psi. \tag{8}$$

The key point about this is that if we adopt the gauge principle as a hypothesis, then from the nature of the symmetry we can infer the form of the interaction. Whether the gauge principle is actually *true* is of course debatable. There is certainly as yet no proof that global symmetries are impossible. What is undeniable is that the principle has been amazingly fruitful.

4 The Intermediate Vector Boson and Parity Violation

Yang & Mills were of course looking for a theory of strong interactions. But the problem with any such efforts was that no one knew how to calculate

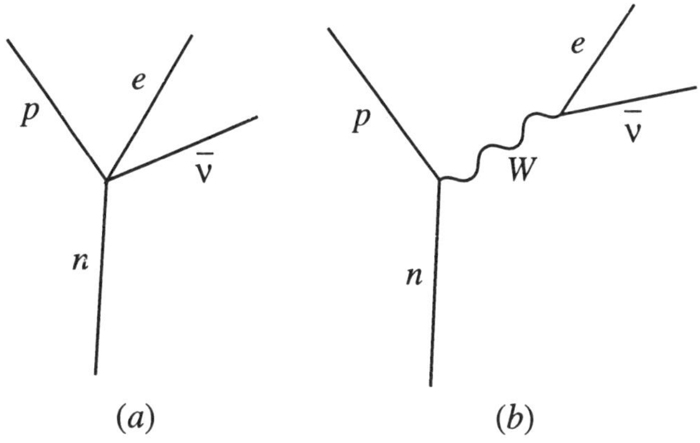

Figure 1: Decay of the neutron (*a*) in Fermi theory (*b*) in intermediate vector boson theory

anything in a theory with a coupling constant of order 1. So gradually the weak interactions began to seem a better bet. Salam, especially with his collaborator John Ward, spent a great deal of time and effort over the next decade searching for a gauge theory of weak interactions — or better still a unified theory of weak and electromagnetic interactions.

There were promising hints of similarities between the two. The phenomenologically successful theory of weak interactions was of course the four-fermion interaction theory of Fermi. For example, neutron decay is represented by the diagram of Figure 1(a). By analogy with QED, it is natural to suppose that the interaction is mediated by the exchange of a vector particle, as in Figure 1(b).

One of the most important advances in the theory of weak interactions was the proposal by Lee and Yang in 1956 [12] of parity non-conservation, and its experimental verification by Wu *et al.* in 1957.[13] In the same year, Salam [14] made a very important contribution, in which he pointed out the relation between a new chiral (γ_5) symmetry and the vanishing of the neutrino mass. As a matter of fact he might he might have published this *before* the appearance of the paper by Wu *et al.*. He actually wrote it immediately after hearing Lee and Yang's proposal. But he then sent it for comments to the two physicists whose opinions he most respected — Peierls and Pauli. The reactions of both were so negative that, rather uncharacteristically, he decided not to publish. He did however receive a handsome public apology from Pauli at a meeting the following year!

These developments led on to the $V - A$ theory of weak interactions, proposed by Marshak & Sudarshan [15,16] and Feynman & Gell-Mann.[17] The $V - A$ structure gave strong support to the idea that weak interactions are mediated by an "intermediate vector boson", and hence to the possibility of a unified gauge theory of weak and electromagnetic interactions.

There were, however, serious problems to be overcome. There were two obvious ways in which the weak interactions differed from the electromagnetic. First, the short range and weakness of the weak interactions require that the mass of the intermediate vector boson W be large: if the interactions of the W and the photon are of equal strength, one requires

$$m_W \approx 40 m_p, \tag{9}$$

whereas a gauge theory would naturally suggest that its mass should vanish, like that of the photon.

Secondly, the weak interactions exhibit parity violation, while simple analogues of the electromagnetic interaction are parity conserving.

In fact, immediately after the $V - A$ theory was proposed, Salam & Ward [18] proposed a "unified gauge theory" of weak and electromagnetic interactions. In this theory the electron, positron and neutrino were combined in an SO(3) triplet, as were the $W^{\pm}$ and the photon:

$$\begin{pmatrix} e^+ \\ \nu \\ e^- \end{pmatrix} \quad \begin{pmatrix} W^+ \\ A \\ W^- \end{pmatrix}. \tag{10}$$

But in fact the theory only generated the parity-conserving part of the weak interaction, and of course had $m_W = 0$. Salam & Ward had to cure these problems by force: they first formed the gauge theory, then imposed γ_5 symmetry on it to give parity violation, and finally put in m_W by hand, in the process spoiling the desired symmetry.

In 1961 they developed [19] an even more ambitious unified theory of weak, electromagnetic *and strong* interactions, based on the group SO(8). This was a remarkable attempt, well ahead of it time. But as Salam and Ward were themselves very well aware, these theories didn't really work. The key question was how to break the symmetry between the photon and the $W^{\pm}$.

5 Folk Theorems

As often happens, progress were delayed by what I have previously called "folk theorems", results that everybody knows to be true, but that turn out on closer inspection to be not quite true, or not quite applicable.

The first of these was the belief that gauge symmetry implies the vanishing of the photon mass. Indeed this was often quoted as one of the great successes of QED! But Schwinger pointed out in 1962[20] that this is not necessarily true: if the interaction between the gauge field A and the current is made strong enough, m_A may become non-zero. However, this formulation itself led some people to suppose that it could not be relevant to the weak interactions.

The key problem, as I have said, was how to break the symmetry. But there again, one encountered an apparent obstacle: the Goldstone theorem.[21,22,23] This suggested that spontaneous symmetry breaking would always lead to the appearance of massless scalar particles, the "Goldstone bosons". No such particles were known. There were certainly conterexamples in condensed matter; indeed, the proof of the theorem requires an assumption that there are no long-range forces. However, it was generally believed that it must hold in a relativistic theory.

To understand why, it may be useful to recall a simple proof of the theorem in a relativistic context. A symmetry implies the existence of a conserved Noether current j^μ:

$$\partial_\mu j^\mu = 0. \tag{11}$$

That the symmetry is broken means that there is some field ϕ whose expectation value $\langle\phi\rangle$ is not invariant under the transformation generated by the integral of j^0, i.e.,

$$i\langle 0|[Q, \phi(0)]|0\rangle \neq 0, \tag{12}$$

where

$$Q = \int d^3\mathbf{x}\, j^0(x). \tag{13}$$

Now define the Fourier transform

$$f^\mu(k) = i \int d^4x\, e^{ik.x} \langle 0|[j^\mu(x), \phi(0)]|0\rangle. \tag{14}$$

Relativistic invariance implies that f^μ is a vector function of the only available vector, k^μ, whence

$$f^\mu = k^\mu \{g(k^2) + \epsilon(k^0)h(k^2)\}. \tag{15}$$

Then (11) implies that

$$0 = k_\mu f^\mu = k^2 g(k^2) + \epsilon(k^0)k^2 h(k^2), \tag{16}$$

whence both g and h must be proportional to $\delta(k^2)$. On the other hand (12) requires

$$\int \frac{dk^0}{2\pi} f(k^0, \mathbf{0}) \neq 0, \tag{17}$$

whence h *cannot* vanish. It follows there must exist massless intermediate states that can couple to the vacuum through the field ϕ. These are the Goldstone bosons.

At the time this proof seemed pretty robust, though of course we now understand that the covariance assumption has to be modified in a gauge theory.

6 Electro-weak Unification

In 1963, Anderson [24] pointed out that the plasmon mass is an example of Schwinger's mechanism. In a plasma, the photon acquires a mass, $m_A = \hbar\omega_{\text{plasma}}/c^2$, where the plasma frequency is given by

$$\omega_{\text{plasma}} = \sqrt{\frac{e^2 n_e}{m_e}}. \tag{18}$$

Moreover, he noted that the example of superconductivity suggests that the zero-mass Yang-Mills bosons and the zero-mass Goldstone bosons can "cancel each other out".

When I re-read that paper now, the argument seems perfectly clear, but at the time many field theorists remained sceptical; it was widely believed that this kind of thing couldn't happen in a relativistic theory — until explicitly relativistic versions of the mechanism were constructed by Englert & Brout,[25] by Higgs,[26] and later by my colleagues and myself.[27,28]

The "Higgs mechanism" provided a means of giving a non-zero mass to the W, while leaving the photon massless. But there was still a serious problem: if the photon were identified with the W^0, how could one reconcile parity conservation for the electromagnetic interactions with parity violation for the weak?

The solution is of course well known: we have to extend the symmetry from SU(2) to SU(2)×U(1). The photon is not the W^0 but a mixture of W^0 and the U(1) gauge boson B:

$$\begin{aligned} A &= B\cos\theta + W^0 \sin\theta, \\ Z &= -B\sin\theta + W^0\cos\theta. \end{aligned} \tag{19}$$

This idea had already been proposed by Sheldon Glashow [29] in 1961, though without the crucial element of spontaneous symmetry breaking. Indeed Salam & Ward [30] wrote down a very similar model in 1964, in which the electron was combined in a triplet with the positive muon:

$$\begin{pmatrix} \mu^+ \\ \nu \\ e^- \end{pmatrix}. \tag{20}$$

So it is perhaps surprising that it took another three years to bring these two ideas together.

The presently accepted unified theory of weak and electromagnetic interactions was first written down by Steven Weinberg [31] in 1967. Salam had been working along similar lines, and presented essentially the same theory in research lectures he gave at Imperial College in the Autumn of 1967. However, he didn't publish it until the following Spring,[32] when it appeared in the Proceedings of a Nobel Symposium.

This was by no means the end of the story of the electro-weak theory. The next important milestones were 't Hooft's proof of renormalizability in 1971 [33] and the verification of the existence of neutral current interactions at CERN in 1973,[34] which together persuaded people to take the theory seriously, and led to the award of the Nobel Prize to Glashow, Salam and Weinberg in 1979. Then of course there was the discovery of the W and Z particles, also at CERN in 1983.[35]

7 Wider Activities

Nor was this the end of the story so far as Salam's contributions to physics are concerned. He went on to make many more key advances, notably in his work with Strathdee on the superspace–superfield formalism,[36] which underlies much later work on supergravity and superstring theory, and the development of the Pati-Salam model,[37] which for the first time suggested in the context of a specific theory that the proton might be unstable.

But after 1970, the centre of gravity of Salam's work shifted increasingly away from Imperial College and towards Trieste. He remained on our staff of course, on a part-time basis, until his retirement in 1993, and played an important role there, but we saw him only occasionally for relatively short periods. So I shall not talk about his later work.

Looking back now on Salam's time at Imperial College, what I find extraordinary is to realise how many other things he was doing at the same time. To us, his colleagues, he seemed to be a very full-time theoretical physicist. He was writing half a dozen papers a year, giving numerous research lectures and looking after six or more research students. His students had a tough time of it. He expected to hear of daily progress, and was impatient of stupidity.

Quite a few were led to seek a less demanding supervisor; indeed, one of our number more or less specialized in taking Salam's rejects! But he was also very supportive of those students who met his high standards.

While he was doing all this, however, Salam was also acting as a Science Advisor to the Pakistani Government, and of course working to set up the ICTP.

When he was appointed to the Chair at Imperial College in 1957, Salam was the first Asian to hold such a post in any science faculty in any British university, indeed I believe only the second in *any* faculty. This gave him a high profile in Pakistan. He was appointed Science Advisor by President Ayub Khan. He worked on developing atomic energy in Pakistan, set up a Pakistani space research organization, arranged for studies of waterlogging and salinity of the soil, and much else.

Salam must also have spent a lot of time working towards the ICTP. He was very unhappy at having to leave his native country to pursue his chosen career, and determined to do what he could to help others in similar positions to avoid having to make this agonizing choice. He conceived the idea of a centre of excellence to which scientists from the Third World would come on a regular basis to maintain contact with science at the frontier. His opportunity came when he was appointed Pakistani delegate to the General Council of the International Atomic Energy Agency. Despite the opposition of the major powers, he sold the idea to his fellow delegates. The opportune offer of a site in Trieste with generous support from the Italian Government and the City of Trieste made the idea a reality. Despite the initial scepticism of many in the science community, it has proved an astonishing success, and is a fitting living memorial to Abdus Salam.

I shall always look back on the early years after I came to Imperial College as the most exciting of my life. There was a wonderful atmosphere at the College; I cannot think of anywhere I would rather have been. I shall always be immensely grateful to Abdus Salam for making it possible for me to participate in these memorable developments.

References

1. W. Heisenberg, *Z. f. Phys.* **77**, (1932) 1.
2. B. Cassen and E.U. Condon, *Phys. Rev.* **50**, (1936) 846.
3. N. Kemmer, *Phys. Rev.* **52**, (1937) 906.
4. S. Sakata, *Prog. Theor. Phys.* **16**, (1956) 686.
5. M. Ikeda, S. Ogawa and Y. Ohnuki, *Prog. Theor. Phys.* **22**, (1959) 715.
6. A. Salam and J.C. Ward, *Il Nuovo Cimento*, ser. X, **20**, (1961) 419.

7. M. Gell-Mann, *California Institute of Technology Synchrotron Laboratory Report* CTSL-20 (1961, unpublished).
8. Y. Ne'eman, *Nucl. Phys.* **26**, (1961) 222.
9. H. Weyl, *Z. f. Phys.* **56**, (1929) 330.
10. C.N. Yang and R.L. Mills, *Phys. Rev.* **96**, (1954) 191.
11. R. Shaw, PhD Dissertation, Cambridge University (1954, unpublished).
12. T.D. Lee and C.N. Yang, *Phys. Rev.* **104**, (1956) 254.
13. C.S. Wu *et al.*, *Phys. Rev.* **104**, (1957) 1413.
14. A. Salam, *Il Nuovo Cimento*, ser. X, **5**, (1957) 299.
15. R.E. Marshak and G. Sudarshan, In *Proceedings of the Padua-Venice Conference on Mesons and Recently Discovered Particles* (1957).
16. R.E. Marshak and G. Sudarshan, *Phys. Rev.* **109**, (1958) 1860.
17. R.P. Feynman and M. Gell-Mann, *Phys. Rev.* **109**, (1958) 193.
18. A. Salam and J.C. Ward, *Il Nuovo Cimento*, ser. X, **11**, (1958) 568.
19. A. Salam and J.C. Ward, *Il Nuovo Cimento*, ser. X, **19**, (1961) 165.
20. J. Schwinger, *Phys. Rev.* **125**, (1962) 397.
21. Y. Nambu and G. Jona-Lasinio, *Phys. Rev.* **122**, (1961) 345.
22. J. Goldstone, *Il Nuovo Cimento*, ser. X, **19**, (1961) 154.
23. J. Goldstone, A. Salam and S. Weinberg, *Phys. Rev.* **127**, (1962) 965.
24. P.W. Anderson, *Phys. Rev.* **130**, (1963) 439.
25. R. Brout and F. Englert, *Phys. Rev. Lett.* **13**, (1964) 321.
26. P.W. Higgs, *Phys. Lett.* **12**, (1964) 132.
27. G.S. Guralnik, C.R. Hagen and T.W.B. Kibble, *Phys. Rev. Lett.* **13**, (1964) 585.
28. T.W.B. Kibble, *Phys. Rev.* **155**, (1966) 1554.
29. S. Glashow, *Nucl. Phys* **22**, (1961) 579.
30. A. Salam and J.C. Ward, *Phys. Lett.* **13**, (1964) 168.
31. S. Weinberg, *Phys. Rev. Lett.* **19**, (1967) 1264.
32. A. Salam, in *Elementary Particle Theory*, ed. N. Svartholm, pp. 367. Stockholm: Almqvist & Wiksell (1968).
33. G. 't Hooft, *Nucl. Phys.* **B35**, (1971) 167.
34. F.J. Hasert *et al.*, *Phys. Lett.* **46B**, (1973) 138.
35. G. Arnison *et al.*, *Phys. Lett.* **122B**, (1983) 1102; **126B**, (1983) 398.
36. A. Salam and J. Strathdee, *Nucl. Phys.* **B76**, (1974) 477; *Phys. Rev. D* **11**, (1974) 1521.
37. J. Pati and A. Salam, *Phys. Rev. Lett.* **31**, (1973) 661; *Phys. Rev. D* **8**, (1973) 1554.

THE STANDARD MODEL:
ABDUS SALAM'S LASTING LEGACY

John ELLIS
Theoretical Physics Division, CERN
CH - 1211 Geneva 23, Switzerland

The present phenomenological status of the Glashow-Salam-Weinberg electroweak model is reviewed. Precision electroweak data indicate that the Higgs boson weighs around 100 GeV. This and measurements of $\sin^2\theta_W$ are consistent with supersymmetric predictions. LEP data indicate that the lightest supersymmetric particle weighs at least 42 GeV. The LHC should be able to detect the Higgs boson and supersymmetric particles.

1 Testing the Standard Model

The fundamental forces described by the Standard Model are those of electromagnetism, the weak interactions and the strong interactions. The bosons carrying these forces are in many respects similar: they have spin 1 and are gauge bosons. However, they differ widely in their masses: the photon and gluon are believed to be massless, whereas the $W^{\pm}$ and Z^0 of the weak interactions weigh around 80 and 91 GeV respectively. How and why does this difference arise? The answer proposed by Abdus Salam [1] and Steven Weinberg [1] was the Higgs mechanism, to which we return shortly. However, it is salutary to recall that there is no *direct* experimental evidence for its validity, despite the many successful tests of the Standard Model described below.

The gauge bosons interact with the elementary matter particles. We know from LEP that there exist only three light neutrino species, corresponding in the Standard Model to the lepton species e, μ, τ, which are complemented by 6 quarks, that ensure the quantum consistency of the theory by cancelling anomalies. The roster of elementary matter particles was recently completed by the discovery of the top quark [2], with mass around 170 GeV as predicted by fitting the precision electroweak data from LEP and elsewhere, which have verified the Standard model with 1‰ precision.

Electroweak measurements range in energy scale from parity violation in atoms, through νe and $\nu N, eN, \mu N$ scattering to tests at momentum transfers $q^2 \sim 10^4$ GeV^2 at $\bar{p}p$ colliders, the e^+e^- colliders LEP and SLC, and the ep collider HERA. Abdus Salam was among those instrumental in securing approval for the construction of LEP, which is the largest particle accelerator to date, with a circumference of about 27 km. Starting in 1989, most measurements there have been performed at centre-of-mass energies $E_{cm} \sim M_{Z^0} \sim 91$ GeV, with about 2×10^7 Z^0 decays detected by the four major experiments ALEPH, DELPHI, L3 and OPAL. Higher-energy measurements have been underway since 1995.

Basic measurements at the Z^0 peak include the total cross-section

$$\sigma_h = \frac{12\pi}{m_{Z^0}^2} \frac{\Gamma_{ee}\Gamma_{had}}{\Gamma_Z^2} \tag{1}$$

which is reduced by about 30% by radiative corrections, and the total decay rate

$$\Gamma_Z = \Gamma_{ee} + \Gamma_{\mu\mu} + \Gamma_{\tau\tau} + N_\nu \Gamma_{\nu\nu} + \Gamma_{had} \tag{2}$$

The LEP experiments have also measured many partial decay rates, checking lepton universality: $\Gamma_{ee} = \Gamma_{\mu\mu} = \Gamma_{\tau\tau} \equiv \Gamma_{\ell\ell}$ and the rates for hadronic decay modes such as $\Gamma_{bb} = \Gamma_{had} \times R_b$. The total rate for invisible Z^0 decays is parametrized in (2) as $\Gamma_{inv} \equiv N_\nu \Gamma_{\nu\nu}$, and the current result is

$$N_\nu = 2.993 \pm 0.011 \tag{3}$$

How exciting life would have been if N_ν had turned out inconsistent with an integer!

In order to measure m_{Z^0} and Γ_Z with high precision, it is essential to calibrate the LEP beam energy as accurately as possible. Many subtle effects need to be taken into account in achieving the quoted precision of about 2 MeV in m_{Z^0} [3]. Tides in the rock in which LEP is embedded cause the ring to expand and contract. The tuning of the accelerating RF cavities then forces the beam trajectories to move relative to the bending magnets, resulting in beam energy variations of about ± 5 MeV. Similar effects are caused by heavy rainfall and seasonal variations in the surface level of Lake Geneva. These effects are all accounted for by the energy calibration.

More recently, it was realized that the LEP beam energy varies during a fill, particularly during the daylight hours. This effect was finally traced to the passage of trains along a nearby railway line, as seen in Fig. 1. They generate stray currents in the ground, which like to run around the LEP ring, because it is a relatively good conductor, modifying the fields of the magnets. After this TGV effect is taken into account, LEP measurements yield [3]

$$m_{Z^0} = 91,186.7 \pm 2.0\,\mathrm{MeV}, \tag{4}$$

a precision greater than that for the muon decay constant G_μ.

What use is this measurement? At tree level in the Standard Model, one has

$$m_{Z^0} = \sqrt{\frac{\pi\alpha}{\sqrt{2}G_\mu}} \quad \frac{1}{\sin\theta_W \cos\theta_W} \tag{5}$$

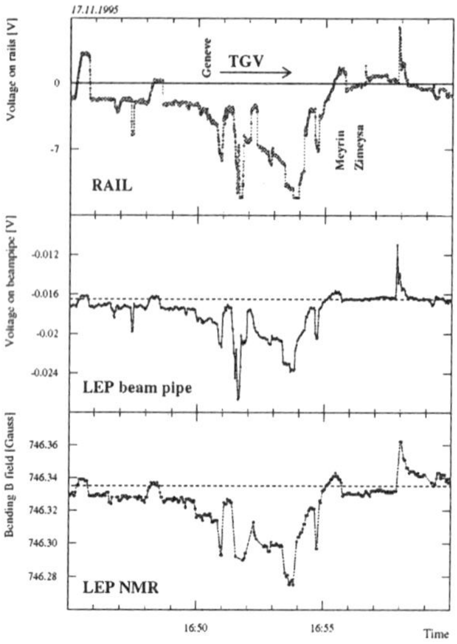

Figure 1: Passing TGV trains generate a leakage current which passes through the LEP beam pipe, modifying the bending magnetic field and hence the beam energy [3].

where θ_W is related to the ratio of electromagnetic and weak couplings. There are by now many independent high-energy determinations of $\sin^2\theta_W$ which are in good agreement [4], as shown in Fig. 2. In past years, there have been nagging discrepancies in the measurements of R_b at LEP and the left-right polarization asymmetry A_{LR} at the SLC. However, these both now seem to be going away. The best-fit value

$$\sin^2\theta_W = 0.23149 \pm 0.00021 \tag{6}$$

is a key test of the Grand Unified Theories of strong and electroweak interactions, much loved by Abdus Salam, to which we return later.

2 Theoretical Interpretation of the Data

As already seen in Fig. 2, the inferred values of $\sin^2\theta_W$ depend on the masses of the top quark m_t and the Higgs boson m_H. This is because relations such as (5) are subject to radiative corrections, for example

$$m_W^2 \sin^2\theta_W = m_Z^2 \sin^2\theta_W \cos^2\theta_W = \frac{\pi\alpha}{\sqrt{2}G_\mu}(1+\Delta r) \tag{7}$$

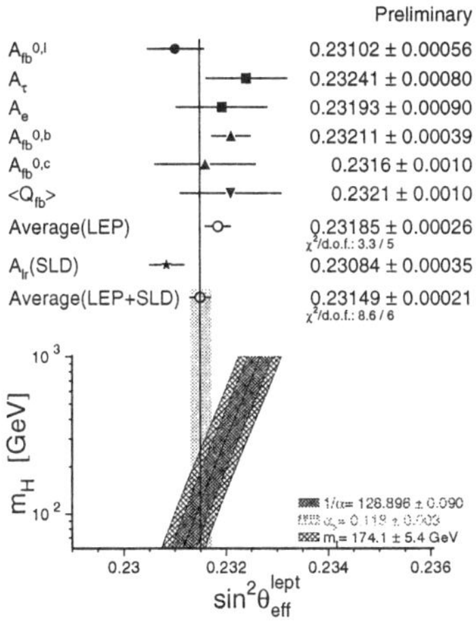

Figure 2: The sensity of different precision measurements of $\sin^2\theta_W$ to m_H and other input quantities [4].

at the one-loop level. The quantum correction Δr in (7) receives important contributions from top and bottom loops in vacuum polarization diagrams, that yield [5]

$$\Delta r \ni \frac{3G_\mu}{8\pi^2\sqrt{2}} \quad m_t^2 \qquad \text{for} \qquad m_t \gg m_b, \tag{8}$$

reflecting the fact that the Standard Model would have been non-renormalizable in the absence of the top quark. Analogously, there are also contributions from loops involving the Higgs boson [6]:

$$\Delta r \ni \frac{\sqrt{2}G_\mu}{8\pi^2\sqrt{2}} \quad m_W^2 \quad \left\{\frac{11}{3}\,\ell n\,\frac{m_H^2}{m_W^2} + \cdots\right\} \quad \text{for} \quad m_H \gg m_W. \tag{9}$$

Again, the divergence as $m_H \to \infty$ reflects the non-renormalizability of the Standard Model in the absence of a Higgs boson, though the logarithmic m_H dependence in (9) is more subtle than the quadratic m_t dependence in (8).

The latter enabled an upper limit $m_t \lesssim 170$ GeV to be given already before the start-up of LEP [7], and it was pointed out that the advent of a precision measurement of m_Z would also allow a lower limit on m_t to be set [8]. Using the latest available electroweak data, and leaving m_H as a free parameter, a global fit now yields [4]

$$m_t = 158^{+14}_{-11}\,\text{GeV}, \tag{10}$$

to be compared with the latest FNAL $\bar{p}p$ collider measurement

$$m_t = 173.8 \pm 5.2\,\mathrm{GeV}. \tag{11}$$

The consistency between the prediction (10) and the measurement (11) permits them to be combined in a global fit to determine m_H. For some time now, such fits have indicated a relatively low range for m_H. The latest fit yields [4]

$$m_H \simeq 70^{+70}_{-40}\,\mathrm{GeV} \tag{12}$$

if one takes all the data at their face values.

Now is the time to assess more carefully the status of the Higgs mechanism. It is well known that a massless vector boson such as the photon or a massless $W^\pm/Z^0$ has just two polarization states, whereas a massive vector boson such as the ρ or massive $W^\pm/Z^0$ must have three polarization states, corresponding to $m = 1, 0, -1$ for $j = 1$. To provide the third polarization state, we need a field that has zero angular momentum ($m = 0$) in any frame of reference, namely a scalar field, as in the Higgs mechanism. The ratio of $W^\pm$ and Z^0 masses, as well as the existence of quark and lepton masses, require this field to have weak isospin 1/2. A complex isospin doublet, such as the Higgs doublet in the Standard Model, contains four states, three of which are "eaten" by the $W^\pm$ and Z^0 to acquire their masses, leaving behind one physical Higgs boson H. There is no direct evidence for its existence, and searches at LEP require [9]

$$m_H \gtrsim 88\,\mathrm{GeV}. \tag{13}$$

Comparing with the predicted range (12), we can see that the Higgs boson may not lie far beyond our current reach. It may appear in future higher-energy runs of LEP or at the LHC, as discussed later.

What is the significance of the range indicated indirectly by the precision electroweak data? It has long been argued that unitarity provides an upper bound on the mass of the Higgs boson. At the tree level, the bound was 1 TeV, and this was reduced by loop corrections. The best way to formulate this upper bound is to require that the couplings of the Standard Model remain finite at all energies up to some cut-off Λ. If one takes $\Lambda = m_H$ itself, one finds $m_H \lesssim 650\,\mathrm{GeV}$, but this is strengthened to $m_H \lesssim 200\,\mathrm{GeV}$ if one takes $\Lambda = m_{GUT}$ or m_P.

There is also a lower bound on m_H imposed by the (meta)stability of the electroweak vacuum against transitions to any rival vacuum with $|H| \leq \Lambda$. The resulting range of m_H is shown in Fig. 3 as a function of Λ [10]. The indirectly indicated range (12) is (depressingly) consistent with the validity of the Standard Model all the way up to $\Lambda = m_p$. However, it is also consistent with the supersymmetry much loved by Abdus Salam, as we review later.

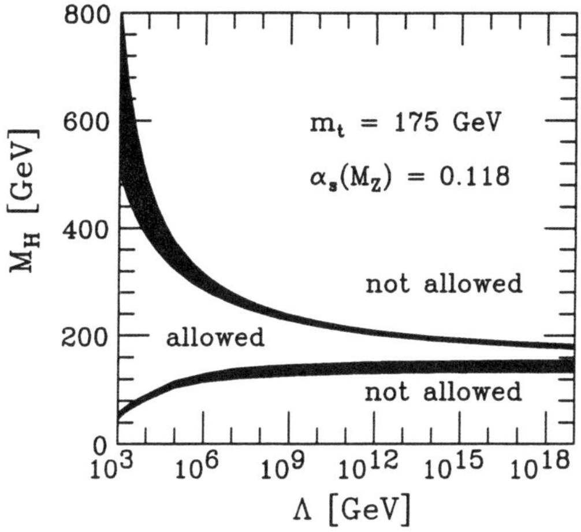

Figure 3: The range of m_H allowed as a function of the cutoff Λ up to which the Standard Model is assumed to remain valid [10].

3 Physics at LEP 2

The first runs of LEP significantly above m_{Z^0} were in Autumn 1995 at centre-of-mass energies between 130 and 140 GeV, around 1.5 times m_{Z^0}, and hence called LEP 1.5. True LEP 2 running started in Summer 1996, with a first run just above the $e^+e^- \to W^+W^-$ threshold. This was followed later in 1996 by a run at 170/172 GeV, and another in 1997 at 180/183 GeV. It is planned to run at about 189 GeV in 1998, to continue running in 1999, and (hopefully) to run also in the year 2000, possibly at a centre-of-mass energy close to 200 GeV - if enough money can be found and a higher accelerating field squeezed out of the LEP RF cavities.

Fig. 4 shows some principal cross sections at LEP 2. The previous focus of interest, $e^+e^- \to f\bar{f}$, now becomes a background to more interesting measurements. The centre of interest is provided by the $e^+e^- \to W^+W^-$ reaction, which will enable m_W to be measured and the gauge nature of the γW^+W^- and $Z^0W^+W^-$ couplings to be established. Somewhat lower is the cross-section for $e^+e^- \to Z^0Z^0$, which is not particularly interesting in its own right, but may be a serious background to the search for $e^+e^- \to Z^0H$ [11], particularly if $m_H \sim 90$ GeV.

Of primary interest for testing the Standard Model is the measurement of

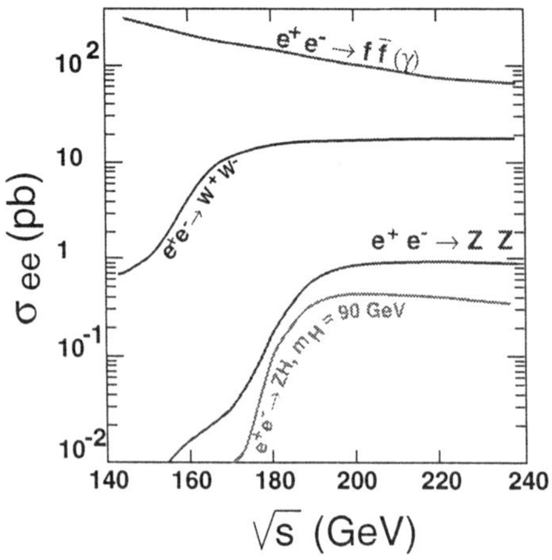

Figure 4: The principal cross sections at LEP 2.

m_W. Current $\bar{p}p$ collider measurements yield [4]

$$m_W = 80.40 \pm 0.090 \text{ GeV} \tag{14}$$

whilst the Standard Model prediction derived from precision electroweak data is [4]

$$m_W = 80.333 \pm 0.040 \text{ GeV} \tag{15}$$

The principal uncertainty in the prediction (15) is provided by the top and Higgs masses. Conversely, a measurement more precise than (14) could constrain indirectly m_H, and also constrain extensions of the Standard Model such as supersymmetry. For example, for $m_t = 180 \pm 5$ GeV, the following would be the implications for m_H of measuring m_W with the stated accuracy [12]:

$$\begin{array}{lclll} \Delta m_W & = & 25, & 50 & \text{MeV} \\ m_W & = & \left\{ \begin{array}{l} 100(+86,-54) \\ 300(+196,-126) \end{array} \right. & \begin{array}{l} (+140,-72) \\ (+323,-168) \end{array} & \text{GeV} \end{array} \tag{16}$$

even without the aid of any other precision electroweak measurement.

There are two principal ways to measure m_W at LEP 2. One is by measuring the cross-section for $e^+e^- \to W^+W^-$ close to threshold. It has been estimated [12] that

$$\Delta m_W \geq 91\text{MeV}\sqrt{\frac{100pb^{-1}}{\mathcal{L}}} \tag{17}$$

at $E_{cm} \simeq 161$ GeV, with the lower bound being attainable in the limit of 100 % detection efficiency, no background and negligible uncertainty in E_{cm}. The second method is by direct reconstruction of $W^\pm$ decays, which has been estimated [12] to have a sensitivity

$$\Delta m_W \geq 50 \text{ MeV} \sqrt{\frac{100pb^{-1}}{\mathcal{L}}} \tag{18}$$

at any centre-of-mass energy above 170 GeV, with the same assumptions as before, and even then only if the detector resolution effects are negligible. There are two classes of W^+W^- final states that can be used for this analysis: $(\bar{q}q)(l^\pm\nu)$ and $(\bar{q}q)(\bar{q}q)$. The former are less constrained, but the latter may have additional uncertainties associated with the possibility of interference between the hadronic decay products of the $W^\pm$. This could arise from colour reconnection effects [13], exogamous hadronic cluster formation [14] and Bose-Einstein correlations between final-state hadrons [15]. None of those effects has yet been established in the data, but they may occur close to the present experimental sensitivity.

Fig. 5 shows the current status of measurements of $\sigma(e^+e^- \to W^+W^-)$. The expected threshold rise is seen, and both the γW^+W^- and $Z^0W^+W^-$ vertices must be present. The measurements close to threshold, in particular, yield [4]

$$m_W = 80.34 \pm 0.10 \text{ GeV} \tag{19}$$

The sensitivity of this measurement is mainly limited by the low statistics accumulated in the run at 161 GeV, which are not likely to be increased by future runs.

The search for $(\bar{q}q)(\bar{q}q)$ final states has significantly more background than $(\bar{q}q)(^\pm_\nu)$, limiting the sensitivity that can be achieved. The current LEP averages are [4]

$$m_W = 80.27 \pm 0.11 \text{ GeV } (\bar{q}q)(l^\pm\nu), \; 80.40 \pm 0.15 \text{ GeV } \; (\bar{q}q)(\bar{q}q) \tag{20}$$

from $(\bar{q}q)(l^\pm\nu)$ and $(\bar{q}q)(\bar{q}q)$ final states, respectively. There is no significant difference between them, so they may be combined to yield [4]

$$m_W = 80.35 \pm 0.09 \text{ GeV} \tag{21}$$

from reconstructed $W^\pm$ decays. Combining this in turn with the threshold measurement (19), we find [4]

$$m_W = 80.35 \pm 0.09 \text{ GeV} \tag{22}$$

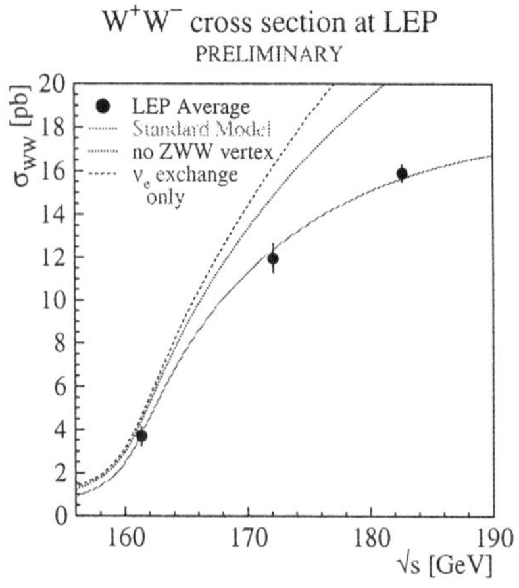

Figure 5: Measurements of $\sigma(e^+e^- \to W^+W^-)$ at LEP 2 [4]

from LEP 2. Combining this in turn with the $\bar{p}p$ collider measurements (14), the current world average is [4]

$$m_W = 80.375 \pm 0.064 \text{ GeV} \tag{23}$$

to be compared with the Standard Model prediction (15). So far, so good for the Standard Model! Fig. 6 illustrates the implications of these and possible future measurements for constraining m_H. Eventually, the LEP error Δm_W should fall below 50 MeV, which will be very significant.

The dominant Higgs production mechanism at LEP 2 is the $e^+e^- \to Z^0H$ [11] mentioned earlier, with $e^+e^- \to \bar{\nu}H\nu$ becoming important at higher energies [16]. Since the Higgs boson provides the masses for other particles, its couplings $g_{H\bar{f}f} \propto m_f$, and it prefers to decay into the heaviest $\bar{f}f$ pairs that are kinematically accessible. In the range of interest to LEP, this means that $H \to \bar{b}b$ should dominate, followed by $H \to \tau^+\tau^-$. The LEP experiments are therefore searching primarily for $e^+e^- \to (Z^0 \to \bar{q}q, \bar{\nu}\nu)(H \to \bar{b}b)$.

Each individual experiment has an expected sensitivity to $m_H \sim 83$ GeV, and their combined lower limit is $m_H \simeq 88$ GeV [9]. Beyond this mass, each is now encountering backgrounds, e.g., from $e^+e^- \to (Z^0 \to \bar{b}b)(Z^0 \to l^+l^-)$, that are irreducible when $m_H \simeq m_Z$. We can expect to hear more about candidate events in this mass range, and only time will tell whether there is a genuine Higgs signal lurking beneath the background. The likely future sensitivity of LEP depends on the E_{cm} and integrated luminosity $\mathcal{L}$ that can

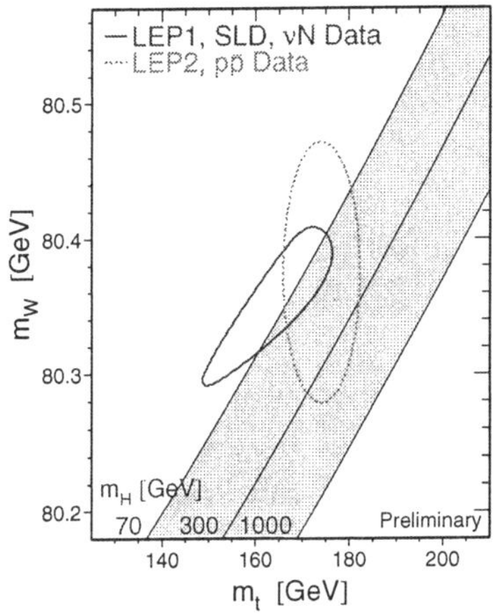

Figure 6: Measurements of m_t and m_W (dotted ellipse) compared with predictions based on precision electroweak data in the context of the Standard Model (solid lines), showing the sensitivity to the value of m_H [4].

be achieved. With $150pb^{-1}$ at $E_{cm} = 200$ GeV, which is optimistic, the Higgs boson could be discovered if $m_H \lesssim 100$ GeV and excluded if $M_H \lesssim 106$ GeV[12].

4 Beyond the Standard Model

The open problems left unsolved by the Standard Model can be grouped into three categories: the problems of unification, flavour and mass. Abdus Salam was a pioneer in studying the *Problem of Unification*: can one find a simple formalism for all gauge forces within a Grand Unified Theory (GUT)? If this occurs, the three Standard Model gauge couplings for SU(3), SU(2) and U(1) should all run logarithmically to become equal at some high energy scale $m_{GUT} \sim 10^{15}$ GeV. New interactions should appear near this energy scale that might lead to Abdus Salam's favourite prediction of proton decay - for which there is no experimental evidence - and/or neutrino masses - for which there are now tantalizing experimental hints. The *Problem of Flavour* is that of understanding the proliferation of quark and lepton flavours, the ratios of their masses, and their mixed couplings to the $W^{\pm}$. Finally, the *Problem of Mass* is that of understanding the origin of all the particle masses. As we have discussed, this is currently believed to be a Higgs boson. However, an elementary Higgs boson in isolation would raise further problems[17].

These originate from the effort to understand why the known particle masses, in particular m_W, are so much smaller than $m_P \simeq 10^{19}$ GeV, the only candidate we have for a fundamental mass scale in physics. This question can be rephrased as: why is $G_F(\sim 1/m_W^2) \gg G_N = 1/m_P^2$? or even why the Coulomb potential in an atom $V_C(\sim e^2/r)$ is so much greater than the Newtonian potential $V_N(\sim G_N m_1 m_2/r)$? It is not enough simply to suppress $m_W << m_P$ by hand, as loop corrections immediately raise it again:

$$\delta m_W^2 \simeq O\left(\frac{\alpha}{\pi}\right)\Lambda^2 \gg m_W^2 \tag{24}$$

if the mass scale Λ up to which the Standard Model applies is $O(m_{GUT})$ or $O(m_P)$. However, the corrections (24) can be made naturally small if one postulates equal numbers of bosons B and fermions F with identical couplings:

$$\delta m_W^2 \simeq O\left(\frac{\alpha}{\pi}\right)\left(m_B^2 - m_F^2\right) \tag{25}$$

as in models with approximate supersymmetry [18], as also much loved by Abdus Salam. One has $\delta m_W^2 \lesssim m_W^2$ if

$$|m_B^2 - m_F^2| \lesssim 1 \text{ TeV}^2 \tag{26}$$

which is the only theoretical motivation for supersymmetry at accessible energies.

All of the above problems are eventually to be resolved within a *Theory of Everything* (TOE) that should also include gravity and reconcile it with quantum mechanics, as well as explain the origin of space-time, why there are three spatial dimensions and one time, and make coffee. The only candidate we have for a TOE is string theory, presumably in the non-perturbative formulation currently termed M (or F) theory, which is the subject of many other talks at this meeting [19].

Supersymmetry predicts that all the known particles should be accompanied by partners with identical internal quantum numbers but spins differing by half a unit. None of these have been seen by experiments, which establish the lower limits [20]:

$$m_{\tilde{q},\tilde{g}} \gtrsim 200 \text{ GeV},\ m_{\tilde{l}} \gtrsim 80 \text{ GeV},\ m_{\chi^\pm} \gtrsim 90 \text{ GeV} \tag{27}$$

for squarks, gluinos, sleptons and charginos, respectively. The minimal supersymmetric extension of the Standard Model (MSSM) also predicts a richer spectrum of physical Higgs bosons than the Standard Model. There should be three neutral bosons h, H, A and two charged bosons $H^\pm$. The mass of the lightest neutral boson can be calculated [21]:

$$m_h \lesssim 130 \text{ GeV} \tag{28}$$

The prediction (28) provides one of the tantalizing indirect experimental hints for supersymmetry, since it is in perfect agreement with the indirect LEP estimate (12). A second hint is provided by the agreement of the experimental determination of $\sin^2\theta_W$ (6) with the prediction $\sin^2\theta_W \sim 0.231$ of minimal supersymmetric GUTs[22], to be contrasted with the non-supersymmetric GUT prediction $\sin^2\theta_W \sim 0.21$ to 0.22. With the experimental value (6), the couplings unify perfectly at $m_{GUT} \sim 10^{16}$ GeV if supersymmetry is postulated, but not if only the Standard Model particles are included when calculating the evolution of the gauge couplings to higher energies. But there is still no "smoking gluino"!

Among the most interesting LEP limits on supersymmetric particles are those on charginos and neutralinos. Within the context of the MSSM, these together provide a lower limit on the lightest neutralino χ, which is commonly thought to be the lightest supersymmetric particle and an excellent candidate for the dark matter in the Universe[23].

The direct experimental searches imply that $m_\chi \gtrsim 30$ GeV if $m_{\tilde{\nu}} \gtrsim 200$ GeV, but smaller values are possible if the sneutrino $\tilde{\nu}$ is lighter. Conversely, the lightest neutralino χ is forced to be heavier if supplementary experimental, theoretical or cosmological inputs are used, as seen in Fig. 7. In particular, taking into account the LEP lower limit on the Higgs mass, assuming universal masses for the spin-zero sparticles at the supersymmetric GUT scale, and requiring that the relic neutralino density fall within the range $0.1 \lesssim \Omega h^2 \lesssim 0.3$ preferred by astrophysics and cosmological inflation, we find[24]

$$m_\chi \gtrsim 42 \text{ GeV} \tag{29}$$

which should be borne in mind when designing experiments to look for neutralino dark matter.

5 LHC Physics

CERN is currently constructing the next large energy step in accelerator particle physics: the LHC, which will explore for the first time the TeV energy range, as well as explore the quark-gluon plasma and look at CP violation in the decays of hadrons containing bottom quarks. In pp mode, the LHC has a design $E_{cm} = 14$ TeV and luminosity $\mathcal{L} = 10^{34} cm^{-2} s^{-1}$, and in $Pb - Pb$ collision mode $E_{cm} = 1.2$ PeV and $\mathcal{L} = 10^{27} cm^{-2} s^{-1}$. Five experiments are currently planned: two large detectors ATLAS and CMS oriented towards discovery physics such as the Higgs boson and supersymmetry, ALICE for heavy-ion physics, LHC-B for CP-violation studies, and TOTEM to measure the elastic and total pp cross sections.

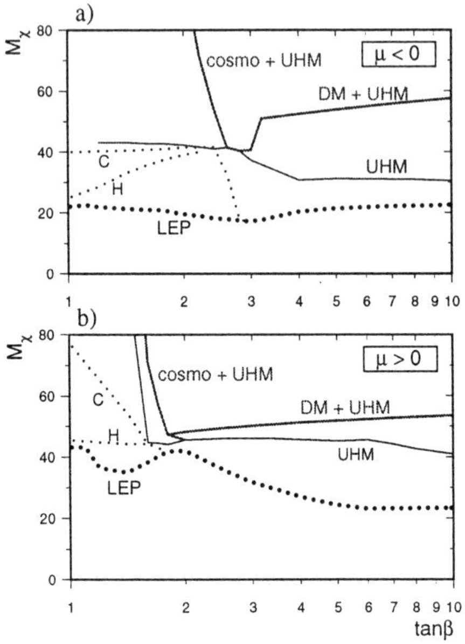

Figure 7: Lower limits on the mass of the lightest neutralino χ, assumed to be stable. The heavy dotted line is the direct LEP limit, the thin dotted lines incorporate limits on Higgs masses and cosmological assumptions. The solid lines labelled UHM incorporate the theoretical assumption that input Higgs, slepton and squark masses are universal. The parts of these lines labelled DM (cosmo) are obtained assuming additionally that $\Omega_\chi h^2 > 0.1 (< 0.3)$ [24].

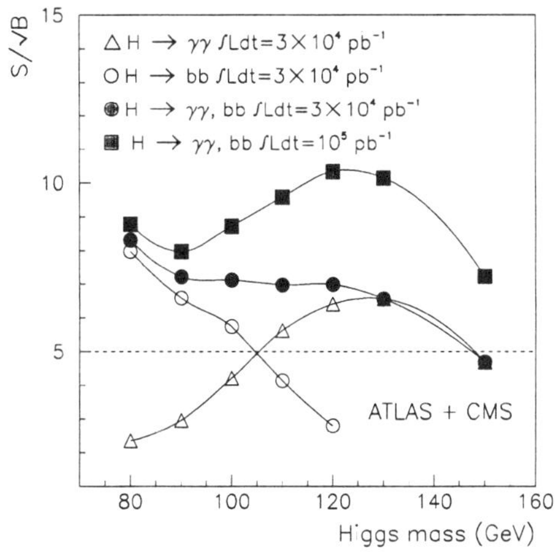

Figure 8: Significances of prospective LHC searches for light Higgs bosons by ATLAS and CMS, expressed as the ratio of the signal S to the statistical fluctuations in the background B.

Fig. 8 illustrates the discovery potential for the Higgs boson of the Standard Model. It should be detectable at the many-standard-deviation level for any mass 90 GeV $\lesssim m_H \lesssim$ 1 TeV. Fig. 9 illustrates the discovery potential for Higgs bosons in the MSSM [25]. Everywhere in the parameter plane, at least one production and decay signature is detectable, and in many regions of the plane there are more than one detectable channel.

Fig. 10 illustrates the ability to pick out the classic missing-transverse-energy signature of supersymmetry at the LHC. Studies indicate that it should be possible to reach $m_{\tilde{q},\tilde{g}} \lesssim 2$ TeV, covering comfortably the expected range (25). Indeed, the region of supersymmetric parameter space favoured if the lightest neutralino is to be the dark matter in the Universe may be covered many times over by the LHC, using searches for leptons, missing energy and two or more jets. Moreover, many detailed measurements of supersymmetric spectroscopy would be possible at the LHC [26].

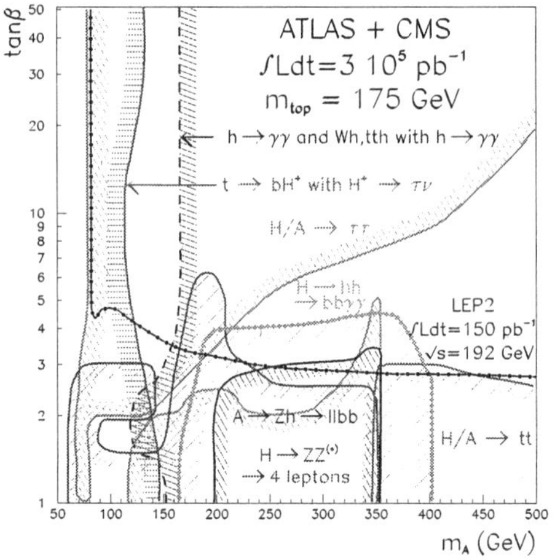

Figure 9: Regions of the $(m_A, \tan\beta)$ plane in which MSSM Higgs bosons may be detected via the indicated signatures, under the assumptions described in Ref. 25.

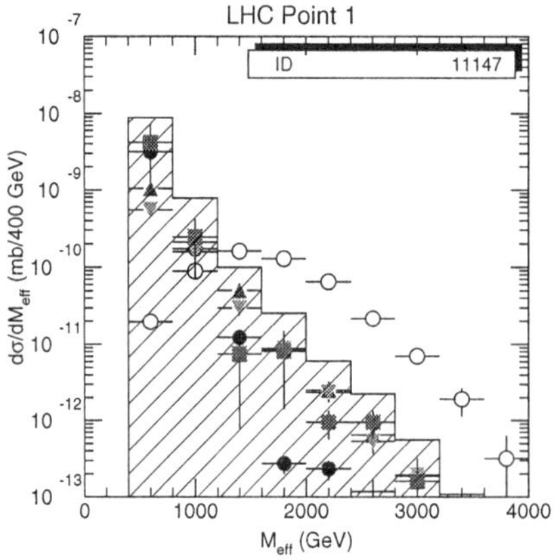

Figure 10: The prospective missing-energy signal (open circles) for supersymmetry at the LHC, compared with the physics and detector backgrounds, for one choice of MSSM parameters [26].

6 Conclusions

Abdus Salam's legacy remains as a monument to his memory. The Standard Model he proposed together with Sheldon Glashow and Steven Weinberg has so far survived all tests by accelerator experiments. However, its experimental verification remains incomplete, and we are still searching for the Higgs boson. There are tantalizing hints from precision experiments at LEP and elsewhere for many of the ideas that Abdus Salam held dear, in particular supersymmetry and grand unification. These are provided by the measurement of $\sin^2\theta_W$ (6) and the estimate (12) of m_H. If the Higgs boson and supersymmetry do not show up in the remaining runs of LEP, they shall surely be discovered by experiments at the LHC. It is sad that Abdus Salam cannot be among us when these dreams come true.

References

1. A. Salam, *Proc. Eighth Nobel Symposium*, Stockholm 1968, ed. N. Svartholm (Almqvist and Wiksells, Stockholm, 1968), p. 367;
   S. Weinberg, *Phys. Rev. Lett.* **19** (1967) 1264.
2. R. Raja, for the CDF and D0 Collaborations, `hep-ex/9706011`;
   B. Klima, for the D0 Collaboration, `hep-ex/980417`.
3. R. Assmann et al., CERN preprint EP/98-40, SL/98-12.
4. LEP Electroweak Working Group, CERN preprint PPE/97-154, and `http://www.cern.ch/LEPEWWG/Welcome.html`.
5. M. Veltman, *Nucl.Phys.* **B123** (1977) 89;
   M.S. Chanowitz, M. Furman and I. Hinchliffe, *Phys.Lett.* **78B** (1978) 285.
6. M. Veltman, *Acta Phys.Pol.* **8** (1977) 475.
7. U. Amaldi, A. Bohm, L.S. Durkin, P. Langacker, A.K. Mann, W.J. Marciano, A. Sirlin, H.H. Williams, *Phys. Rev.* **D36** (1987) 1385;
   G. Costa, J. Ellis, G.L. Fogli, D.V. Nanopoulos, F. Zwirner, *Nucl. Phys.* **B297** (1988) 244;
   J. Ellis and G.L. Fogli, *Phys. Lett.* **B213** (1988) 526.
8. J. Ellis and G.L. Fogli, *Phys.Lett.* **B231** (1989) 189.
9. LEP working group for Higgs boson searches, CERN preprint EP/98-046;
   L3 Collaboration, M. Acciarri et al., CERN preprint EP/98-52, and references therein.
10. T. Hambye and K. Riesselmann, `hep-ph/9708416`.

11. J. Ellis, M.K. Gaillard and D.V. Nanopoulos, *Nucl.Phys.* **B106** (1976) 292;
B.L. Ioffe and V.A. Khoze, *Sov.J.Part.Nucl.* **9** (1978) 50;
B.W. Lee, C. Quigg and H. Thacker, *Phys.Rev.Lett.* **38** (1977) 883 and *Phys.Rev.* **D16** (1977) 1519.
12. *Proceedings of the Workshop on Physics at LEP 2*, eds. G. Altarelli, T. Sjöstrand and F. Zwirner, CERN Report 96-01.
13. Yu.L. Dokshitzer, V.A. Khoze, L.H. Orr and W.J. Stirling, *Nucl.Phys.* **B403** (1993) 65;
V.S. Fadin, V.A. Khoze and A.D. Martin, *Phys.Lett.* **B320** (1994) 141; *Phys.Rev.* **D49** (1994) 2247;
G. Gustafson, U.Pettersson and P. Zerwas, *Phys.Lett.* **B209** (1988) 90;
T. Sjöstrand and V.A. Khoze, *Phys.Rev.Lett.* **72** (1994) 28; *Z.Phys.* **C62** (1994) 281;
G. Gustafson and J. Häkkinen, *Z.Phys.* **C64** (1994) 659.
14. J. Ellis and K. Geiger, *Phys.Rev.* **D54** (1996) 1755;
J. Ellis and K. Geiger, *Phys. Lett.* **B404** (1997) 230.
15. L. Lönnblad and T. Sjöstrand, *Phys.Lett.* **B351** (1995) 293.
16. D.R.T. Jones and S.T. Petcov, *Phys.Lett.* **84B** (1979) 440.
17. L. Maiani, *Proceedings of the Summer School on Particle Physics*, Gif-sur-Yvette, 1979 (IN2P3, Paris, 1980), p. 3;
G. 't Hooft, in *Recent Developments in Field Theories*, eds. G. 't Hooft et al. (Plenum Press, New York, 1980);
E. Witten, *Nucl.Phys.* **B188** (1981) 513;
R.K. Kaul, *Phys.Lett.* **109B** (1982) 19.
18. Y.A. Gol'fand and E.P. Likhtman, *Pis'ma Zh.E.T.F.* **13** (1971) 323;
D. Volkov and V.P. Akulov, *Phys.Lett.* **46B** (1973) 109;
J. Wess and B. Zumino, *Nucl.Phys.* **B70** (1974) 39;
for a review, see P. Fayet and S. Ferrara, *Physics Reports* **32C** (1977) 249.
19. J. Ellis, CERN preprint TH/98-139, `hep-ph/9804440`.
20. LEP 2 supersymmetry working group, `http://www.cern.ch/LEPSUSY/`.
21. Y. Okada, M. Yamaguchi and T. Yanagida, *Progr.Theor.Phys.* **85** (1991) 1;
J. Ellis, G. Ridolfi and F. Zwirner, *Phys.Lett.* **B257** (1991) 83; *Phys.Lett.* **B262** (1991) 477;
H.E. Haber and R. Hempfling, *Phys.Rev.Lett.* **66** (1991) 1815;
R. Barbieri, M. Frigeni and F. Caravaglios, *Phys.Lett.* **B258** (1991) 167;
Y. Okada, M. Yamaguchi and T. Yanagida, *Phys.Lett.* **B262** (1991) 54.

22. S. Dimopoulos, S. Raby and F. Wilczek, *Phys.Rev.* **D24** (1981) 1681; W.J. Marciano and G. Senjanovic, *Phys.Rev.* **D25** (1982) 3092; L.E. Ibàñez and G.G. Ross, *Phys.Lett.* **105B** (1982) 439; M.B. Einhorn and D.R.T. Jones, *Nucl.Phys.* **B196** (1982) 475; J. Ellis, S. Kelley and D.V. Nanopoulos, *Phys.Lett.* **B249** (1990) 441 and **B260** (1991) 131; P. Langacker and M. Luo, *Phys.Rev.* **D44** (1991) 817; U. Amaldi, W. de Boer and H. Furstenau, *Phys.Lett.* **B260** (1991) 447; F. Anselmo, L. Cifarelli, A. Petermann and A. Zichichi, *Nuovo Cimento* **104A** (1991) 1817.
23. J. Ellis, J.S. Hagelin, D.V. Nanopoulos, K.A. Olive and M. Srednicki, *Nucl.Phys.* **B238** (1984) 453.
24. J. Ellis, T. Falk, G. Ganis, K.A. Olive, M. Schmitt, CERN preprint TH/98-32, hep-ph/9801445.
25. E. Richter-Was, D. Froidevaux, F. Gianotti, L. Poggioli, D. Cavalli and S. Resconi, CERN Preprint TH/96-111 (1996).
26. I. Hinchliffe, F.E. Paige, M.D. Shapiro, J. Soderqvist and W.Yao, *Phys.Rev.* **D55** (1997) 5520.

GOING BEYOND THE SALAM-WEINBERG STANDARD MODEL WITH DECAYING NEUTRINOS

D.W. SCIAMA
SISSA and ICTP, Trieste,
Department of Physics, University of Oxford.

It has not yet been possible to use experiments to go beyond the standard model, although one would expect this model to be superseded by a more comprehensive one. Here I show that the decaying neutrino theory for the ionisation of the interstellar medium (Sciama 1990, 1993) requires a decay lifetime at least 10^{13} times smaller than the standard model value. The required short lifetime can be achieved, for example, in supersymmetric theories with broken R parity.
The present status of the decaying neutrino theory is reviewed, and a variety of observational tests, from both astronomy and cosmology, are described. Finally an account is given of a potentially decisive experiment being conducted on a satellite currently orbiting the earth.

1 Introduction

I first met Abdus Salam in 1950, when we were both research students at Cambridge. He was beginning his famous work on renormalisation theory, and it was obvious that he was a man of great brilliance and drive. Of course we did not then know what lay in the future.

One of the things which lay in the future was the Salam-Weinberg standard model. We have heard from John Ellis how difficult it is to go beyond this model using laboratory experiments, yet everyone expects that one day it will be superseded. In this talk I would like to describe my speculative proposal (Sciama 1990, 1993) for using astronomy and cosmology to go beyond the standard model. The idea is to explain the otherwise puzzling ionisation of hydrogen in individual galaxies and in the intergalactic medium in terms of photons emitted by radiatively decaying dark matter neutrinos. This theory requires the decay lifetime τ of the neutrino to be at least 10^{13} times less than the value expected on the standard model – a gross deviation indeed!

But what chance is there that this neutrino theory is correct? In this talk I will argue that it has three advantages supporting it:

1) If correct it would possess a wide variety of strong implications for elementary particle physics, for astrophysics and for cosmology.

2) Its domain of validity is extremely small, so that some of its predictions are very specific and therefore very vulnerable.

3) It leads to many observational tests, some of which already seem rather successful. I will conclude by describing an observational test which is currently

being conducted from outer space and which could yield a decisive result by the time this book is published.

2 The Motivation of the Decaying Neutrino Theory

The original motivation of the decaying neutrino theory arose from the discovery by Reynolds (1989) of the large scale height of the free electrons in the interstellar medium of our Galaxy. Using recently performed observations of the dispersion of radio signals from pulsars lying in globular clusters situated well beyond the electron layer of the Galaxy he derived a scale height for this layer of about 1 kiloparsec. By contrast, the scale height of the stars whose ultra-violet photons were thought to be responsible for the ionisation of hydrogen in the interstellar medium is only about 100 parsecs. Moreover, interstellar space contains many neutral hydrogen clouds which are opaque to the ionising photons. Despite recent attempts by Miller and Cox (1993) and by Dove and Shull (1994) to study the percolation of the photons between these clouds, it is widely regarded as problematic to attribute the ionisation to stellar photons, especially for the ionisation observed within a kiloparsec of the sun, where the clouds are directly seen to constitute a highly opaque screen.

By contrast, if the dark matter needed to account for the extended flat rotation curve of the Galaxy is assumed to consist mainly of neutrinos with non-zero rest mass, then these neutrinos would be expected to pervade the Galaxy in an essentially continuous and large-scale distribution. There would then be many neutrinos within one optical depth of any point where there are free electrons observed, and all that is required to account for those electrons, using photons from decaying neutrinos, are suitable values for the decay lifetime τ and the decay photon energy E_γ.

3 The Radiative Decay of the Neutrino

A neutrino of non-zero rest mass m_1 would be expected to decay into a photon and a neutrino of lower rest mass m_2 via a virtual transition to a charged W boson and a lepton. The decay lifetime τ was first calculated for the gauge theory of the Salam-Weinberg standard model by Marciano and Sanda (1977) and by Petcov (1977). They obtained

$$\tau \geq 10^{36} \left(\frac{30eV}{m_1} \right)^5 \text{sec} . \tag{1}$$

In the same year Cowsik (1977) suggested that the photons emitted by "dark matter" neutrinos might lead to an observable radiation background, the large number of neutrinos along a line of sight outweighing a possibly long decay lifetime.

The typical energy of a photon in this background would be mainly determined by the energy E_γ of a decay photon in the rest frame of the decaying neutrino. Conservation of energy and momentum in the decay yields the relation

$$E_\gamma = \frac{1}{2} m_1 \left(1 - \frac{m_2^2}{m_1^2}\right) .$$

For simplicity we shall assume hereafter that $m_2 \ll m_1$ (as would be true, for example, in the see-saw model for neutrino masses). We would then have

$$E_\gamma = \frac{1}{2} m_1 .$$

This is one of the fundamental relations of our decaying neutrino theory. Because of it, the only parameters to be determined are E_γ and τ.

Of course if a decay photon is to be able to ionise hydrogen we must have

$$E_\gamma \geq 13.6 eV ,$$

and so

$$m_1 \geq 27.2 eV .$$

We immediately make contact with the cosmological dark matter problem, because we already know that to just close the universe with neutrinos we require that $m_1 \sim 30$ eV. We shall say more about this later. For the moment we warn the reader that the closeness of m_1 to twice the ionisation potential of hydrogen will be an important element in the structure of our theory.

4 The Required Value of the Decay Lifetime τ

4.1 The Ionisation of the Interstellar Medium

In this subsection we derive an upper limit on τ from recent estimates of the maximum permitted density of dark matter near the sun and of the free electron density n_e which we are here attributing to the ionisation of hydrogen by decay photons.

In subsection 4.2 a lower limit on τ will be determined from recent observational upper limits on the extragalactic diffuse background at 1500 A, which in turn impose an upper limit on the flux of red shifted decay photons emitted by the cosmological distribution of neutrinos.

Our derived upper and lower limits on τ are only just consistent with one another, and so lead to a highly constrained value for this quantity.

To derive the upper limit on τ consider a region near the sun whose atomic hydrogen density makes it opaque to ionising decay photons. In ionisation equilibrium (which we would expect to hold) every decay photon in the region would ionise a nearby hydrogen atom which would then recombine, and

$$\frac{n_\nu}{\tau} = \alpha n_e^2 \ , \tag{2}$$

where n_ν is the local number density of neutrinos and α is the recombination coefficient (excluding recombinations directly to the ground state which simply produce additional ionising photons). The value of n_e which our theory attributes to ionisation by decay photons has recently been rediscussed by Sciama (1977a) who obtained $n_e = 0.05 \pm 0.01\text{cm}^{-3}$.

The value of α depends on the temperature T of the recombining gas. For the regions which we are considering (both clouds and the intercloud medium), one has $T < 2 \times 10^4$K (Reynolds 1985, Spitzer and Fitzpatrick 1993). Hence $\alpha > 1.3 \times 10^{-13}$ cm^3 sec^{-1} (Ferland et al. 1992).

Finally, to derive an upper limit on τ an upper limit on the number density n_ν of neutrinos near the sun must be established. This can be obtained from an upper limit on the local mass density ρ_ν of neutrinos near the sun, since an independent estimate of m_1, namely $\sim$ 28eV, will be obtained in Section 5. An upper limit on ρ_ν near the sun was recently derived by Sciama (1977b) in connection with values for the largest permitted flattening of the neutrino halo and for the rotational velocity of the Galaxy at the sun's position.

By taking into account various estimates for column densities of material near the sun, an upper limit for ρ_ν of 0.03 $M_\odot$pc^{-3} was obtained. This upper limit is associated with the largest permitted flattening of the dark halo (Dehnen and Binney 1997) corresponding to an axial ratio of 0.2. We return briefly to the observational validity of this extreme requirement in subsection 6.1. If we now suppose that $m_1 = 27.4$eV (see later), our upper limit on n_ν near the sun is $4.2 \times 10^7\text{cm}^{-3}$. Our limits on the values of n_ν, α and n_e (≥ 0.04 cm^{-3}) now lead in conjunction with (2), to the conclusion that

$$\tau \leq 2 \times 10^{23} \text{ sec.}$$

4.2 The Extragalactic Background at 1500 A

Some of the earliest lower limits on τ (for a decaying neutrino not then related to the ionisation of the interstellar medium) were based on observational estimates of the cosmic background in the far ultra-violet, which was compared with the red shifted decay flux produced by the cosmological distribution of neutrinos (Stecker 1980, Kimble, Bowyer and Jakobsen 1981). Estimates of this background are still controversial (compare Bowyer 1991 with Henry 1991). Recent contributions to this discussion have been made by Witt, Friedmann and Sasseen (1997) and by Korpela, Bowyer and Edelstein (1997).

Theoretical aspects of the decay flux have been rediscussed by Sciama (1991), Overduin, Wesson and Bowyer (1993), Dodelson and Jubas (1994) and Overduin and Wesson (1997). It would not be appropriate to enter into these detailed matters here. We believe that a fair summary of the present situation can be expressed in terms of a lower limit for τ given by

$$\tau \geq 2 \times 10^{-23} \text{ sec} ,$$

(cf. Bowyer et al. 1995b).

4.3 The Required Value of τ

Our overall conclusion from this discussion is that, if the decaying neutrino theory for the ionisation of the interstellar medium is correct, the decay lifetime τ is determined as

$$\tau \sim 2 \times 10^{23} \text{sec} ,$$

and the neutrino halo of our Galaxy has to possess its maximum permitted degree of flattening. The uncertainty in our estimate for τ is difficult to assess. We provisionally adopt

$$\tau = 2 \pm 1 \times 10^{23} \text{sec} . \qquad (3)$$

This is the quantity that is at least about 10^{13} times less than the value of τ derived from the standard model and given in equation (1). On the other hand, our required value of τ is about a million times longer than the age of the universe – strong testimony to Cowsik's point that such a long lifetime can still lead to easily observable effects.

Is the discrepancy with the standard model fatal for our decaying neutrino theory or can we claim that this theory forces us to seek a different particle physics model which is compatible with (3)? In the rest of this article we will argue that various astronomical and cosmological considerations strongly favour the decaying neutrino theory. It is fortunate, therefore, that particle physics models do exist which are compatible with (3). In particular, it has been shown by Roulet and Tommasini (1991) and by Gabbiani, Masiero and Sciama (1991) that supersymmetric theories with broken R parity can lead to (3) (see also Bowyer et al. 1995a). It is comforting to know that Salam himself was particularly fond of supersymmetric theories!

5 The Required Values of E_γ and m_1

Existing attempts to observe the hydrogen recombination line H_α from opaque intergalactic neutral hydrogen clouds at essentially zero red shift lead to an upper limit on the hydrogen-ionising flux F incident on the clouds arising from the cosmological distribution of neutrinos. This upper limit on F then leads to an upper limit on E_γ because of the role of the red shift in reducing the energy of a decay photon to below 13.6 eV. Write

$$E_\gamma = 13.6 + \varepsilon \ eV \ .$$

Then, if $\frac{\varepsilon}{13.6} \ll 1$, F at zero red shift $F(0)$ is given by

$$F(0) = \frac{n_\nu(0)}{\tau} \frac{c}{H} \frac{\varepsilon}{13.6} \ , \tag{4}$$

where $n_\nu(0)$ is the cosmological number density of neutrinos at $z = 0$, c is the velocity of light, and H_0 is the present value of the Hubble constant. As we shall see in Sec.6.3 H_0 in the decaying neutrino theory is 55 ± 0.3 km sec^{-1} Mpc^{-1}.

The H_α upper limit observations of Vogel et al. (1995) and of Donahue, Aldering and Stocke (1995) imply that $F(0) \leq 10^5$ photons cm^{-2}sec^{-1}. Since $n_\nu(0) = 112.6 \pm 0.5$ cm^{-3} (Sciama 1997c) it follows that

$$\frac{\varepsilon}{\tau} \leq 7 \times 10^{-25} eV \ \text{sec}^{-1} \ .$$

Combining this inequality with our value for τ one obtains

$$\varepsilon \leq 0.2 eV \ ,$$

so that indeed $\frac{\varepsilon}{13.6} \ll 1$. It follows that

$$E_\gamma \leq 13.8 eV \ .$$

We need not be too surprised that E_γ is required to be so close to the ionisation potential of hydrogen since we have already seen that a cosmologically based estimate of m_1 leads to a value of $\frac{1}{2}m_1$ within a few per cent of 13.6 eV.

In Sec. 6.4 we shall find that considerations involving the metagalactic ultra-violet background in the red shift range 2 to 5 suggest that the actual value of E_γ is close to the upper limit given in (4). To be conservative we shall here assume that

$$E_\gamma = 13.7 \pm 0.1 eV \ ,$$

so that

$$m_1 = 27.4 \pm 0.2eV \ .$$

These quantities are thus determined with a precision of 1%. Together with

$$\tau = 2 \pm 1 \times 10^{23}\text{sec} \ ,$$

which is less precise, they provide the numerical values of the fundamental parameters of our theory.

6 Observational Tests of the Decaying Neutrino Theory

6.1 The Extreme Flattening of the Neutrino Halo of our Galaxy

The high value for the required local mass density ρ_ν can be tested by comparing dynamical estimates of the total density ρ_0 near the sun (the Oort limit) with the densities of known stars and gas. The value of ρ_0 has been controversial; an estimate will be used here which is based on recent Hipparcos observations of F stars, and so may lead to a more reliable result. This result is $\rho_0 = 0.11 M_\odot \text{pc}^{-3}$ (Pham 1996, 1997).

However, a significantly smaller value for ρ_0 has also been obtained from Hipparcos data by Crézé et al. (1998), namely $0.076 \pm 0.015 M_\odot \text{pc}^{-3}$. This result seems rather low when compared to the density of known matter (see below) and the disagreement remains to be clarified. Perhaps the errors have been underestimated. Here we provisionally adopt Pham's value.

The contribution of known stars and gas to ρ_0 is itself somewhat uncertain. Often quoted values for each are 0.04 $M_\odot \text{pc}^{-3}$ (e.g. Bienaymé, Robin and Crézé 1987, Crézé et al. 1998). The stellar contribution may have to be increased to allow for the existence of faint stars, but the Hubble Space Telescope deep survey suggests that this additional contribution may be small (e.g. Gould et al. 1996). If this is correct, one again obtains an upper limit for ρ_ν of 0.03 $\text{M}_\odot$ pc^{-3}. This upper limit was also previously derived by Bienaymé et al. (1987). However if the Crézé et al. (1998) value for ρ_0 is correct the decaying neutrino theory would probably be ruled out.

Recently it has been argued (Olling and Merrifield 1997) that analysis of the hydrostatic equilibrium of the gaseous disk of the Galaxy shows that the flattening of its dark halo must be much less than we are requiring here. However, their analysis assumes that the contribution of cosmic rays and magnetic fields to the vertical pressure gradient in the outer regions of the Galaxy are negligible. This assumption neglects the observational evidence from gamma ray emission data (Bloemen 1989) that in fact the radial gradient of the cosmic ray density in the outer regions of the Galaxy is rather small. If a reasonable contribution to the vertical pressure gradient from cosmic rays and magnetic fields is included in the hydrostatic balance a dark halo as flattened as we require is readily achieved.

In this connection it is somewhat reassuring that some other galaxies do seem to exhibit the same large flattening as we require for our Galaxy (Sackett et al. 1994, Olling 1996, Becquaert and Combs 1997).

6.2 The Ionisation of Warm Opaque Interstellar Clouds near the Sun

At the time of writing the most decisive observational test in favour of the decaying neutrino theory concerns the density of free electrons in warm opaque interstellar clouds near the sun (where n_ν is sensibly constant). Because of their opacity the ionising photons produced by stars cannot penetrate these clouds and in general one would not expect other ionising agencies such as cosmic rays to produce a density of free electrons in their interiors as large as ~ 0.05 cm^{-3} (Sciama 1993).

By contrast, one sees from equation (2) that decay photons would produce an appreciable electron density in the interiors of these opaque clouds. Moreover this electron density would be the same in each cloud, so long as the temperatures, and so the recombination coefficients, of these clouds are similar. In particular n_e would be independent of the neutral hydrogen density n_{HI}, so long as this density is sufficient to render the clouds opaque.

The decaying neutrino theory therefore makes two strong and unambiguous predictions for the value of n_e in warm opaque clouds near the sun:

(a) Despite the opacity, n_e should be appreciable in the interiors of these clouds (~ 0.05 cm^{-3}).

(b) n_e should have essentially the same value in each cloud.

These two predictions are often difficult to test. When free electrons are detected in a cloud, one then has to decide whether these electrons are located in the skins of the cloud, where they might be produced by photons from distant stars, or truly in their interiors where the opacity is high. Secondly, the derived value of n_e is usually rather inaccurate, so that its claimed constancy cannot be adequately tested.

Fortunately there is one case where both these difficulties can be avoided. This involves Hubble Space Telescope observations of the absorption spectrum of the halo star HD 93521 made by Spitzer and Fitzpatrick (1993). They observed several opaque clouds along the line of sight to this star, and inferred n_e in each cloud from the measured excitation effect of electron collisions with singly ionised carbon. Their analysis of their data, further amplified by Sciama (1997a) provides strong evidence that for the slowly moving clouds which are presumably undisturbed by shock waves

(a) the free electrons are located in the interiors of the clouds,

(b) n_e is the same in each cloud with a precision of about 10%.

Both the predictions of the decaying neutrino theory are thus clearly verified. Moreover it would be difficult to understand (a) and (b) in conventional theories for the origin of the ionisation.

6.3 The Hubble Constant and the Age of the Universe

We shall now see that our precise value for m_1 leads to precise predictions for the values of the Hubble constant H_0 and the age of the universe t_0, which should soon be subject to reasonably accurate observational tests.

The reasoning involved has been recently updated by Sciama (1997c), which can be consulted for the detailed arguments. We first derive the contribution ρ_ν of neutrinos to the cosmological density. Since n_ν is $3/11 n_\gamma$, where n_γ is the photon density in the cosmic microwave background we have $n_\nu(0) = 112.6 \pm 0.5$ cm^{-3}. We multiply this by our value for m_1 to obtain ρ_ν. It is conventional to express this density in terms of $\Omega = \rho/\rho_{\text{crit}}$, and $H_0 = 100$ h km sec^{-1} Mpc^{-1}, where ρ_{crit} is the critical density given by $8\pi/3$ $G\rho_{\text{crit}} = H_0^2$.

We then find that

$$\Omega_\nu h^2 = 0.293 + 0.003 \ .$$

Since we need to know the total density Ωh^2 we must include the (small) contribution Ω_b from baryons. This contribution is still uncertain, depending on whether one adopts a low or a high value for the abundance of deuterium at large red shifts. The high value, in conjunction with considerations of big bang nucleosynthesis, leads to a low value of the baryon density, namely $\Omega_b h^2 = 7 \pm 1 \times 10^{-3}$. Fortunately the uncertainty in $\Omega_b h^2$ will make only a small difference to our final result. We provisionally adopt the low value of $\Omega_b h^2$ and so obtain

$$\Omega h^2 = 0.300 \pm 0.003 \ .$$

To derive H_0 and t_0 from our value of Ωh^2 we proceed as follows. First we assume from the observations that

(a) $h \geq 0.5$

and

(b) $t_0 \geq 12$ Gyr.

From (a) and $\Omega h^2 = 0.3$ we derive that $\Omega < 1.2$ and so (for a zero cosmological constant λ) that $t_0 \leq 13.4$ Gyr.

From (b) and $\lambda = 0$ we derive that $\Omega \geq 1$ and $h \leq 0.55$. Combining these results we obtain

$$H_0 = 52.5 \pm 2.5 \text{ km sec}^{-1}\text{Mpc}^{-1} ,$$

$$t_0 = 12.7 \pm 0.7 \text{ Gyr} ,$$

and

$$\Omega = 1.1 \pm 0.1 .$$

Since Ω is required to be so close to 1 it is tempting to assume that it is exactly 1 (or at least 1 to within the level of uncertainty to which we are working here). In that case uncertainties in H_0 and t_0 are controlled by our 1 percent uncertainty in Ωh^2, and so we find that for $\lambda = 0$

$$H_0 = 54.8 \pm 0.3 \text{ km.sec}^{-1}\text{Mpc}^{-1} ,$$

and

$$t_0 = 11.96 \pm 0.06 \text{ Gyr} ,$$

(where 1 yr is defined to be 3.15×10^7 sec).

The observational values of H_0 and t_0 are still controversial. Our precise predictions lie within the presently acceptable range, and changes to H_0 and t_0 enforced by the recently announced Hipparcos data have been generally helpful. Astronomers are hopeful that within a few years it should be possible to determine H_0 and t_0 with a precision of 10%.

6.4 The Metagalactic Ultra-violet Background at Red Shifts 2-5

In order to test the decaying neutrino theory at non-zero red shifts we need to know how the ionising decay flux from the cosmological distribution of neutrinos depends on z. This dependence follows immediately from equation (4), since $n_\nu(z) = (1+z)^3 n_\nu(0)$ and $H(z) = (1+z)^{3/2} H_0$. This last relation follows from our conclusion that $\Omega \sim 1$. If $\Omega = 1$ exactly (and $\lambda = 0$) the universe conforms to the Einstein-de Sitter model which gives the z dependence of H as above. Hence we conclude that

$$F(z) = (1+z)^{3/2} F(0) .$$

Absorption of decay photons by intergalactic clouds does not have to be taken into account in deriving $F(z)$ even at $z \sim 5$, where in principle the absorption would be much larger than at $z \sim 0$ (e.g. Haardt and Madan 1996) because with our small value for ε the red shift reduces the energy of a decay photon to below 13.6 ev in a much shorter distance than the absorption free path even at large z.

There now exist observational estimates of the metagalactic ionising flux at the HI, HeI and HeII Lyman edges. These estimates can be derived from the Gunn-Peterson effect and from the proximity effect. We consider first the Gunn-Peterson effect. As soon as quasars of red shift ~ 2 were identified by Schmidt various authors pointed out that they could be used to probe the density of intergalactic neutral hydrogen. This density could then be used in turn to constrain the intergalactic flux of ionising photons. Consider a layer of neutral hydrogen lying at a red shift z along the line of sight to a quasar of greater red shift Z. Photons emitted by the quasar and reaching the layer with the wavelength of Lyman α would be able to excite neutral hydrogen in the layer to its first excited state. When the excited atom decays to its ground state it would re-emit a Lyman α photon in some general direction. This scattering process would deplete the photon beam propagating towards the observer. Thus these would be an apparent absorption trough in the spectrum of the quasar lying between 1216 Å and 1216 $(1 + Z)$ Å. For $Z \sim 2$ the long wavelength end of this trough would be accessible to ground-based telescopes.

The first attempt to detect this absorption trough was made by Gunn and Peterson (1965) and the effect is now named after them. Up to the present time no such trough has been found either for HI or HeI but very recently one has been found for HeII (Zheng, Davidson and Kriss 1997) based on a detailed numerical analysis of data from telescopes in space. These results constrain the spectrum of the metagalactic ionising flux between the HI edge and the HeII edge at $z \sim 2 - 3$.

The proximity effect is a property of the so-called Lyman α clouds. If one considers a discrete cloud of neutral hydrogen on the line of sight to a quasar the Gunn-Peterson absorption trough would collapse into a Lyman α absorption line with a red shift that of the cloud itself. Many such Lyman α absorption lines have been identified in the spectra of quasars.

The proximity effect is then the reduction in the number of Lyman α clouds near a quasar (Murdoch et al. 1986). This effect was first used by Bajtlik, Duncan and Ostriker (1988) to derive the background ionisation rate Γ per H atom for $1.7 < z < 3.8$ by attributing this proximity effect to the additional and known direct ionising influence of the quasar on nearby clouds. They obtained $\Gamma \sim 3 \times 10^{-12}\text{sec}^{-1}$, with an uncertainty of a factor 3 either way. They also found that, within their uncertainty, Γ did not vary significantly with z. Of the many later attempts to determine Γ and its possible variation with z the most recent are due to Giallongo et al. (1996), Lu et al. (1996), Savaglio et al. (1997) and Cooke, Epsey and Carswell (1997). Despite the considerable remaining uncertainties there is a consensus that $\Gamma \sim 2 \times 10^{-12}\text{sec}^{-1}$, with no significant z variation out to $z = 4.5$.

There has been much discussion in the literature as to whether the population of quasars alone can provide a sufficient ionising flux Γ_Q to account for Γ, and for the Gunn-Peterson type results. According to Haardt and Madau (1996) $\Gamma_Q \sim 10^{-12}\mathrm{sec}^{-1}$ at $z \sim 2$ but only 2×10^{-13} sec^{-1} at $z \sim 4.5$. The discrepancy of a factor ~ 10 at $z \sim 4.5$ appears to be significant, even if the one at $z \sim 2$ lies within the uncertainties.

It has often been suggested that any discrepancy between Γ and Γ_Q could be resolved by appealing to ionising radiation emitted by hot stars in galaxies (e.g. Miralda-Escudé and Ostriker 1990). Recent discussions of this possibility have been given by Giroux and Shapiro (1996), Madau and Shull (1996), Haiman and Loeb (1997), and Gnedin and Ostriker (1997). They used metal-enrichment arguments to show that if $\sim 25\%$ of the Lyman continuum photons emitted by hot stars escape from their parent galaxies, then these photons would be responsible for an ionisation rate $\sim 10^{-12}\mathrm{sec}^{-1}$ per H atom at $z \sim 3$. The actual escape fraction is not known at high z, but at $z \sim 0$ it is less than 1% (Deharveng et al. 1997). It is therefore suggestive that the missing ionising photons at $z \sim 4.5$ are produced by the cosmological distribution of decaying neutrinos and not by hot stars.

This assumption would enable us to understand in a general way why the proximity effect leads to a value of Γ which is approximately independent of z. If at $z \sim 2$ Γ_ν were of the same general order as Γ_Q then, as z increases, Γ_ν would increase as $(1+z)^{3/2}$ while Γ_Q decreases considerably, leaving Γ approximately constant.

A second method of estimating Γ_ν at $z \sim 2-3$ has become available thanks to the recent observations and analyses of HeII absorption at these red shifts (Jakobsen et al. 1994, Tytler et al. 1995, Jakobsen 1996, Davidsen, Kriss and Zheng 1996, Hogan , Anderson and Rugers 1997, Reimers et al. 1997, Zheng, Davidsen and Kriss 1997). This work shows that $N_{HeII}/N_{HI} \sim 100$ in Lyman α clouds at $z \sim 2-3$. This ratio can be related to the emission spectrum of quasars if one allows for the absorption due to intervening clouds. In particular it has been shown by Haardt and Madau (1998) that if quasars have the Zheng et al. (1997) power law spectrum with index $\alpha = -1.8$ between the HI and HeII edges, then the resulting radiation background at $z \sim 2-3$ would lead to the observed value of 100 for N_{HeII}/N_{HI}.

However, it has recently been pointed out that such an emission spectrum would not account (via excitation effects) for the observed strengths of the HeII emission lines in quasar spectra (Korista, Ferland and Baldwin 1997). These authors suggest that either the quasars have a suitably complicated geometry, or that their emission spectrum contains a significant bump in the vicinity of the HeII ionisation edge at 54.4 eV.

The existence of such a bump had previously been proposed by Sciama (1994) for a different reason and related to a Guilbert - Rees (1988) reprocessing of harder x-rays from the central regions of quasars by optically thick cold material. Since 1994 further observational evidence has accumulated for the prevalence of a soft x-ray excess in the spectra of quasars. This evidence has been reviewed by Gondhalekar, Rouillon-Foley and Kellett (1996). It is also noteworthy that a 50 eV bump would fit nicely in the gap (due to galactic absorption) in the composite spectrum shown in Fig. 6 of Laor et al. (1997).

On the theoretical side it has been found recently that a slim accretion disk around a black hole at the centre of a quasar would produce a soft x-ray excess without any Guilbert-Rees reprocessing (Shimura and Takahara 1995, Szuszkiewicz, Malkan and Abramowicz 1996, Hubeney and Hubeney 1997).

If we accept provisionally that the emissivity of quasars beyond the HeII edge has to be increased beyond the Zheng et al. (1997) value, perhaps by a factor ~ 3, then in order to avoid disturbing the Haardt-Madau (1998) analysis of $N_{\text{HeII}}/N_{\text{HI}}$ we must increase Γ at the HI edge by a similar factor. If we attribute this increase to Γ_ν, we find that $\Gamma_\nu \sim 2\Gamma_Q$ at $z \sim 2$, (which is compatible with our previous estimate for Γ_ν based on the observed approximate constancy of Γ with z).

If this value of Γ_ν at $z \sim 2$ is correct it implies a value of Γ_ν at $z \sim 0$ of 4×10^{-13} sec^{-1}. Since the decay photons have an energy close to the Lyman limit this ionisation rate converts to a photon flux by using the photoionisation cross-section at this limit, which is 6×10^{-18} cm^2. Hence one obtains $F(0) \sim 7 \times 10^4 \text{cm}^{-2}\, \text{sec}^{-1}$, which is compatible with the upper limit $\sim 10^5 \text{cm}^{-2}\, \text{sec}^{-1}$, derived by Vogel et al. (1995) and by Donahue, Aldering and Stock (1996).

We may summarise this discussion by saying that there is already some evidence of a need for additional hydrogen-ionising photons at $z \sim 2-5$, which could be due to decaying neutrinos. Future observations, particularly of HeII absorption in this red shift range, would be expected to clarify the situation.

6.5 The Angular Distribution of the Cosmic Microwave Background

One of the striking predictions of the decaying neutrino theory is that it should lead to suppression of the so-called Doppler peaks (White, Scott and Silk 1994) which are expected in the anisotropy of the cosmic microwave background when it is expressed as a function of the spherical harmonic order ℓ of the angular correlation function. This suppression arises from the early reionisation of the universe by decay photons after its hydrogen combined at red shift $\sim$ 1000. This reionisation gives rise to a thick Thomson scattering screen of free electrons stretching from $z \sim 500$ to $z \sim 0$. It is this scattering screen which would suppress the Doppler peaks.

The first and highest Doppler peak would occur in the conventional theory at $\ell \sim 200$, and there are preliminary signs in the existing data that this first peak does exist. If this is confirmed by future measurements and analysis the decaying neutrino theory would have to be abandoned. However the present indication of a peak is not yet statistically significant, and the problem of the subtraction of contributions from foreground radio sources remains to be completely analysed. The planned launches of the MAP and Planck surveyors, together with further ground-based observations, should lead to a solution of this problem.

6.6 The Transition Magnetic Moment of the Decaying Neutrino

In addition to radiative decay, there is another manifestation of an electromagnetic connection between two neutrino flavours, namely, the transition magnetic moment μ_{12}, which also has direct astronomical significance. To understand the meaning of this quantity it is helpful to recall that when a particle has an ordinary magnetic moment its spin will precess in the presence of a magnetic field B. If the particle ν_1 possesses a transition magnetic moment, then while it precesses in the magnetic field it picks up a component of $\bar{\nu}_2$ (Schechter and Valle 1981).

This process, called flavour changing spin precession, is reminiscent of oscillations between two neutrino flavours and has a similar formal structure. In particular it can be resonantly enhanced by the presence of matter (Akhmedov 1988, Lim and Marciano 1988) in a manner analogous to the MSW effect.

There is a simple relation between τ and μ_{12} which holds quite generally, and does not depend on the particular particle model used to calculate them (Marciano and Sanda 1977). This relation arises because both processes depend on the same first-order Feynman diagram, The relation is

$$\tau = \frac{8\pi}{\mu_{12}^2} \frac{1}{m_1^3} \left[1 - \frac{m_2^2}{m_1^2} \right]^{-3} .$$

In terms of the Bohr magneton $\mu_B = eh/2m_e c$ one can write this relation (assuming that $m_2/m_1 \ll 1$)

$$\tau = 10^{23} \left(\frac{30eV}{m_1} \right)^3 \left(\frac{10^{-14}\mu_B}{\mu_{12}} \right)^2 \text{sec} .$$

Thus a lifetime $\tau \sim 10^{23}$sec would be associated with a transition magnetic moment $\mu_{12} \sim 10^{-14}\mu_B$ if $m_1 \sim 30$ eV.

At least four astronomical applications have been made of a possible transition moment of ν_1.

(i) Cooling rates of stars. Plasma oscillations in the central regions of certain kinds of stars would create neutrino-anti-neutrino pairs via the transition moment. These neutrinos would then escape from the star leading to excess cooling.

To avoid too rapid a cooling rate one must impose the constraint

$$\mu_{12} \leq 3 \times 10^{-12} \mu_B \ .$$

(Raffelt 1997).

This constraint would be satisfied by the decaying neutrino theory.

(ii) Resonant spin flavour precession has been proposed as an alternative to the MSW effect to solve the solar neutrino problem (Akhmedov, Lanza and Petcov 1993). Whether a transition moment $\sim 10^{-14}\mu_B$ would suffice depends on as yet unknown details of the magnetic field distribution in the sun.

(iii) The regeneration of the stalled shock wave in supernova explosions. It was suggested by Fuller et al. (1992) that the MSW effect could give rise to a heating process in the environs of the shock which could lead to its regeneration. Again an alternative possibilty is to appeal to resonantly enhanced spin flavour precession (Akhmedov, Lanza, Petcov and Sciama 1997). One conclusion of this analysis was that for this mechanism to work with $\mu_{12} \sim 10^{-14}\mu_B$, a rather large but not unacceptable supernova magnetic field must be chosen.

(iv) The peculiar velocities of pulsars. Pulsars are observed to possess peculiar velocities of several hundred kilometres per second. This has been attributed to recoil of the pulsar during a nonspherically symmetric supernova explosion which led to the formation of the pulsar. Most of the momentum flux in the explosion is carried by neutrinos, and a non-zero transition moment of the neutrinos would lead to an asymmetric explosion (Akhmedov, Lanza and Sciama 1997).

6.7 The Search from Space for a Decay Line from Neutrinos near the Sun

On 21 April 1997 the first Spanish minisatellite (MINISAT-01) was launched with an ultra-violet detector (EURD) on board, one of whose aims is to search for the predicted decay line from neutrinos near the sun. Details of the detector are given by Bowyer et al. (1995a), Morales et al. (1997) and Bowyer, Edelstein and Lampton (1997).

In this sub-section we predict the intensity, the wavelength and the width of the decay line on the basis of the parameters which we have derived earlier in this article.

To determine the intensity of the line one must know the opacity of the medium surrounding the sun for photons just beyond the Lyman limit. In fact the sun is known to be immersed in a partially neutral hydrogen cloud, which is the central part of what is called the local interstellar medium (Cox and Reynolds 1987). The best determinations of the volume density n_{HI} of HI near the sun have been derived from Hubble Space Telescope observations of the Lyman α absorption line in the spectra of Procyon ($\ell = 214°$, $b = 13°$, $d = 3.5$pc) (Linsky et al. 1995) and of $\in$ Ind($\ell = 336°$, $b = -48°$, $d = 3.46$ pc) (Wood, Alexander and Linsky 1996). They obtained for the lines of sight to these two stars $n_{HI} = 0.1065 \pm 0.0028$ cm^{-3} and 0.094 ± 0.022 cm^{-3} respectively. The two stars lie in rather different directions, so it is comforting that they lead to the same value of n_{HI}, namely 0.1 cm^{-3}. The corresponding mean free path for a photon at the Lyman limit $\sim \frac{1}{2}$ pc, which is substantially less than the distances to the two stars. Accordingly the flux in the line at the sun, which is $n_\nu \ell/\tau$, would be 350^{+350}_{-117} cm^{-2} sec^{-1}, since $n_\nu = 4.2 \times 10^7$ cm^{-3} and $\tau = 2 \pm 1 \times 10^{23}$ sec. A possible variation of this flux with direction has been discussed by Bowyer et al. (1995b).

The wavelength λ of the line for our value for E_γ of 13.7 ± 0.1 eV is

$$\lambda = 905 \pm 7\text{Å} \ .$$

The error quoted is the uncertainty in the central wavelength, not the linewidth, which as we shall see is about 1Å. Our predicted wavelength falls close to the position of a much stronger airglow emission feature from the Earth's outer atmosphere (Chakrabarty 1984, Chakrabarty, Kimble and Bowyer 1984), but it may be possible nonetheless to distinguish the decay line if it is present (Bowyer 1997, personal communication).

Finally, we consider the expected width of the decay line. This width is due to the velocity dispersion v of the neutrinos producing the line. For a simple isotropic isothermal sphere model for the neutrino halo of our Galaxy one would have $c = \sqrt{\frac{3}{2}}\, v_{\rm rot}$, where the asymptotic rotation velocity $v_{\rm rot}$ of the Galaxy can be taken to be about 220 km sec^{-1} (Binney and Tremaine 1987). Thus $v \sim 270$ kmsec^{-1} and so $\Delta\lambda < 1$Å, which is much less than the wavelength resolution ($\sim$ 5Å) of EURD. However, recently Cowsik, Ratnam and Bhattacharjee (1996) have claimed to have constructed a self-consistent model of the dark matter halo of our Galaxy which requires v to lie between 600 and 900 km sec^{-1}. This claim has been challenged by Evans (1997), Gates, Kamionkowski and Turner (1997) and Bienaymé and Pichon (1997), and Cowsik et al. (1997) have replied to the first two criticisms.

We do not wish to enter into this controversy here, and merely note that if the decay line could be detected and its width measured, one would be able to deduce directly the velocity dispersion of the neutrinos. In this connection, it should be noted that in our required strongly flattened model for the neutrino halo, the velocity dispersion, and so the linewidth, would depend appreciably on direction. One could imagine measuring this anisotropic effect in a future mission with adequate wavelength resolution. In addition, if the neutrino halo itself has little or no rotation, one might be able to observe the Doppler effect associated with the sun's rotation in the Galaxy.

Acknowledgments

The author is grateful to Marek Abramowicz, Stuart Bowyer, Walter Dehnen, Piero Madau, Ron Reynolds and Ewa Szuszkiewicz for their crucial help. This work has been supported by the Italian Ministry of Universities and Scientific and Technological Research.

References

1. E.K. Akhmedov, *Phys. Lett.* B **213**, 64 (1988).
2. E.K. Akhmedov, A. Lanza and S.T. Petcov, *Phys. Lett.* B **303**, 85 (1993).
3. E.K. Akhmedov, A. Lanza, S.T. Petcov and D.W. Sciama, *Phys. Rev.* D **55**, 515 (1997).
4. E.K. Akhmedov, A. Lanza and D.W. Sciama, *Phys. Rev.* D **56**, 6117 (1997).
5. S.Bajtlik, R.C. Duncan and J.P Ostriker, *ApJ* **327**, 570 (1988).
6. J.F. Becquaert and F. Combes, *Astron. and Astrophys.* **325**, 41 (1977).
7. O. Bienaymé, A.C. Robin and M. Crézé, *Astron. and Astrophys.* **180**, 94 (1987).
8. O. Bienaymé and C. Pichon, astro-phys/9705210 (1997).
9. J.J. Binney and S. Tremaine, in *Galactic Dynamics* (Princeton University Press, Princeton, 1987).
10. H. Bloemen, *Ann. Rev. Astron. and Astrophys.* **27**, 469 (1989).
11. S. Bowyer, *Ann. Rev. Astron. and Astrophys.* 29, 59 (1991).
12. S. Bowyer, J. Edelstein, M. Lampton, L. Morales, J. Peres-Mercader and A. Gimenez, in *Astrophysics in the EUV* (S. Bowyer, R.F. Malina eds) Dordrecht Kluwer p. 611 (1995a).
13. S. Bowyer, M. Lampton, J.T. Peltoniemi and M. Roos, *Phys. Rev.* D **52**, 3214 (1995b).
14. S. Bowyer, J. Edelstein and M. Lampton, *ApJ* **485**, 523 (1997).
15. S. Chakrabarti, *Geophys. Res. Lett.* **11**, 565 (1984).

16. S. Chakrabarti, R. Kimble and S. Bowyer, *J. Geophys. Res.* **89**, 5660 (1984).
17. A.J. Cooke, B. Espey and R.F. Carswell, *MNRAS* **284**, 552 (1997).
18. R. Cowsik, *Phys. Rev. Lett.* **39**, 784 (1977).
19. R. Cowsik, C. Ratnam and P. Bhattacharjee, *Phys. Rev. Lett.* **76**, 3886 (1996).
20. R. Cowsik, C. Ratnam and P. Bhattacharjee, *Phys. Rev. Lett.* **78**, 2262 (1977).
21. D.P. Cox and R.J. Reynolds, *Ann. Rev. Astron. and Astrophys.* **25**, 303 (1987).
22. M. Crézé, E. Chereul, O. Bienaymé and C. Pichon, *Astron. and Astrophys.* **329**, 920 (1998).
23. J.M. Deharveng, S. Faisse, B. Milliard and V. Le Brun, astro-ph/9704146 (1997).
24. W. Dehnen and J.J. Binney, *MNRAS* in press (1997).
25. S. Dodelson and J.M. Jubas, *MNRAS* **266**, 886 (1994).
26. M. Donahue, G. Aldering and J.T. Stocke, *ApJ* **450**, L45 (1995).
27. J.B. Dove, J.M. Shull, *ApJ* **430**, 222 (1994).
28. N.W. Evans, *Phys. Rev. Lett.* **78**, 2260 (1997).
29. G.J. Ferland, B.M. Peterson, K. Horne, W.F. Welsh and S.N. Naher, *ApJ* **387**, 95 (1992).
30. G.M. Fuller, R. Mayle, B.S. Mayer and J.R. Wilson, *ApJ* **389**, 517 (1992).
31. F. Gabbiani, A. Masiero and D.W. Sciama, *Phys. Lett.* B **259**, 323 (1991).
32. E. Gates, M. Kamionkowski and M.S. Turner, *Phys. Rev. Lett.* **78**, 2261 (1997).
33. E. Giallongo, S. Cristiani, S. D'Odorico, A. Fontana and S. Savaglio, *ApJ* **466**, 46 (1996).
34. M.L. Giroux and P.R. Shapiro, *ApJ* **202**, 191 (1996).
35. N.Y. Gnedin and J.P. Ostriker, *ApJ* **486**, 581 (1997).
36. P.M. Gondhalekar, C. Rouillon-Foley and B.J. Kellett, *MNRAS* **282**, 17 (1996).
37. A. Gould, J.N. Bahcall and C. Flynn, *ApJ* **465**, 759 (1996).
38. P.W. Guilbert and M.J. Rees, *MNRAS* **233**, 475 (1988).
39. J.E. Gunn and B.A. Peterson, *ApJ* **142**, 1633 (1965).
40. F. Haardt and P. Madau, *ApJ* **461**, 20 (1996).
41. F. Haardt and P. Madau, in preparation (1998).
42. Z. Haiman and A. Loeb, *ApJ* **483**, 21 (1997).
43. R.C. Henry, *Ann. Rev. Astron and Astrophys.* **29**, 89 (1991).

44. C.J. Hogan, S.F. Anderson and M.H. Rugers, *Astron. J.* **113**, 1495 (1997).
45. I. Hubeney and V. Hubeney, *ApJ* **484**, L37 (1997).
46. P. Jakobsen, in *Science with the Hubble Space Telescope II*, Proc. of STScI/ST-ECF Workshop (P. Benvenuti, F.D. Macchetto, E.I. Schreier eds.) p. 153 (1996).
47. R. Kimble, S. Bowyer and P. Jakobsen, *Phys. Rev. Lett.* **46**, 80 (1981).
48. K. Korista, G. Ferland and J. Baldwin, *ApJ* **487**, 555 (1997).
49. E.J. Korpela, S. Bowyer and J. Edelstein, astroph/9709291 (1997).
50. A. Laor, F. Fiore, M. Elvis, B.J. Wilkes and J.C. McDowell, *ApJ* **477**, 93 (1997).
51. C.S. Lim and W.J. Marciano, *Phys. Rev.* D **37**, 1368 (1988).
52. J.L. Linsky, A. Diplas, A. Wood, B.E. Brown, T.R. Ayres and B.D. Savage, *ApJ* **451**, 335 (1995).
53. L. Lu, W.L.W. Sargent, D.S. Womble and M. Takad-Hidei, *ApJ* **472**, 509 (1996).
54. P. Madau and J.M. Shull, *ApJ* **457**, 551 (1996).
55. W. Marciano and A.I. Sanda, *Phys. Lett.* B **67**, 303 (1977).
56. W.W. Miller and D.P. Cox, *ApJ* **417**, 579 (1993).
57. J. Miralda-Escudé and J.P. Ostriker, *ApJ* **350**, 1 (1990).
58. C. Morales, J. Trapero, J.F Gomez., S. Bowyer, J. Edelstein and M. Lampton, in *The Local Bubble and Beyond* (D. Brechtschwardt, M. Freyberg eds.) Berlin, Springer in press (1997).
59. H.S. Murdoch, R.W. Hunstead, M. Pettini and J.C. Blades, *ApJ* **309**, 19 (1986).
60. R.P. Olling, *Astron. J.* **112**, 481 (1996).
61. R.P. Olling and M. Merrifield, astro-ph/9710224 (1997).
62. J.M. Overduin, P.S. Wesson and S. Bowyer, *ApJ* **404**, 1 (1993).
63. J.M. Overduin and P.S. Wesson, *ApJ* **483**, 77 (1997).
64. S.T. Petcov, *Sov. J. Nucl. Phys.* **25**, 340 (1997).
65. H.A. Pham, *PhD thesis*, Paris (1996).
66. H.A. Pham, in *Hipparcos Conference*, Venice 97 ESA SP-402 (M.A.C. Perryman ed.) (1997).
67. G. Raffelt, Stars as Laboratories for Fundamental Physics, Chicago University Press (1996).
68. D. Reimers, S. Köhler, L. Wisotzki, D. Groote, P. Rodriguez-Pascal and W. Wamsteker, *Astron. and Astrophys.* **327**, 890 (1997).
69. R.J. Reynolds, *ApJ* **294**, 256 (1985).
70. R.J. Reynolds, *ApJ* **330**, L29 (1989).
71. E. Roulet and D. Tommasini, *Phys. Lett.* B **256**, 218 (1991).

72. P.D. Sackett, H.W. Rix, B.J. Jarvis and K.C. Freeman, *ApJ* **436**, 629 (1994).
73. S. Savaglio, S. Cristiani, S. D'Odorico, A. Fontana and P. Molaro, *Astron. and Astrophys.* **318**, 347 (1997).
74. J. Schechter, J.W.F. Valle, *Phys. Rev.* D **24**, 1883 (1981).
75. D.W. Sciama, *ApJ* **364**, 549 (1990).
76. D.W. Sciama, in *The Early Observable Universe from Diffuse Backgrounds* (G. Rocca-Volmerange, J.M. Deharveng J. Tran Thanh Van eds) Editions Frontieres p. 127 (1991).
77. D.W. Sciama, Modern Cosmology and the Dark Matter Problem, Cambridge Univerity Press, Cambridge (1993).
78. D.W. Sciama, *ApJ* **426**, L65 (1994).
79. D.W. Sciama, *ApJ* **488**, 234 (1997a).
80. D.W. Sciama, astro-ph/9704081 (1997b).
81. D.W Sciama., *MNRAS* **289**, 945 (1997).
82. D. Scott, M.J. Rees and D.W. Sciama, *Astron. and Astrophys.* **250**, 295 (1991).
83. T. Shimura and F. Takahara, *ApJ* **440**, 610 (1995).
84. L. Spitzer and E.L. Fitzpatrick, *ApJ* **409**, 299 (1993).
85. F.W. Stecker, *Phys. Rev. Lett.* **45**, 1460 (1980).
86. E. Szuszkiewicz, in *Physics of Accretion Disks-Advances in Astronomy and Astrophysics* Vol. 2 (S. Kato, S. Minishige, J. Fukue eds.) Gordon and Breach Amsterdam p. 79 (1996).
87. E. Szuszkiewicz, M.A. Malkan and M. Abramowicz, *ApJ* **458**, 474 (1996).
88. D. Tytler, X.M. Fan, S. Burles, L. Cottrell, C. Davis, D. Kirkman and L. Zuo, in *QSO Absorption Lines* (G. Meylan ed.) Springer, Berlin p. 289 (1995).
89. S.N. Vogel, R. Weymann, M. Rauch and T Hamilton., *ApJ* **441**, 162 (1995).
90. M. White, D. Scott and J. Silk, *Ann. Rev. Astron and Astrophys.* **32**, 319 (1994).
91. A.N. Witt, B.C. Friedman and T.P. Sasseen, *ApJ* **481**, 809 (1997).
92. B.E. Wood, W.R. Alexander and J.L. Linsky, *ApJ* **470**, 1157 (1996).
93. W. Zheng, G.A. Kriss, R.C. Telfer, J.P. Grimes and A.F. Davidsen, *ApJ* **475**, 469 (1997).
94. W. Zheng, A.F. Davidsen and G.A. Kriss, astro-ph/9711143 (1997).

THE SUPERWORLD

NATHAN SEIBERG
School of Natural Sciences
Institute for Advanced Study
Princeton, NJ 08540, USA
E-mail: seiberg@sns.ias.edu

An overview of supersymmetry and its different applications is presented. We motivate supersymmetry in particle physics. We then explain how supersymmetry helps us analyze field theories exactly, and what dynamical lessons these solutions teach us. Finally, we describe how supersymmetry is used to derive exact results in string theory. These results have led to a revolution in our understanding of the theory.

1 Introduction

In this talk we will go on a guided tour through the superworld – the world of supersymmetric phenomena. We will explain what supersymmetry is and why many physicists expect to find it in the next generation of experiments. We will also show how powerful it is in leading to exact results in field theory and in string theory and how these results have revolutionized our understanding of string theory.

In section 2 we review the status of the standard model of particle physics, its underlying principles and its success. We also review some of the flaws in the standard model, which lead us to look for extensions of it. In section 3 we introduce supersymmetry as an extension of our ideas about the structure of space and time. In section 4 we explain why many physicist believe that supersymmetry is likely to exist at low energies (around TeV) and to be discovered soon.

In section 5 we turn to a different application of supersymmetry. It turns out that some aspects of supersymmetric field theories can be analyzed exactly. These are extremely complicated systems and the fact that they can be analyzed exactly is by itself surprising. More importantly, the exact solutions exhibit interesting phenomena with new lessons; among them is the crucial role played by electric-magnetic duality in the dynamics. The important applications of supersymmetry to mathematics will not be reviewed here.

In the final section we turn to string theory and show how using the magic of supersymmetry some nonperturbative information can be derived in string theory. This nonperturbative information has taught us many new facts, completely changing our perspective on the theory.

It is logically possible that string theory does not describe Nature and that supersymmetry will not be found in the TeV range. Then, only the applications in section 4 will survive. It is also possible that string theory describes Nature but supersymmetry is not present at low energies. Alternatively, it is also possible that string theory does not describe Nature but supersymmetry is found soon. My personal prejudice is that we should get the whole "package deal" including string theory and low energy supersymmetry. Clearly, the fact that supersymmetry naturally appears in one context makes it more likely that it also appears in another. It will be a shame if Nature does not use a beautiful and powerful idea like supersymmetry. However, as physicists, we should never forget that only experiments are the final judge about what constitutes a correct theory of Nature.

Many people have made crucial contributions to the developments of the subject. In order not to have a list of references longer than the text, we omit all references.

2 Review of the Standard Model

The standard model of particle physics is based on the following ingredients:

- The theory respects special relativity. In other words, space-time is invariant under the 3+1 dimensional Poincaré symmetry.

- The theory is based on the principles of quantum mechanics. It is generally believed that the only quantum theory which respects special relativity and is local (no action at a distance) and causal is local quantum field theory.

- The theory has local gauge symmetry. Unlike ordinary global symmetries (like isospin) gauge symmetry allows arbitrary symmetry transformation at different points in space-time. Therefore the symmetry group is really an infinite product of groups at different space-time points. Such a large symmetry group with arbitrary group element at different space-time points is familiar from general relativity and electrodynamics. The specific gauge group of the standard model is

  $$SU(3) \times SU(2) \times U(1).$$

  The gauge symmetry leads to gauge interactions which are mediated by gauge particles. For example, the electro-magnetic interactions are mediated by photons. Similarly, in the standard model we also have gluons and W and Z gauge bosons.

- The matter particles in the standard model are in a representation of $SU(3) \times SU(2) \times U(1)$. These include the quarks, leptons and Higgs boson. Of these only the Higgs boson has not yet been experimentally found (ν_τ has been "observed" only indirectly).

- The final ingredient of the standard model is its set of parameters. These include the masses of the various particles, the fine structure constant and a few others. Of these only one parameter, the Higgs mass, has not yet been measured.

The success of the standard model is unprecedented. It is a fully consistent theoretical theory. Furthermore, there are many experimental confirmations of the theory and there is no experiment which is manifestly in contradiction with it.

Despite this spectacular success, it cannot be over stressed that this is not "The End of Science." In particular, all the ingredients of the standard model are problematic:

- We included special relativity but did not include general relativity or gravity. In Nature space-time is dynamical and can be curved. In the standard model, which does not include gravity, space-time is static and provides a passive arena for the interactions.

- Trying to add gravity to the standard model and in particular to combine general relativity with quantum mechanics leads to contradictions. Therefore, we must go beyond the framework of local quantum field theory.

- Regarding the other ingredients of the standard model, we would like to understand why the standard model is as it is. Why is this the gauge symmetry? Why is this the particle spectrum? Why are these the values of the parameters?

- All the experimental tests of the standard model have been performed at energies smaller than a few hundred GeVs. Therefore, the standard model should be viewed as an effective field theory valid up to that energy scale. At higher energies it can be extended to another theory. What is this theory?

- The last point allows for the possibility that there is no new physics in the TeV range and new degrees of freedom show up only at much higher energy, say $M_{\text{Planck}} \sim 10^{19}_{\text{GeV}}$, where gravitational effects cannot

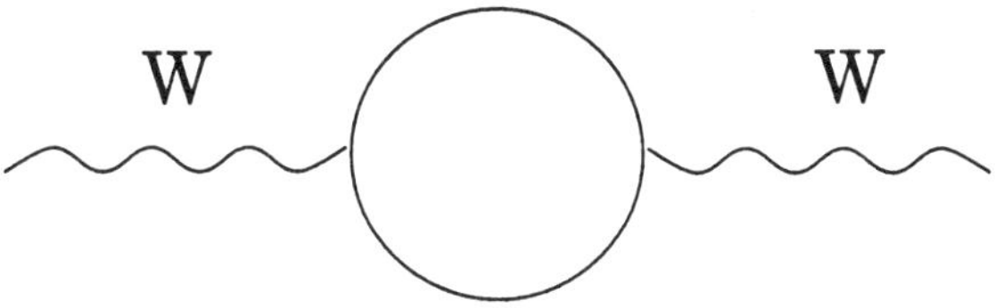

Figure 1: Loop diagram contributing to the mass of the W boson

be ignored and the theory must be modified. If this is the case, we face the hierarchy problem. This is essentially a problem of dimensional analysis. Why is the characteristic energy of the standard model, which is given by the mass of the W boson $M_{\mathrm{W}} \sim 100\mathrm{GeV}$ so much smaller than the next scale, M_{Planck}? It should be stressed that in quantum field theory this problem is not merely an aesthetic problem, but it is also a serious technical problem. Even if such a hierarchy is present in some approximation, radiative corrections tend to destroy it. More explicitly, loop diagrams like the loop of figure 1 restore dimensional analysis and move $M_{\mathrm{W}} \to M_{\mathrm{Planck}}$.

3 What is Supersymmetry?

Supersymmetry is a new kind of symmetry relating bosons and fermions. According to supersymmetry every fermion is accompanied by a bosonic superpartner. For example, the quarks which are fermions are accompanied by squarks which are bosons. Similarly, the gluons which are bosons are accompanied by gluinos which are fermions.

Another presentation of supersymmetry is based on the notion of *superspace.* We do not change the structure of space-time but we add structure to it. We start with four coordinates $X = t, x, y, z$ and add four odd directions θ_α $(\alpha = 1, \cdots, 4)$.

These odd directions are fermions

$$\theta_\alpha \theta_\beta = -\theta_\beta \theta_\alpha;$$

i.e. they are quantum dimensions and have no classical analog. Therefore, it is difficult to visualize or to understand them intuitively. However, they can be treated formally.

The fact that the odd directions are anticommuting has important consequences. Consider a function of superspace

$$\Phi(X, \theta) = \phi(X) + \theta_\alpha \psi_\alpha(X) + \cdots + \theta^4 F(X).$$

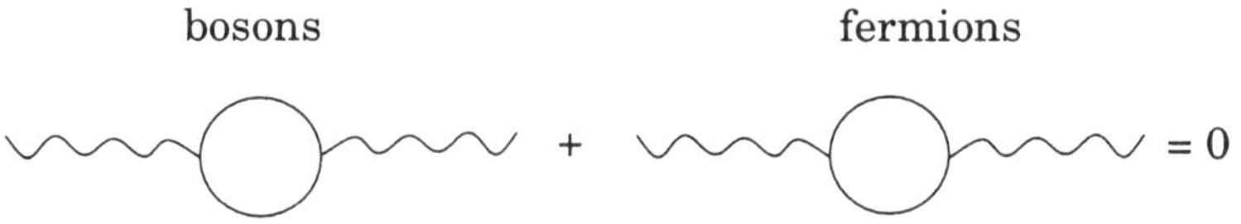

Figure 2: Boson-fermion cancellation in some loop diagrams

Since $\theta_\alpha \theta_\alpha = 0$, and since there are only four different θs, the expansion in powers of θ terminates at the fourth order. Therefore, a function of superspace includes only a finite number of functions of X.

Hence, we can replace any function of superspace $\Phi(X, \theta)$ with the component functions $\phi(X), \psi(X) \cdots$. These include bosons $\phi(X), \cdots$ and fermions $\psi(X), \cdots$. This facts relates this presentation of supersymmetry, which is based on superspace, and the one at the beginning of this section, which is based on pairing between bosons and fermions.

A supersymmetric theory looks like an ordinary theory with degrees of freedom and interactions satisfying some symmetry requirements. A supersymmetric theory is a special case of more generic theory rather than being a totally different kind of theory.

The fact that bosons and fermions come in pairs in supersymmetric theories has important consequences. In some loop diagrams, like in figure 2, the bosons and the fermions cancel each other.

This boson-fermion cancellation is at the heart of most of the applications of supersymmetry. If supersymmetry is present in the TeV range, this cancellation solves the gauge hierarchy problem. Also, this cancellation is one of the underlying reasons for being able to analyze supersymmetric theories exactly.

4 Supersymmetry in the TeV Range?

There are several motivations for assuming that supersymmetry is realized in the TeV range. That means that the superpartners of all the particles of the standard model have masses of the order of a few TeV or less.

- The main motivation is a solution of the hierarchy problem. As we mentioned in the previous section in supersymmetric theories some loop diagrams vanish due to cancellations between bosons and fermions. In particular the loop diagram restoring dimensional analysis (figure 1) is cancelled as in figure 2. Therefore, in its simplest form supersymmetry solves the technical aspects of the hierarchy problem. More sophisticated ideas, known as dynamical supersymmetry breaking, also solve the aesthetic problem.

- The second motivation for low energy supersymmetry comes from the idea of gauge unification. Recent experiments have yielded precise determinations of the strength of the $SU(3) \times SU(2) \times U(1)$ gauge interactions – the analog of the fine structure constant for these interactions. They are usually denoted by α_3, α_2 and α_1 for the three factors in $SU(3) \times SU(2) \times U(1)$. In quantum field theory these values depend on the energy at which they are measured; i.e. these coupling constants run. The rates of change of these coupling constants depend on the particle content of the theory. Using the measured values of the coupling constants and the particle content of the standard model, we can extrapolate to higher energies and determine the coupling constants there. The result is that the three coupling constants do not meet at the same point. However, repeating this extrapolation with the particles of the standard model and their superpartners the three gauge coupling constants meet at a point (see figure 3). How much weight should we assign to this result? Two lines must meet at a point. Therefore, there are only two surprises here. First, the third line meets them at the same point – there is only one non-trivial number here. Second, which is more qualitative, the meeting point, the unification energy, is at a "reasonable value of the energy." My personal view is that this is far from a proof of low energy supersymmetry but it is certainly encouraging circumstantial evidence.

- The next generation of experiments at Fermilab and CERN will explore the energy range where at least some of the superpartners are expected to be found. Therefore, in a few years we will know whether supersymmetry exists at low energies. If supersymmetry is discovered in the TeV range, the parameters of the superpartners like their masses and coupling constants will also be measured. These numbers will be extremely interesting as they will give us a window into the physics of higher energies.

- Finally we should point out that some of these superpartners might also be relevant for the dark matter of the Universe.

If supersymmetry is indeed discovered in the TeV range, this will amount to the discovery of the new odd dimensions. This will be a major change in our view of space and time, comparable to and perhaps bigger than the discovery of parity violation. It should be stressed that at the moment supersymmetry does not have a solid experimental motivation. If it is discovered, this will

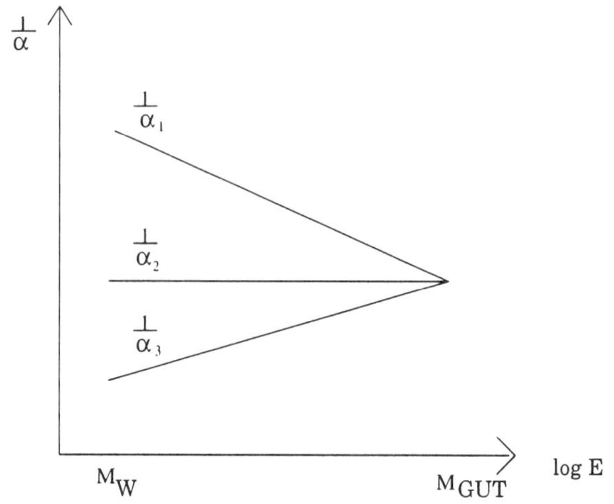

Figure 3: Coupling constant unification in supersymmetric theories

be one of the biggest successes of theoretical physics – predicting such a deep notion without any experimental input!

5 Exact Results in Quantum Field Theory

Quantum field theory is notoriously complicated. It is a non-linear system of an infinite number of coupled degrees of freedom. Therefore, (except in two dimensions) there are only a few exact results in quantum field theory. However, the special quantum field theories which are supersymmetric can be analyzed exactly!

The main point is that these systems are very constrained. The dependence of some observables on the parameters of the problem is so constrained that there is only one solution which satisfies all the consistency conditions. More technically, some observables vary holomorphically (complex analytically) with the coupling constants which are complex numbers. Due to Cauchy's theorem, such analytic functions are determined in terms of very little data: the singularities and the asymptotic behavior. Therefore, if supersymmetry requires an observable to depend holomorphically on the parameters and we know the singularities and the asymptotic behavior, we can determine the exact answer. The boson-fermion cancellation, which we mentioned above in the context of the hierarchy problem, can also be understood as a consequence of a constraint following from holomorphy.

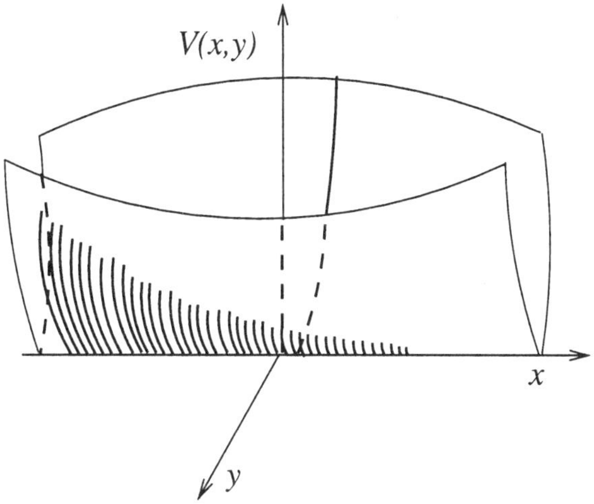

Figure 4: Typical potential in supersymmetric theories exhibiting "accidental vacuum degeneracy"

Another property of supersymmetric theories makes them tractable. They have a family of inequivalent vacua. To understand this fact recall first the situation in a magnet. It has a continuum of vacua, labeled by the orientation of the spins. These vacua are all equivalent; i.e. the physical observables in one of these vacua are exactly the same as in any other. The reason is that these vacua are related by a symmetry and the phenomenon of many vacua leads to spontaneous symmetry breaking.

We now study a situation with inequivalent vacua. Consider the case of degrees of freedom with the potential $V(x, y)$ in figure 4. The vacua of the system correspond to the different points along the valley of the potential, $y = 0$ with arbitrary x. However, as we tried to make clear in figure 4, these points are inequivalent – there is no symmetry which relates them. More explicitly, the potential is shallow around the origin but becomes steep for large x. Such "accidental degeneracy" is usually lifted by quantum effects. For example, if the system corresponding to the potential in the figure had no fermions, the zero point fluctuations around the different vacua would have been different. They would have led to a potential along the valley pushing the minimum to the origin.

However, in a supersymmetric theory the zero point energy of the fermions exactly cancels that of the bosons and the degeneracy is not lifted. The valleys persist in the full quantum theory. Again, we see the power of the boson-fermion cancellation.

We see that a supersymmetric system typically has a continuous family of vacua. It is referred to as *moduli space of vacua*, and the modes of the system corresponding to motion along the valleys are called moduli.

The analysis of the systems is usually simplified by the presence of these manifolds of vacua. Asymptotically, far along these flat directions of the potential the analysis of the system is simple and various approximate techniques are applicable. Then by using the asymptotic behavior along several such asymptotics as well as the constraints from holomorphy the solution is unique.

This is a rather unusual situation in physics. We perform approximate calculations which are valid only in some regime and this gives us the exact answer. *This is a theorist's heaven – exact results with approximate methods!*

Once we know how to solve these system, we can analyze many examples. The main lesson which was learned is the fundamental role played by *electric-magnetic duality*. It turns out to be the underlying principle controlling the dynamics of these systems.

When faced with a complicated system with many coupled degrees of freedom it is common in physics to look for weakly coupled variables which capture most of the phenomena. For example, in condensed matter physics we formulate the problem at short distance in terms of interacting electrons and nuclei. The solution is the macroscopic behavior of the matter and its possible phases. It is described by weakly coupled effective degrees of freedom. Usually they are related in a complicated, and in most cases unknown, way to the microscopic variables. Another example is hydrodynamics, where the microscopic degrees of freedom are molecules and the long distance variables are properties of a fluid which are described by partial differential equations.

In one class of supersymmetric field theories the situation is similar to that. The long distance behavior is described by a set of weakly coupled degrees of freedom. As the characteristic length scale becomes longer, the interactions between these effective degrees of freedom become weaker, and the description in terms of them becomes more accurate.

In another class of examples there are no variables in terms of which the long distance theory is simple. The theory remains interacting (it is in a non-trivial fixed point of the renormalization point). In these situations there are two (in some cases more than two) descriptions of the physics leading to identical results for the long distance interacting behavior.

In both classes of examples an explicit relation between the two sets of variables is not known. However, there are several reasons to consider these pairs of descriptions as being electric-magnetic duals of one another. The original variables at short distance are referred to as the electric degrees of freedom. The other set of variables are magnetic.

These two dual descriptions of the same theory give us a way to address strong coupling problems. When the electric variables are strongly coupled, they fluctuate rapidly and their dynamics is complicated. However, the magnetic degrees of freedom are weakly coupled. They do not fluctuate rapidly and their dynamics is simple. In the first class of examples the magnetic degrees of freedom are the macroscopic ones. They are massless bound states of the elementary particles. In the second class of examples there are two valid descriptions of the long distance theory: the electric and the magnetic ones. As one of them becomes more strongly coupled, the other becomes more weakly coupled.

Finally, using this electric-magnetic duality we can find a simple description of complicated phenomena associated with the phase diagram of the theories. For example, as the electric degrees of freedom become strongly coupled, they can lead to confinement. In the magnetic variables, this is simply the Higgs phenomenon (superconductivity) which is easily understood in weak coupling.

We summarize the electric-magnetic relations in the following table:

	electric	magnetic
coupling	strong	weak
fluctuations	large	small
phase	confinement	Higgs

Apart from the "practical" application to solving quantum field theories, the fact that a theory can be described either in terms of electric or magnetic variables has deep consequences:

- In theories of the first class of examples it is natural to describe the magnetic degrees of freedom as composites of the elementary electric ones. The magnetic particles typically include massless gauge particles reflecting a new magnetic gauge symmetry. These massless composite gauge particles are associated with a gauge symmetry which is not present in the fundamental electric theory. This is rather surprising because most people believed that such a phenomenon cannot take place in four dimensions. The lesson from these examples is that *gauge invariance cannot be fundamental.*

- For theories of the second kind the notion of elementary particle breaks down. There is no invariant meaning to which degrees of freedom are elementary and which are composite. The magnetic degrees of freedom are composites of electric ones and vice versa. Again, such behavior is very surprising in four dimensions.

6 The String Revolution

We do not know how to formulate string theory nor do we know its underlying principles. Surprisingly, this fact does not stop us from making progress. In particular, as in field theory, the magic of supersymmetry allows us to obtain some exact results and to control the theory in extreme situations. These results have completely changed our perspective on the theory. In the remainder of this talk we will briefly mention some of the main lessons:

- Just as in supersymmetric field theories, string theory has many inequivalent vacua – a moduli space of vacua. It turns out that the supersymmetric compactifications of all five string theories are connected. A "map" of these vacua is given in figure 5. At different boundaries of the map we find the five known string theories as well as the mysterious eleven-dimensional theory whose low energy limit is eleven-dimensional supergravity. Without the magic of supersymmetry only the vicinity of each boundary could be explored in perturbation theory and there was no way to extrapolate from one boundary to another. Now, with these extrapolations, it is clear that all the vacua are connected. We conclude that instead of five string theories there is only one theory with many solutions. *The theory is unique!*

- As we extrapolate from one boundary to another a phenomenon, which we have already discussed in the previous section, takes place. The "elementary" degrees of freedom at one boundary appear composite elsewhere. There is no universal object which appears elementary everywhere and should therefore be viewed as preferred. Furthermore, in the boundary where the theory becomes eleven-dimensional there are no strings at all. We conclude that *the theory is not a theory of strings.* Therefore it is appropriate to change its name and it is often being referred to as *M-theory* (M stands for magic, mysterious, membrane, mother, ...).

- At various boundaries of the map in figure 5 there is a preferred notion of space-time. However, as we extrapolate from one boundary to another, the underlying space-time becomes ambiguous. The theory can be described either as one kind of strings propagating on one background or as

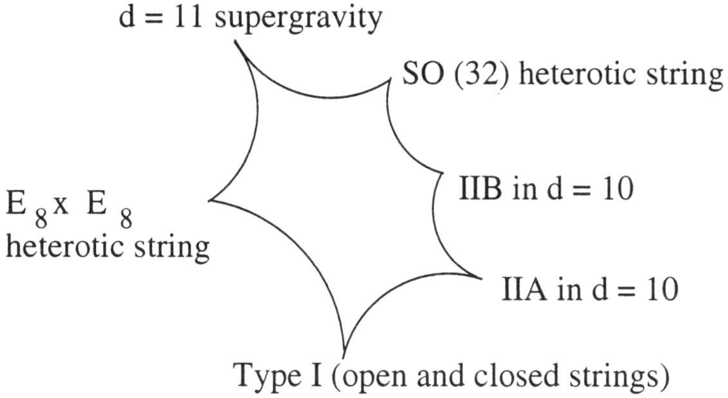

Figure 5: "Map" of some of the vacua of string theory

another kind of strings propagating on another background. This ambiguity is known as *string duality*. It has led to a proposal to formulate the full theory in terms of the dynamics of large matrices – the coordinates of space-time are non-commuting matrices in this approach.

- The map in figure 5 includes the value of a parameter which can loosely be called $\hbar$. As we approach various boundaries we seem to take it to zero or infinity. However, a more careful examination of the theory shows that even as we set $\hbar \to 0$ the theory still includes sectors, which remain quantum mechanical. Furthermore, in the eleven-dimensional vacuum there is no parameter like $\hbar$. We see that there is no classical theory whose quantization leads to string theory. Instead, *the theory is inherently quantum mechanical!*

- Certain black-hole solutions of string theory where examined. Using the magic of supersymmetry an extrapolation from weak coupling to strong coupling can be performed and one can exactly enumerate the black-hole states. It turns out to coincide with the number predicted by the Bekenstein-Hawking entropy formula. Therefore, the black-hole entropy reflects the existence of many microscopic states. This is a crucial step toward resolving the black-hole information paradox. It points in the direction that the full theory is unitary and no information is being lost in Hawking radiation.

Unfortunately, these exciting developments have not yet led to direct comparison with experiment. The situations where exact answers are possible are very idealized and have a lot of supersymmetry – even more than the amount of supersymmetry we expect to find in the TeV range. Even worse, before these developments one could have hoped that the ten or eleven-dimensional vacua are somehow inconsistent. Now, they appear perfectly consistent and are unified into a beautiful picture. Therefore, the question "why don't we live in ten or eleven dimensions?" becomes sharper.

However, these developments are an enormous step toward uncovering the underlying dynamical principles of string theory.

Acknowledgments

This work was supported in part by DOE grants #DE-FG02-90ER40542. IASSNS-HEP-98/16. Based on talks given at various conferences and will appear in various proceedings.

ASPECTS OF SIX DIMENSIONAL SUPERSYMMETRIC THEORIES

S. RANDJBAR-DAEMI
The Abdus Salam International Centre for Theoretical Physics, Trieste, Italy
E-mail: daemi@ictp.trieste.it

In this contribution some aspects of supergravity and super Yang-Mills systems in $D = 6$ are briefly reviewed and, in some cases, are contrasted with the analogous features in $D = 4$. Particular emphasis is laid on the stringy solutions of the $D = 6$ super Yang-Mills systems.

1 Introduction

I was indeed very privileged to be for many years a close associate of Abdus Salam. I have learned many things from him. Salam combined the vigorous western thought in a coherent manner with his oriental culture. He believed deeply that the social life of an individual has a sense and purpose only in relationship with those of others. It becomes richer and more purposeful if its guiding principles are compassion and tolerance. He himself was a proverbially generous person.

Salam deeply appreciated the relevance of Science to the enrichment of human life. He spent a major part of his active life to disseminate scientific knowledge among the less privileged nations. Being a singularity as he was, he also contributed substantially to the advancement of the fundamental sciences. In fact the very best existing theory of Nature, the Standard Model of Particle Physics, bears his name.

My scientific collaboration with Abdus Salam started with a study of theories of Kaluza - Klein type in a space time of six dimensions.[1] I have therefore chosen to review in this memorial contribution some of the recent developments in 6-dimensional theories. The presentation will be mostly, but not always, non technical and elementary.

2 Particles and Strings in D=4 and D=6

Physical theories in a six dimensional manifold of Lorentzian signature differ in many respects from the four dimensional theories. In $D = 6$ in addition to spinors, scalars, vectors, and second rank symmetric tensors, which are the basic objects of 4-dimensional field theories, we also have second rank antisymmetric tensor potentials. Also the fact that the fundamental spinor representation of $SO(1,5)$ is pseudo real, as opposed to the complex Weyl spinors of $SO(1,3)$, has some significance in constructing anomaly free models in D=6. We shall start with a summary of differential forms and the extended objects to which they couple.[2]

In $D = 4$ the only interesting forms are the 1-forms and their exterior derivatives which correspond to, respectively, a Maxwell (or Yang-Mills) potential A and its field strength F. Being a 2 form, F admits a dual, ${}^*F$, which is also a 2 form. Maxwell's equations are essentially symmetric under the exchange of F and ${}^*F$. This is called the electromagnetic duality under which the electric and magnetic charges interchange their roles.[3] This type of duality is the prototype of a larger class of duality symmetries which can occur in space times of higher than 4 dimensions. Note that in D dimensions the dual of a p form is a $D-p$ form. It is thus only for $D = 4$ that the dual of the electromagnetic 2 form F is again a 2 form. To appreciate the physical significance of this simple fact let us recall that the electromagnetic potential A couples to particles through the term $\int_C A$ where C denotes the world line of the charged particle. If ${}^*F$ is derived from a dual potential $\tilde{A}$ then there will be a dual particle which could couple to $\tilde{A}$ through $\int_{\tilde{C}} \tilde{A}$, where $\tilde{C}$ is the world line of the dual particle. In D dimensions a $p+1$ form potential couples naturally to a p dimensional extended object, called a p-brane. This coupling is a direct generalization of the electromagnetic coupling, namely, $\int_{\Sigma_{p+1}} A$, where Σ_{p+1} denotes the $p+1$ dimensional world volume of the p dimensional extended object. Note that a p brane will occupy a p dimensional subspace of the D dimensional space time. For example, we can think of a p dimensional hyperplane extended along p of the $D-1$ space coordinates. Therefore, in the remaining $D-1-p$ space dimensions our object will look like a point, which we call the position of the p brane. A large sphere around this position will have $D-2-p$ dimensions, which equals to the rank of the dual of the $p+1$ form $F = dA$. We can thus integrate ${}^*F$ over this large sphere and call it the electric charge of the p brane. On the other hand the integral of the $p+2$ form F over a $p+2$ dimensional sphere is called the magnetic charge of the considered p brane. These are direct extensions of the well known definitions in $D = 4$. Thus in D space time dimensions the dual of a p brane is an extended object with $D-p-4$ space dimensions, i.e. a $D-p-4$ brane.

With the above definitions, in $D = 6$ the dual of a particle is a 2 brane, while the dual of a 1 brane is again a 1 brane. This indicates that in $D = 6$ one dimensionally extended objects, namely, strings, and the two form potentials to which they couple, play a role analogous to the role of particles and vector potentials in $D = 4$.

To mention other interesting differences let us impose the condition $F ={}^*$

F. This condition is meaningful only for a 2 form F in $D = 4$ and a 3 form F in $D = 6$. Thus in $D = 4$, if the self-duality equation had a real solution, there could exist self dual particles. However, it turns out that the self duality equation in $D = 4$ has interesting solutions only if the 4 dimensional manifold is Euclidean and the gauge group is non Abelian. These solutions are called instantons. They are localised finite action solutions of the 4 dimensional Euclidean Yang-Mills equations. The value of their Euclidean action, appropriately normalised, equals to their topological charge. These solutions do not have particle type interpretation in 4 dimensional Minkowski space time. In contrast with $D = 4$, the self duality equation $F =^* F$ has a solution for an Abelian 3 form F in $D = 6$ Minkowski space-time. These are self-dual strings.[4] Such strings will carry both electric and magnetic charges and their magnitude will be equal. The non Abelian version of higher rank forms are not yet known.

One may wonder what type of interpretation the ordinary Yang-Mills instantons can have in $D = 6$. To answer this question we need to consider a non Abelian gauge theory in a 6-dimensional space time. One can consider a 4-dimensional Euclidean subspace and an instanton configuration localised in this subspace. From the point of view of the 6-dimensional space time this object will look like a string with a thickness given by the size of the instanton. As the size goes to zero it will look more and more like a fundamental string.[5] We shall make use of this interpretation in section four.

3 Supersymmetry and Chiral Anomalies

The possibility of having self dual or anti-self dual gauge fields in $D = 6$ make the six dimensional supersymmetric theories more akin to D=10 theories rather than their D=4 counterparts. The conditions for the cancellation of chiral anomalies are also more stringent in $D = 6$. For example, there are no pure gravitational anomalies in $D = 4$ theories. Such anomalies can exist, however, in $D = 6$. The requirement that the pure gravitational anomalies do cancel imposes restrictions on the supermultiplet structure of the six dimensional supergravity theories.

A model with a minimum number of supersymmetries in $D = 6$ has four independent complex supercharges which can be assembled into a Weyl spinor of $SO(1,5)$. It is customary to double the number of components and impose a symplectic Majorana condition. For this reason such models are sometimes called $(2,0)$. We shall refer to them as $(1,0)$. These models like their extended versions with 16 real supercharges which we shall denote as $(2,0)$ are chiral, while their $(1,1)$ and $(2,2)$ theories, which have respectively 16 and 32 supercharges, are non chiral.

In section 6 we shall give a little more detail about the models with more than 8 real supercharges and explain briefly how they can be obtained from the $D = 10$ superstring theories upon compactification on a four manifold. In the rest of this section we shall exclusively discuss theories with $(1,0)$ supersymmetry.

The $(1,0)$ models admit the following type of super multiplets:

1) gravity: $E_a^m, \Psi_\mu, B_{ab}^-$
2) hypermatter : ψ^r, ϕ^α
3) Yang-Mills: λ, A_a, Y
4) Tensor : χ, B_{ab}^+, σ

The spinors in the gravity and the Yang-Mills multiplets are left handed, while those in the tensor and hypermultiplets are right handed with respect to $SO(1,5)$. Furthermore, the spinors in the gravity, Yang-Mills and the tensor multiplets are doublets of an automorphism $Sp(1) = SU(2)$, and they are Majorana symplectic in the sense that they satisfy a constraint of the type $\psi = \Omega\psi_c$ where Ω is the $Sp(1)$ invariant metric and ψ_c is the charge conjugate of ψ. The superscripts $\pm$ on the antisymmetric tensor potentials indicate that their field strengths are self dual (+) or anti-self dual (-). The scalars in the hypermultiplets are the $2n$ complex coordinates of a quaternionic manifold. Thus $\alpha = 1, ..., 2n$ and the index r in the hypermatter fermions, $r = 1, ..., n$, where, n counts the number of the hypermatter multiplets. Y in the Yang-Mills multiplet is an auxiliary field. It is a triplet of the automorphism $Sp(1)$.

In any $D = 6$ coupled Yang-Mills supergravity theory, the condition for the cancellation of pure gravitational anomalies imposes a restriction on the number of multiplets listed above. This condition is [6]

$$n = m + 273 - 29k$$

where m, n and k are respectively the number of Yang-Mills, hyper and tensor multiplets.

Anomaly free supergravity models with $(1,0)$ supersymmetry in $D = 6$ can be obtained from the compactifications of the ten dimensional heterotic strings on a particular class of complex manifolds called K_3.[7]

Having cancelled the pure gravitational anomalies the remaining anomalies can be cancelled with the help of a mechanism discovered by Green and Schwarz in the context of ten dimensional superstring models [8] and extended to the six dimensional models in refs.6,7.

Let us consider models with $k = 1$. For such models one can construct invariant supergravity actions. The anomaly condition simplifies to $n = m + 244$. The Green-Schwarz anomaly cancellation mechanism starts from an 8 form $P_8 = X_4.\tilde{X}_4$, where "." indicates a wedge product. The 4 forms X_4 and $\tilde{X}_4$ have the following general structures

$$X_4 = trR^2 - \Sigma v_\alpha trF_\alpha^2$$

$$\tilde{X}_4 = trR^2 - \Sigma \tilde{v}_\alpha trF_\alpha^2$$

where v_α and $\tilde{v}_\alpha$ are numerical constants and F_α is the field strength 2 form associated with the gauge group G_α. Note that both X_4 and $\tilde{X}_4$ are closed. Locally we can write $X_4 = d\Omega_3$, where Ω_3 is a Chern-Simons three form. It is not gauge invariant. Since X_4 is gauge invariant, under a gauge transformation, we need to have $\Omega_3 \to \Omega_3 + d\alpha$. The 2 -form α can be constructed explicitly.[9] Up to total derivatives, the gauge and gravitational chiral anomaly is proportional to the integral of the 6 form $\alpha.\tilde{X}_4$ over the Euclideanised six-dimensional space time.

The GS mechanism requires the addition of a local counter term of the form $B.\tilde{X}_4$ to the effective Lagrangian. If we demand that the two form potential B undergoes a gauge transformation of the form $B \to B - \alpha$ then the one loop effective action will be gauge invariant. A gauge invariant field strength associated with the 2 form potential B should thus be defined by $H = dB + \Omega_3$.

All this is very similar to the application of the Green-Schwarz mechanism to the anomaly problem in $D = 10$ heterotic string theory. In that case the anomaly polynomial is a 12 form which factorises as $P_12 = X_4.X_8$, where X_4 and X_8 are closed forms and furthermore $X_4 = \frac{1}{30}TrF^2 - trR^2$. Here Tr and tr refer, respectively, to the adjoint and the fundamental representations.

The modified definition of H has very interesting consequences for K_3 compactifications. First note that

$$dH = trR^2 - \Sigma v_\alpha trF_\alpha^2$$

Integrate this expression on K_3. If dH has no δ-function type singularities on K_3 its integral will vanish. We then obtain $24 = \int \Sigma v_\alpha trF_\alpha^2$, where 24 is the Euler number of K_3. The integral on the right-hand side of this expression is the Chern number (instanton number) associated with the background gauge field configuration. For the $E_8 \times E_8$ heterotic string it will equal to the sum of instanton numbers embedded in each E_8 factor. We thus obtain $24 = n_1 + n_2$.

If dH has singularities they can be interpreted as 5 branes and the result of integration becomes $24 = n_1 + n_2 + n_5$ for the $E_8 \times E_8$ and $24 = n + n_5$ for the $SO(32)$ heterotic strings, where n_5 indicates the number of 5 branes.

In the above discussion we considered only models with one tensor multiplet. A more general framework has been developed in ref.10 for models with $k > 1$.

In addition to anomalies in local symmetries there is also the possibility of global anomalies in the supergravity and super Yang-Mills theories in $D = 6$. Such anomalies exist in $D = 4$ theories if the fourth homotopy group, Π_4, of the gauge group is non trivial.[11] If Π_6 of the gauge group is non trivial, a super Yang-Mills theory in $D = 6$ can be inconsistent due to the presence of global gauge anomalies. This happens for the groups $SU(2)$, $SU(3)$ and G_2. The requirement of the absence of such anomalies imposes further restrictions on the structure of the consistent six dimensional models. We shall return to this issue in section 5.

Very interesting physics can be extracted by studying the moduli space of scalars. We saw above that there are two types of scalars in the spectrum of $(1, 0)$ models, namely, those in the hypermultiplets and those in the tensor multiplets. Seiberg and Witten parametrise the Coulomb branch of the $(1, 0)$ theories by the expectation values of the scalars in the tensor multiplets.[12] The reason for naming the tensor moduli space as the Coulomb branch is that upon compactification to lower dimensions the tensor multiplets become vector multiplets of the lower dimensional theories.

The metric in the hypermultiplet moduli space is independent of the scalars in the tensor multiplets. Likewise the metric in the Coulomb branch and the kinetic energy terms for the vector and the tensor fields are independent of the scalars in the hypermultiplets. In the infrared region, where the physics is described by classical field theory, there is no supersymmetric coupling which could lead to masses for massless particles by wondering in the moduli space of scalars in the tensor multiplets. Conversely, it is impossible for massive particles to become massless at a particular point on the Coulomb branch, by any mechanism that can be described at low energies by a free field theory. Seiberg and Witten then conclude that the singularities in the Coulomb branch necessarily involve non infrared freephysics which is associated with non-critical tensionless strings, where non-critical means that gravity is not a mode of the string. Thus the non trivial infrared physics associated with the Coulomb branch singularities is occurring in flat six dimensional Minkowski space time. In the next section we shall study a flat space model which exhibits some of these features.

4 Super Yang-Mills coupled to tensors in D=6 and non critical Strings

As it was argued in the preceding section the singularities in the Coulomb branch of $(1, 0)$ models are conjectured to be related to non critical strings which do not couple to gravity and therefore they can be studied in a flat

Minkowski space time. Supersymmetric models in flat $D = 6$ which involve only the Yang-Mills and the tensor multiplets have been constructed in ref.13 In addition to the argument given at the end of the last section, the introduction of tensor multiplets is called for also by the requirement of anomaly cancellation, because, it is the presence of such tensor fields in six dimensions which enable us to make use of the Green-Schwarz anomaly cancellation mechanism. The tensor multiplet by itself does not have nonsingular solitonic solutions. This is one more reason for considering the tensor multiplet coupled to Yang-Mills fields. [a]

The supersymmetry transformations of the fields in our system are as follows.[13]

$$\begin{aligned} \delta A_a &= -\bar{\epsilon}_i\, \Gamma_a \lambda^i \\ \delta\lambda^i &= \frac{1}{8}\Gamma^{ab}F_{ab}\epsilon^i - \frac{1}{2}Y^{ij}\epsilon_j \\ \delta Y^{ij} &= -\bar{\epsilon}^{(i}\, \Gamma^a D_a \lambda^{j)} \end{aligned} \tag{1}$$

where the index i is a doublet index of the automorphism $Sp(1)$. The corresponding rules for the tensor multiplet coupled to Yang–Mills are given by

$$\begin{aligned} \delta\sigma &= \bar{\epsilon}\chi \\ \delta\chi^i &= \frac{1}{48}\Gamma^{abc}H^{+}_{abc}\epsilon^i + \frac{1}{4}\Gamma^a\partial_a\sigma\epsilon^i - \frac{\alpha'}{4}\mathrm{Tr}\,\Gamma^a\lambda^i\bar{\epsilon}\Gamma_a\lambda \\ \delta B_{ab} &= -\bar{\epsilon}\Gamma_{ab}\chi - \alpha'\mathrm{Tr}\,A_{[a}\bar{\epsilon}\Gamma_{b]}\lambda \end{aligned} \tag{2}$$

where

$$\begin{aligned} H_{abc} &= 3\partial_{[a}B_{bc]} + 3\alpha'\mathrm{Tr}\left(A_{[a}\partial_b A_{c]} + \tfrac{2}{3}A_a A_b A_c\right) \\ H^{\pm}_{abc} &= \frac{1}{2}\left(H_{abc} \pm \tilde{H}_{abc}\right). \end{aligned} \tag{3}$$

The closure of the supersymmetry algebra leads to the following field equations for various fields,

$$H^{-}_{abc} = -\frac{\alpha'}{2}\mathrm{Tr}\,(\bar{\lambda}\Gamma_{abc}\lambda) \tag{4a}$$

$$\Gamma^a\partial_a\chi^i = \alpha'\mathrm{Tr}\left(\frac{1}{4}\Gamma^{ab}F_{ab}\lambda^i + Y^{ij}\lambda_j\right) \tag{4b}$$

$$\partial^2\sigma = \alpha'\mathrm{Tr}\left(-\frac{1}{4}F^{ab}F_{ab} - 2\bar{\lambda}\Gamma^a D_a\lambda + Y^{ij}Y_{ij}\right). \tag{4c}$$

---

[a]This section and section 5 follow closely ref.14. Similar results have been obtained from the flat space limit of a $D = 6$ supergravity model in ref.15.

Further, by virtue of its definition, H_{abc} satisfies the identity

$$\partial_{[a} H^{+}_{bcd]} = \alpha' \mathrm{tr} \left(\frac{3}{4} F_{[ab} F_{cd]} - \bar{\lambda} \Gamma_{[abc} D_{d]} \lambda \right) \tag{5}$$

We shall look for a bosonic background configuration in which all the fermions as well as the auxiliary field Y^{ij} will vanish. The six-dimensional coordinates will be chosen as x^+, x^- and x^μ where $\mu = 1, ...4$. We shall consider a multi instanton-type configuration in the $\mathbf{R}^4$ spanned by x^μ. It will be shown that that the moduli of this instanton can depend on x^+. This will require that the A_+-component of the vector potential is different from zero. In this sense the solution looks like a static monopole configuration in the six-dimensional spacetime in which x^- is taken to be the time coordinate. This configuration will preserve half the six-dimensional $N = 1$ supersymmetry.

It follows from (4a) that if $\lambda = 0$, then H is self dual.[b]. Now setting $\delta\lambda$ and $\delta\chi$ equal to zero we obtain

$$\Gamma^{ab} F_{ab} \epsilon = 0, \qquad \left(\Gamma^a \partial_a \sigma + \frac{1}{12} \Gamma^{abc} H_{abc} \right) \epsilon = 0 \tag{6}$$

To satisfy these equations, we can choose $\epsilon = \begin{pmatrix} \varepsilon \\ 0 \end{pmatrix}$, where, $\gamma_5 \varepsilon = \pm \varepsilon$, and γ_5 gives the four-dimensional chirality. We shall first discuss the case of positive chirality; the case of negative chirality can be obtained by essentially trivial change of some self duality conditions. With this choice the fields must obey the equations

$$H_{05\mu} = -\partial_\mu \sigma \tag{7a}$$
$$H_{0\mu\nu} = \tilde{H}_{0\mu\nu} = \tfrac{1}{2} \epsilon_{\mu\nu\alpha\beta} H_{0\alpha\beta} \tag{7b}$$
$$F_{\mu\nu} = \tilde{F}_{\mu\nu} = \tfrac{1}{2} \epsilon_{\mu\nu\alpha\beta} F_{\alpha\beta} \tag{7c}$$

together with $F_{+-} = 0$, $F_{-\mu} = 0$, $\partial_- \sigma = 0$. Choosing the gauge $A_- = 0$, these reduce to $\partial_- A_\mu = \partial_- A_+ = \partial_- \sigma = 0$.

The constraint (5) for H_{abc}, expressing its coupling to the Yang-Mills fields via the Chern-Simons 3-form, now gives the following conditions,

$$\partial_- H_{+\mu\nu} = 0 \tag{8a}$$
$$\partial_\lambda H_{+\lambda\alpha} = \partial_+ \partial_\alpha \sigma - 2c \mathrm{Tr}(\mathrm{F}_{\lambda\alpha} \mathrm{F}_{+\lambda}) \tag{8b}$$
$$\partial_\mu \partial_\mu \sigma = -\frac{c}{2} \mathrm{Tr}(\mathrm{F}_{\mu\nu} \tilde{\mathrm{F}}_{\mu\nu}) \tag{8c}$$

[b]We shall henceforth drop the superscript + from H.

where $c = 3\alpha'/4$. Further, $H_{-\mu\nu} = 0$ and $H_{+\mu\nu} = \tilde{H}_{+\mu\nu}$. Setting the auxiliary field Y^{ij} to zero implies $D_a F^{ab} = 0$.[13] The only nontrivial surviving component of this equation is

$$D_\lambda(D_\lambda A_+ - \partial_+ A_\lambda) = 0 \tag{9}$$

where $D_\lambda A_+ = \partial_\lambda A_+ + [A_\lambda, A_+]$.

The strategy for solving these equations is as follows. We first choose $F_{\mu\nu}$ to be a multi-instanton configuration in $\mathbf{R}^4$. Then equation (8c) gives σ, and (7a) gives $H_{05\mu}$. Since $D_\lambda D_\lambda$ is invertible in the instanton background, (9) can be uniquely solved for A_+. Finally, equation (8b) can be solved, consistently with its self duality, to get $H_{+\mu\nu}$. As a consequence of $\partial_- A_\mu = \partial_- A_+ = 0$, the instanton parameters, collectively denoted by ξ, obey the condition $\partial_- \xi = 0$, but they can, of course, depend on x^+. (They are thus left-moving modes in the (x^0, x^5)-subspace.)

Using the self duality of $H_{+\mu\nu}$, we can rewrite (8b) as

$$\partial_\lambda \partial_\lambda H_{+\mu\nu} = (\partial_\mu J_\nu - \partial_\nu J_\mu) + \tfrac{1}{2}\epsilon_{\mu\nu\alpha\beta}(\partial_\alpha J_\beta - \partial_\beta J_\alpha) \tag{10}$$

where $J_\alpha = \partial_+(\partial_\alpha \sigma) - 2c\mathrm{Tr}(\mathrm{F}_{\lambda\alpha}\mathrm{F}_{+\lambda})$. It is easy to see that $\partial_\alpha J_\alpha = 0$, as required by the consistency of the equations. Since the four-dimensional Laplacian is invertible, the above equation can easily be solved, once we have J_α. For gauge group $SU(2)$, A_+ is given by

$$A_+^a = \int d^4y\ \Delta^{ab}(x,y)\epsilon^{bkl}(A_\lambda^k \partial_+ A_\lambda^l)(y) \tag{11}$$

where the Green's function $\Delta^{ab}(x,y)$ for $D_\lambda D_\lambda$ in the instanton background is given in ref.16. To make the above solutions explicit, we can, for example, take the 't Hooft ansatz for instantons, viz., $A_\mu^a = \bar{\eta}_{\mu\nu}^a \partial_\nu(\log\phi)$ where $\phi = 1 + \sum_1^N \rho_i^2/(x-a_i)^2$ and insert it in various equations above. In this case, σ, for example, becomes $2c\partial_\mu\phi\partial_\mu\phi/\phi^2$.

5 String Interpretation

To see the stringy interpretation of our solution, we need to analyze its moduli or zero mode structure. From the above equations, we see that, given the gauge field $F_{\mu\nu}$, all the fields are uniquely determined up to the addition of the freely propagating six-dimensional waves for the tensor multiplet. [c] Therefore the only zero modes correspond to the moduli of the instantons.

In order for our models to be mathematically meaningful they should be free from local and global gauge anomalies. In the absence of hypermatter, the

[c]Note the soliton does not modify their propagation.

gauge groups $SU(2)$, $SU(3)$, G_2, F_4, E_6, E_7 and E_8 can be made perturbatively anomaly-free with the help of the Green-Schwarz prescription . However, since the homotopy group Π_6 of the first three groups in this list are nontrivial, these theories will harbour global gauge anomalies. To make them consistent we need to introduce hypermatter for these theories.[17] The allowed matter contents for the cancellation of the global [17] as well as the local [18] anomalies in the presence of one tensor multiplet are $n_2 = 4, 10$ for $SU(2)$, $n_3 = 0, 6, 12$ for $SU(3)$ and $n_7 = 1, 4, 7$ for G_2, where n_2, n_3 and n_7 represent the number of the doublets for $SU(2)$, triplets for $SU(3)$ and 7-dimensional representation of G_7, respectively. All other gauge groups are free from global anomalies and they can be made free from perturbative anomalies (using the Green-Schwarz prescription) if an appropriate amount of hypermultiplets are taken together with the gauge and the tensor multiplets.[17,18]

For the gauge group $SU(2)$, for the four-dimensional space being $\mathbf{R}^4$ and for instanton number k, we have $8k$ bosonic moduli corresponding to the instanton positions, scale sizes and group orientations. (The equations of motion, despite the appearance of the dimensional parameter c, have scale invariance and give the scale size parameter in the solutions.) These moduli appear in the solution for the fields B_{ab} as well.

The surviving supersymmetry has $\gamma_5 \varepsilon = \varepsilon$, i.e., left-chirality in the four-dimensional sense corresponding to a $(4,0)$ world-sheet supersymmetry for the solitonic string. There must necessarily be fermionic zero modes. For the gauginos, we have $4k$ zero modes for the gauge group $SU(2)$, which are of right-chirality in the four-dimensional sense and are in the right-moving sector. The Dirac equation for the gauginos along with the half-supersymmetry condition shows that the gaugino zero mode parameters are constants; the bosonic parameters are constant as well, by supersymmetry. The fermionic zero mode parameters are complex, i.e., we have $8k$ real Grassman parameters which balance the $8k$ bosonic parameters. Some of the fermionic zero modes correspond to the supersymmetries which are broken by the background and can be obtained by such supersymmetry variations. With hypermatter, there are also hyperino zero modes, which are in the left-moving sector. There is no supersymmetry for these modes and generically there are no hyperscalar zero modes.

For higher gauge groups, there will be more moduli. Thus, for example, for $SU(3)$, with the standard embedding of the instanton and $n_3 = 0$, we have $12k$ bosonic parameters and $6k$ fermionic parameters. It is easy to see that the number of moduli for all of the anomaly-free gauge groups listed above is always a multiple of 4. We may thus interpret these solutions as six-dimensional strings with 4 transverse coordinates corresponding to the zero modes for the broken translational symmetries. The remaining zero modes can be regarded as additional world-sheet degrees of freedom. In this way for instanton number k, we have k strings with $(4,0)$ world-sheet supersymmetry.

As an example, consider an $SU(2)$ theory with 10 hypermatter doublets.[19] In this case, for instanton number equal to one, we have eight instanton moduli, eight gaugino zero modes for the right-moving sector and 20 hypermatter zero modes for the left-moving sector. The $SU(2)$ symmetry can be spontaneously broken by vacuum expectation values of the scalars originating from the moduli corresponding to the global $SU(2)$ rotations and the scale size of the instanton. By supersymmetry this should remove four of the gaugino zero modes from the right moving sector by giving them a non zero mass, which will also eat up four hyperino zero modes in the left moving sector. One is left with four moduli for the instanton, four gaugino modes in the right-moving sector and 16 hyperino zero modes in the left-moving sector. These 16 hyperino zero modes presumably generate a left moving E_8 current algebra. This looks like the spectrum of the non critical string which lives in the boundary of a membrane joining a 5-brane to a 9-brane in M-theory and which becomes tensionless as the 5-brane approaches the 9-brane.[20] It has been argued in ref.21 that the same model corresponds to one of the phases of the F-theory.

There are also independent solutions with the opposite chirality. The choice $\gamma_5\varepsilon = -\varepsilon$ leads to anti-self dual $H_{+\mu\nu}$, $F_{\mu\nu}$ with $A_+ = 0$ and $\partial_+\xi = 0$.

The solution we have obtained is a static one. The choice of four-dimensional chirality as $\gamma_5\epsilon = \pm\epsilon$ leads to static solitons. By Lorentz boosts, it is possible to obtain a solution whose center of mass is moving at a constant velocity. For a moving soliton, the condition $\gamma_5\epsilon = \pm\epsilon$ must be modified. Consider, for example, the one-soliton (one-instanton) solution. We choose the supersymmetry parameters ε as $S\varepsilon_{(0)}$ where $S = \exp(-\frac{1}{2}\omega^\mu\gamma_\mu) \approx 1 - \frac{1}{2}\omega^\mu\gamma_\mu$ and $\varepsilon_{(0)}$ obeys $\gamma_5\varepsilon_{(0)} = \varepsilon_{(0)}$. (For small velocities, the parameter $\omega^\mu \approx v^\mu$, the velocity.) The vanishing of the gaugino variation, viz., $\Gamma^{ab}F_{ab}\epsilon = 0$, now gives, to first order in v^μ,

$$\begin{aligned} F_{\mu\nu} - \tilde{F}_{\mu\nu} &= 0 \\ F_{-\mu} + \frac{1}{\sqrt{2}}F_{\mu\nu}v^\nu &= 0 \\ F_{+-} - \frac{1}{\sqrt{2}}F_{+\nu}v^\nu &= 0 \end{aligned} \tag{12}$$

To this order, $F_{\mu\nu}$ is still self dual. The other two equations are seen to be satisfied if we take the instanton position a^α to move with velocity v^α, i.e., $\partial_0 a^\alpha = v^\alpha$. (We can make a gauge transformation $A_- \to A_- - (1/\sqrt{2})A_\mu v^\mu$ to restore the $A_- = 0$ gauge.) There is a similar set of statements for the vanishing of the tensorino variation. What we have shown is that a soliton whose center of mass is moving at a constant velocity v^α is also a supersymmetric solution with supersymmetry parameters being $S\varepsilon_{(0)}$, $\varepsilon_{(0)}$ having definite four-dimensional chirality.

6 $D = 6$ Models with Extended Supersymmetries

In the foregoing sections we discussed only the $D = 6$ models with a minimum number of supersymmetries. Apart from the $(1,0)$ type supersymmetry in $D = 6$ there are also models with $(1,1)$, $(2,0)$ and $(2,2)$ type supersymmetries. The number of real components of the supercharges are respectively 16, 16 and 32. Out of these three types only the $(2,0)$ models are chiral and therefore can be anomalous. Like the four dimensional theories, if we do not want to have a physical field of spin larger than 2, then the total number of real supercharges should not exceed 32. This is the number of supersymmetries of the $D = 11$ supergravity which is conjectured to be the low energy limit of a unifying theory of all known $D = 10$ string theories and is called the M theory. When we obtain a lower dimensional theory from the $D = 10$ or $D = 11$ some of the super symmetries can be broken. For example the K_3 compcatification which takes us from $D = 10$ to $D = 6$ breaks $1/2$ of supersymmetries. Thus starting from the type IIB theory, which has 32 real chiral supersymmetries in $D = 10$, and compactifying on a K_3 we obtain a $D = 6$ theory with 16 chiral supersymmetries. This is an example of a $(2,0)$ model with 21 tensor multiplets.[22] It is exactly 21 tensor multiplets which is required by the anomaly cancellation in $D = 6$. Although an invariant Lagrangian has not yet been constructed for these models, the field equations with an arbitrary number of tensor multiplets have been kn own for some time.[23]

The $(2,0)$ models involve self dual and anti-self dual tensor fields. One can then contemplate self dual or anti-self dual string like solutions of the type discussed in the previous section. Presumably these strings are also tensionless. An intuitive way of understanding this is to remember that in $D = 10$ the IIB theory has a four form potential whose field strength is self dual. There are also self dual 3-brane solutions.[24] One can imagine that a self dual brane wraps around a 2 cycle of the K_3 to produce an object which will look like a string from the $D = 6$ point of view. In general this string will have some thickness. But as the area of the cycle shrinks to zero the thickness also will decrease. Furthermore, the string will be self dual by construction and its tension will be proportional to the area of the 2 cycle and hence will vanish as the area goes to zero.

The type IIA theory has the same number of supersymmetries in $D = 10$

as the type IIB, however it is a non chiral theory. For that reason upon compactification from $D = 10$ to $D = 6$ on a K_3 one obtains a non chiral theory in $D = 6$ with 16 real supercharges which generate a $(1,1)$ supersymmetry. The same type of model can be obtained from the compactification of the heterotic strings on a T^4. There exist many compelling evidence that the theories obtained from the type IIA compactifications on a K_3 and the $E_8 \times E_8$ compactifications on T^4 are dual in the sense that the strong coupling limit of one can be set in correspondence with the weak coupling limit of the other.[25] At a first glance this looks puzzling, because, although at a generic point on the moduli space of the heterotic compactification the six dimensional gauge group is $U(1)^{24}$ it is known from Narain's work that at some special points the gauge symmetry can be enlarged to a non Abelian group.[26] For the duality to work one needs to find mechanisms for the generation of non Abelian gauge symmetries on the type IIA side. The possibility which has been suggested[27] is that the 2-branes of the type II can wrap around 2 cycles of K_3 and produce, in the limit that the area of the cycles shrink to zero, particle like objects in $D = 6$. The masses of these particles will be proportional to the area of the 2 cycles and will vanish in the limit of the vanishing cycles. These massless particles should then match with the massless particles generated at the special points in the heterotic moduli space at which the gauge symmetries are enhanced.

Finally the compactifications of type IIA or the type IIB theories on a four dimensional torus will produce non chiral $D = 6$ supergravity models with $(2,2)$ supersymmetries.

An important role in the study of these models is played by the moduli space of vacua, i.e. the expectation values of the massless scalars. It has been conjectured that all known compactifications with $(1,1)$, $(0,2)$ and $(2,2)$ supersymmetries belong to the same moduli space of vacua.

For example, we mentioned above that the heterotic string on T^4 is dual to the type IIA compactification on K_3. It is also known that if we compactify type IIA on a S^1 of radius R from $D = 10$ to $D = 9$ then it gives the same theory as the one obtained from the compactification of type IIB from $D = 10$ to $D = 9$ on a circle of radius $1/R$. It then follows that the compactification of the heterotic string on a T^5 should produce a theory dual to the compactification of the type IIB on $S^1 \times K_3$. This duality is a strong weak duality. On the heterotic side the string coupling is given by e^{ϕ}, where ϕ denotes the vev of the dilaton field. On the type IIB side this modulus corresponds to the radius of S^1. This implies that the strong coupling limit of the heterotic string in $D = 5$ is dual to the large radius limit of the type IIB string in $D = 5$. The large radius limit is, of course the same as the decompactification limit which takes us back to $D = 6$ space time again. Thus by moving around in the moduli space of type IIB or heterotic in $D = 5$ we can end up with the $(2,0)$ type IIB in $D = 6$.

References

1. S. Randjbar-Daemi, Abdus Salam and J. Strathdee, *Nucl. Phys.* **B214**, 491 (1983).
2. For excellent reviews see, for example, the contributions by A. Sen, P. K. Townsend and C. Vafa in the Proceedings of the 1996 Trieste Summer School.
3. For a review of the electromagnetic duality and the Olive Mantonen conjecture see D. Olive in ICTP Conference on Recent Developments in Statistical Mechanics and Quantum Field Theory, Trieste, Italy, 10-12 Apr 1995. Published in Nucl. Phys. Proc. Suppl. 45A:88-102, 1996, hep-th/9508089.
4. M.J. Duff and R. Minasian, *Nucl. Phys.* **B436**, 507 (1995).
5. P. K. Townsend, *Phys. Lett.* **B202**, 53 (1988).
6. S. Randjbar-Daemi, Abdus Salam, E. Sezgin and J. Strathdee, *Phys. Lett.* **B151**, 351 (1985).
7. M. Green, J. Schwarz and P. West, *Nucl. Phys.* **B254**, 327 (1985); J. Schwarz, *Phys. Lett.* **B371**, 223 (1996), hep-th 9512053.
8. M. Green and J. Schwarz, *Phys. Lett.* **B149**, 117 (1984).
9. B. Zumino, in *Relativity, Groups and Toplogy II*, Eds. B.S. De Witt and R. Stora (North-Holland, 1984).
10. A. Sagnotti, *Phys. Lett.* **B294**, 196 (1992), hep-th/9210127; Fabio Riccioni and Augusto Sagnotti, hep-th/9806129.
11. E. Witten, *Phys. Lett.* **B117**, 324 (1982); S. Elitzur and V.P. Nair, *Nucl. Phys.* **B243**, 205 (1984).
12. N. Seiberg and E. Witten, hep-th/9603003; N. Seiberg, **B 390**, 169 (1997), hep-th/9609161.
13. E. Bergshoeff, E. Sezgin and E. Sokatchev, *Class. Quantum Grav.*
14. V.P. Nair and S. Randjbar-Daemi, *Phys. Lett.* **B418**, 312-316 (1998), hep-th/9711125.
15. M.J. Duff, J.T. Liu, H. Lu, C.N. Pope, hep-th hep-th/9711089.
16. L.S. Brown *et al*, *Phys. Rev.* **D17**, 1583 (1978).
17. M. Bershadsky and C. Vafa, hep-th 97030167.
18. U.H. Danielsson, G. Ferretti, J. Kalkkinen and P. Stjernberg, hep-th 9703098.
19. P. Mayr and C. Vafa, unpublished.
20. Ori J. Ganor and Amihay Hanany, hep-th 96020120.

21. A. Klemm, P. Mayr and C. Vafa, hep-th 96070139.
22. P. Townsend, *Phys. Lett.* **B139**, 283 (1984).
23. L. Romans, *Nucl. Phys.* **B276**, 71 (1986).
24. M.J. Duff and J.X. Lu, *Phys. Lett.* **B273**, 41 (1991).
25. P. Townsend and C. Hull, *Nucl. Phys.* **B243**, 205 (1984), hep-th/9410167.
26. K.S. Narain, *Phys. Lett.* **B169**, 409 (1986); K.S. Narain, M.H. Sarmadi and E. Witten, *Nucl. Phys.* **B279**, 369 (1987).
27. E. Witten, *Nucl. Phys.* **B443**, 85 (1995), hep-th/9503124; P.S. Aspinwall, *Phys. Lett.* **B357**, 329 (1995)[, hep-th/9507012.

DISCRETE SYMMETRIES AND SUPERSYMMETRIES – POWERFUL TOOLS FOR STUDYING QUANTUM MECHANICAL SYSTEMS

J. NIEDERLE
Institute of Physics, Academy of Sciences, Prague, Czech Republic
E-mail: niederle@fzu.cz

Distinguished colleagues, ladies and gentlemen,

It is a great pleasure and honour for me to give a talk in this Memorial Meeting devoted to Abdus Salam. This is not only due to the fact that Professor Abdus Salam was a man I have admired very much for his brilliance in theoretical physics, his prodigious intuition and many remarkable activities for well-being of the people but also that it was precisely the ICTP – the place of this meeting – where I met him for the first time thirty-three years ago. In the ICTP I had opportunities from time to time to talk to him (and not only about physics!), to keep an eye on development of his ingenious physical ideas from a "short distance" and to admire his outstanding capabilities for handling and solving many problems threatening the world today. I remember, however, that he always respected a hierarchy of values where physics was at the top. This was well reflected in a quotation of Robert Oppenheimer posted on the wall of his office that our only anchor in honesty is to preserve our competence in our professions, to preserve what we know intimately, to preserve our mastery. I have come to appreciate the depth of this view more and more as I grew older and as the number of my commitments increased.

According to Franz Kafka people need three things for their existance: beauty, sense of life and sustainabilities. I am pretty sure if Franz Kafka had known scientists like Abdus Salam he would have added a fourth thing important for people's existence and typical for Abdus Salam – zeal for unveiling the mysteries of Nature.

Ladies and gentlemen, there are and always will be many mysteries in Nature that are not explained by contemporary science, many open hows and whys and it is difficult for us to become reconciled with the fact that Abdus Salam is no longer with us to answer these questions in his own original way. We shall greatly miss him.

In my today's lecture I would like to talk about the subject which was very close to Abdus Salam's heart and which was a common denominator of his many outstanding contributions to contemporary physics – about symmetries. I shall discuss this very fascinating subject mostly based on some recent results of Nikitin and myself (see refs.[1-4]). I shall confine to discrete symmetries and supersymmetries of quantum-mechanical systems and moreover rather to a few typical examples of them since the rest can be found in detail in the published papers [1-4]. First let us make several introductory remarks.

1 Introduction

It is well-known that quantum-mechanical systems are usually described in terms of (ordinary or partial, linear or nonlinear) differential equations and that symmetries of these equations form powerful tools for their studies. They are used to separate variables in these equations (see e.g. [5]), to find their solutions as well as to solve their labelling problem (see e.g. [5-9]), to derive spectra and related complete sets of functions of linear differential operators (see e.g. [5,10,11]), to derive the corresponding conservation laws (see e.g. [12]), to construct of new theories, i.e., to find out differential equations invariant with respect to a given symmetry (see e.g. [13]), and so on.

Let us recall that in quantum mechanics the statement: "The physical system S has a symmetry group G" means that there is a group of transformations leaving both the equation of motion of system S as well as the rules of quantum mechanics invariant. In particular no transformation from symmetry group G is allowed to produce an observable effect. Thus if system S is described by an observable A in states $|\psi>, |\varphi>, \ldots$ and system S' obtained by a symmetry transformation $g \in G, g : S \to S'$, is described by the corresponding observable A' in the states $|\psi'>, |\varphi'>, \ldots$, then the equality

$$|\langle\psi'|A'|\phi'\rangle|^2 = |\langle\psi|A|\phi\rangle|^2 \tag{1}$$

must hold. As shown by E. Wigner [14], equality (1) implies that to any symmetry g there exists a unitary or anti-unitary operator U_g (representing g in the Hilbert space of the states of system S) such that

$$|\psi'> = U_g|\psi> \quad \text{and} \quad A' = U_g A U_g^+ \tag{2}$$

describe the effect of g, i.e., the change $S \to S'$. It is obvious that unitary and anti-unitary operators preserve (1). It is, however, more difficult to show that they exhaust all possibilities.

There are two types of symmetries: *continuous* (e.g. rotations) and *discrete* (e.g. reflections). For continuous symmetries any $g \in G$ is a function of one or more continuous parameters α^i, $i = 1, 2, \ldots, n$, i.e. $g = g(\alpha^1, \alpha^2, \ldots \alpha^n)$ and any U_g can be expressed in terms of Hermitian operators, i.e. observables

$B_1, B_2, \ldots$ via $e^{i\alpha^j B_j}$. Due to continuity of parameters α^j any B_j is a constant of motion, since for a given quantum-mechanical system described by Hamiltonian H

$$\left[e^{i\alpha^j B_j}, H\right] = 0 \Leftrightarrow \sum_{n=0}^{\infty} \frac{(i\alpha^j)^n}{n!} [B_j^n, H] = 0 \Leftrightarrow [B_j, H] = 0 \ . \tag{3}$$

Now, if $g \in G$ is a discrete symmetry, it does not depend on continuous parameters. The corresponding operator U_g can still be written as e^{iB} or Ke^{iB}, where B is an observable and K is an anti-unitary operator, but the equality $[B, H] = 0$ is only a sufficient condition for $\sum\limits_{n=0}^{\infty} \frac{(i)^n}{n!} [B^n, H] = 0$ but not necessary. Consequently, B is not, in general, a constant of motion. However, all discrete symmetries in physics fulfil the condition $U_g^2 = 1$, i.e., they are involutive. Then if U_g is unitary ($U_b U_g^+ = U_g^+ U_g = 1$), it is also Hermitian $U_g^+ = U_g$, and therefore an observable. This is not true for $U_g^2 \neq 1$.

Now we are ready to discuss some of the results derived in [1-4] on a few typical examples.

2 Involutive Symmetries and Reduction of the Physical Systems

First we describe new symmetry algebras of the equations describing various quantum-mechanical systems. As an example we consider the free Dirac equation

$$L_0 \psi = (i\gamma^\mu \partial_\mu - m)\psi = 0 \tag{4}$$

with

$$\gamma_0 = \begin{pmatrix} 0 & 1_2 \\ 1_2 & 0 \end{pmatrix}, \ \gamma_a = \begin{pmatrix} 0 & -\sigma_a \\ \sigma_a & 0 \end{pmatrix}, \ a = 1, 2, 3, \ \gamma_5 = \begin{pmatrix} 1_2 & 0 \\ 0 & 1_2 \end{pmatrix} .$$

It is invariant w.r.t the complete Lorentz group. Discrete involutive symmetries form a finite subgroup of the Lorentz group consisting of 4 reflections of x_μ, 6 reflections of pairs of x_μ, 4 reflections of triplets of x_μ, reflection of all x_μ and the identity transformation.

If the coordinates x_μ in (4) are transformed by these involutive symmetries, function $\psi(x)$ contransforms according to a projective representation of the symmetry group, i.e., either via $\psi(x) \to R_{kl}\psi(x)$ or via $\psi(x) \to B_{kl}\psi(x)$ (for details see [1]). Here R_{kl} and $B_{kl} = CR_{kl}$ are linear and anti-linear operators respectively which commute with L_0 and consequently transform solutions of (4) into themselves. The operators $R_{kl} = -R_{lk}$ form a representation of the algebra $so(6)$ and C is the operator of charge conjugation $C\psi(x) = i\gamma_2\psi^*(x)$.

Among the operators B_{kl} there are six which satisfy the condition that $(B_{kl})^2 = -1$ and nine for which $(B_{kl})^2 = 1$. We shall consider further only B_{kl} fulfilling the last condition (for the reason mentioned in the Introduction and since otherwise B_{kl} cannot be diagonalised to real γ_5 and consequently used for reduction of the system). As shown in [1] the operators R_{kl}, B_{kl} and C form a 25-dimensional Lie algebra. It can be extended to a 64-dimensional real Lie algebra or via non-Lie symmetries (i.e., via linear operators Q of general forms than $ia^\mu \partial_\mu + b$ (with a^μ and b being functions and matrices of x_μ resp.) fulfilling $[Q, L_0] = \alpha_Q L_0$ with L_0 from (4) and α_Q being appropriate operators (for details see [2])), to a 256-dimensional Lie algebra. These algebras do not yield essentially new conservation laws but may be used to reduce considered systems or to find their supersymmetries.

Let us discuss now only one example how to use this new symmetry algebra to reduce physical systems into uncoupled subsystems (for the other examples see [1]). Let our system be a spin $\frac{1}{2}$ particle interacting with a magnetic field A_μ described by the Dirac equation

$$L\psi(x) = (\gamma^\mu(i\partial_\mu - eA_\mu) - m)\psi(x) = 0 \ . \tag{5}$$

Eq. (5) is invariant w.r.t. discrete symmetries provided $A_\mu(x)$ contransforms appropriately. For instance,

$$A_\mu(-x) = -A_\mu(x) \tag{6}$$

for $x \to -x$ and $\psi(x) \to \hat{R}\psi(x) = \gamma_5\hat{\theta}\psi(x) = \gamma_5\psi(-x)$. Then, diagonalising symmetry operator $\hat{R}$ by means of the operator

$$W = \frac{1}{\sqrt{2}}\,(1 + \gamma_5\gamma_0)\,\frac{1}{\sqrt{2}}\,(1 + \gamma_5\gamma_0\hat{\theta})\ , \tag{7}$$

i.e., $W\hat{R}W^+ = \gamma_5$, and applying the same transformation to L, equation (5) is reduced to the block diagonal form

$$\left(-\mu(i\partial_0 - eA_0) - \vec{\sigma}\cdot\left(i\vec{\partial} - e\vec{A}\right)\hat{\theta} - m\right)\psi_\mu(x) = 0\ , \tag{8}$$

where $\mu = \pm 1$ and ψ_μ are two-component spinors satisfying $\gamma_5\psi_\mu = \mu\psi_\mu$.

If equations (8) admit a discrete symmetry again, then they can be further reduced to one-component uncoupled subsystems.

The other treated examples in [1] include that Dirac equation with an anomalous Pauli interaction, the Dirac and Weyl oscillators, and so on.

3 Discrete Symmetries and Supersymmetries

It was shown in [3] that extended, generalized and reduced supersymmetries appear rather frequently in many quantum-mechanical systems. Here I shall illustrate it only on two examples. Let us begin with a familiar case – the Schrödinger equation with a matrix superpotential W:

$$H\psi = \frac{1}{2}\,(p^2 + W^2 + \sigma_3 W')\psi = E\,\psi, \tag{9}$$

where $p = -i\frac{d}{dx}, W' = \frac{d}{dx}\,W(x)$. It is well known that system (9) admits supercharges

$$Q_1 = \frac{1}{\sqrt{2}}\,(\sigma_1 p + \sigma_2 W) \quad \text{and} \quad Q_2 = \frac{1}{\sqrt{2}}\,(\sigma_2 p - \sigma_1 W) \tag{10}$$

satisfying the $N = 2$ superalgebra

$$\{Q_a, Q_b\} = 2\delta_{ab}H, \quad [Q_a, H] = 0, \quad a = 1, 2. \tag{11}$$

There is, however, an additonal supersymmetry of (9) provided superpotential W is an even function of x! Indeed, if $W(-x) = W(x)$, then there exists the third supercharge

$$Q_3 = i\sigma_1 R Q_1 \tag{12}$$

with R defined by

$$R\psi(x) = \psi(-x) \tag{13}$$

such that the relations (11) hold for $a, b, = 1, 2, 3$.

If W is an odd function of x, then Q_1 and Q_2 transform according to a reducible representation of the $N = 2$ superalgebra. This follows from the fact that for $W(-x) = -W(x)$ there exists an invariant operator namely, $I = \sigma_3 R$, which commutes with H and Q_a, $a = 1, 2$, and which can be used to diagonalise H and Q_a to a block diagonal form. Let us remark that the irreducible components of Q_1 and Q_2 are expressed in terms of p, R and W, i.e., without the usual fermionic variables (for details see [3])!

As a second example let us consider a spin $\frac{1}{2}$ particle interacting with a constant and homogeneous magnetic field $\vec{H}$. It is described by the Schrödinger-Pauli equation:

$$H\psi(x) = \left(\left(-i\vec{\partial} - e\vec{A}\right)^2 - \frac{1}{2}\,eg\vec{\sigma}\cdot\vec{H}\right)\psi(x) = 0\,. \tag{14}$$

This system is exactly solvable and admits the $N = 4$ extended supersymmetry (for details see [3]). One of its supercharges is of the standard form

$$Q_1 = \vec{\sigma}\left(-i\vec{\partial} - e\vec{A}\right), \; Q_1^2 = H \; , \tag{15}$$

three additional ones can be constructed by means of space reflections R_a of $x^a, a = 1, 2, 3$ since equation (14) is invariant w.r.t. them. As found in [3], they are of the form:

$$Q_2 = iR_3\vec{\sigma}\cdot\left(-i\vec{\partial} - e\vec{A}\right), \; Q_3 = iCR_4\vec{\sigma}\cdot\left(-i\vec{\partial} - e\vec{A}\right), \; Q_4 = iCR_2\vec{\sigma}\cdot\left(-i\vec{\partial} - e\vec{A}\right) .$$

Supercharges Q_a are integrals of motion for system (14) (notice that without the usual "fermionic" variables) and responsible for the eight-fold degeneracy of the energy spectrum of this system. For details, many other examples and appropriate references see [3].

4 Conclusions

- The new symmetry algebra for the Dirac equation (or the Schrödinger-Pauli equation) describing various quantum-mechanical systems was derived. This algebra is based on discrete involutive symmetries of the considered systems and is of the biggest dimension from all known invariant algebras of the considered systems.

- The new symmetry algebra was used for two purposes: i) to reduce the considered quantum-mechanical system to simpler subsystems and ii) to search for its supersymmetries (extended, reducible or generalized) and to explain degeneracy of its energy spectrum.

- Since the required transformation properties of magnetic field $A_\mu(x)$ for the considered quantum-mechanical systems to be supersymmetric seem to be physically realisable (for details see [3]), these systems give strong indications that supersymmetry is indeed a symmetry realised in Nature.

Thank you for your attention.

References

1. J. Niederle and A.G. Nikitin, *J. Phys. A: Math. Gen.* **30**, 999 (1997).
2. J. Niederle and A.G. Nikitin, *Nonlin. Math. Phys.* **4**, 436 (1997).

3. J. Niederle and A.G. Nikitin, "Extended supersymmetries for the Schrödinger-Pauli equation", *J. Math. Phys.* (1998) (in press).
4. J. Niederle, in the Proc. of the Second International Conference "Symmetries in Nonlinear Mathematical Physics", Kiev, 1997, Vol. 2, pp.495-498;
   A.G. Nikitin, in the Proc. GROUP21 Physical Applications and Mathematical Aspects of Geometry, Groups and Algebras, Vol. 1, Eds. H.-D. Doebner et al. (World Scientific, Singapore, 1997) pp. 509-514.
5. H.Ya. Vilenkin and A.U. Klimyk, *Representations of Lie Groups and Special Functions*, Vol. 1-3. *Recent Advances* (Kluwer, Dordrecht, 1991-1995);
   U. Müller, *Symmetry and Separation of Variables* (Reading, MA: Addison-Wesley, 1997).
6. P. Olver, *Application of Lie Groups to Differential Equations* (Springer, New York, 1986);
   N.Ch. Ibragimov, *Transformation Groups in Mathematical Physics* (Nauka, Moscow, 1983).
7. E. Taflin, *Pacific J. Math.* **108**, 203 (1983);
   M. Flato and J. Simon, *J. Math. Phys.* **21**, 913 (1980); *Phys. Lett.* **94B**, 518 (1980) and references cited therein.
8. W.I. Fushchich, W.M. Shtelen and N.I. Serov, *Symmetry Analysis and Exact Solutions of Equations of Nonlinear Mathematical Physics* (Reidel, Dordrecht, 1993).
9. B.G. Konopelchenko, *Nonlinear Integrable Equations* (Lecture Notes in Physics 270), (Springer, Berlin, 1987).
10. N. Limic, J. Niederle and R. Raczka, *J. Math. Phys.* **7**, 1861, 2026 (1966); *J. Math. Phys.* **8**, 1079 (1967).
11. W.I. Fushchich and A.G. Nikitin, *Symmetries of Equations of Quantum Mechanics* (Allerton, New York, 1994).
12. N.N. Bogolubov, D.V. Shirkov, *An Introduction to the Theory of Quantised Fields* (Nauka, Moscow, 1976);
    N.P. Konoplyova, V.N. Popov, *Gauge Fields* (Atomizdat, Moscow, 1972); see also:
    C. Gardner et al., *Phys. Rev. Lett.* **19**, 109 (1967);
    J. Mickelsson and J. Niederle, *Lett. Math. Phys.* **8**, 195 (1984).
13. I.M. Gel'fand, A.M. Yaglom, *Z. Eksp. Teor. Fiz.* **8**, 703 (1948);
    also in:
    I.M. Gel'fand, R.A. Minlos, Z.Ya. Shapiro, *Representations of the Rotation and Lorentz Groups and their Applications* (Pergamon, New York, 1963);

See also
R. Kotecký, J. Niederle, *Czech. J. Phys.* **B25**, 123 (1975);
R. Kotecký, J. Niederle, *Reports Math. Phys.* **12**, 237 (1977);
J. Mickelsson, J. Niederle, *Ann. Inst. Henri Poincaré* **XXIII A**, 277 (1975).

14. E.P. Wigner, *Group Theory and its Application to the Quantum Mechanics of Atomic Spectra* (Addison-Wesley, New York, 1959).

PRE-BIG BANG COSMOLOGY: A LONG HISTORY OF TIME?

G. Veneziano
Theoretical Physics Division, CERN
CH-1211 Geneva 23

The popular myth according to which the Universe -and time itself- started with/near a big bang singularity is questioned. After claiming that the *two* main puzzles of standard cosmology allow for *two* possible logical answers, I will argue that superstring theory strongly favours the the pre-big bang (PBB) alternative. I will then explain why PBB inflation is as generic as classical gravitational collapse, and why, as a result of symmetries in the latter problem, recent fine-tuning objections to the PBB scenario are unfounded. A hot big bang state naturally results from the powerful amplification of vacuum quantum fluctuations before the big bang, a phenomenon whose observable consequences will be briefly summarized.

1 PREAMBLE

Abdus Salam was not afraid of "crazy" ideas. What follows is quite daring and, most probably, "wrong" (see my conclusions). Yet, I like to think that Abdus would have enjoyed listening to it also because, as we have been reminded by Miguel Virasoro, one of his constant concerns -even till the last days of his life- was expressed in a recurrent question: "What about Gravity?"

Exactly 100 years ago, J.J. Thompson discovered the electron. But, even today, the origin of the small electron mass is clouded in mystery. In the same year, french impressionist Paul Gauguin signed one of his most famous paintings, whose title consists of three questions:

D'ou venons-nous? Que sommes-nous? Ou' allons-nous?

100 years later Gauguin's questions are still very much with us. Modern versions of the second and third question (What is the value of Ω? Will the Universe expand forever?) have been the subject of recent scientific and popular articles [1]. What about Gauguin's first question? In a recent article, E. Witten [2], defending US participation in the LHC project, argues that discovering supersymmetry would "give a huge boost to ...string theory." He then goes on to explain why non-specialists should care about this: it is because strings may hold "the key to answering some of the questions that people who are not scientists most often ask". These modern versions of Gauguin's first question, prompted by experimental evidence that a Big Bang did occur in the early Universe, are (quoting again [2]):

"What happened *before* the Big Bang? What was the beginning of time?"

i.e. precisely the subject of this talk.

2 Two attitudes towards two puzzles

It is commonly believed (see e.g. [3]) that the Universe -and time itself- started some 15 billion years ago from a kind of primordial explosion, the famous Big Bang. Indeed, the experimental observations of the red-shift and of the Cosmic Microwave Background (CMB) lead us quite unequivocally to the conclusion that, as we trace back our history, we encounter epochs of increasingly high temperature, energy density, and curvature. However, as we arrive close to the singularity, our classical equations are known to break down. The earliest time we can think about classically is certainly larger than the so-called Planck time, $t_P = \sqrt{G_N \hbar} \sim 10^{-43}$s ($c = 1$ throughout). Hence, the honest answer to the question: Did the Universe -and time- have a beginning? is: We do not know, since the answer lies in the unexplored domain of quantum gravity.

Besides the initial singularity problem -and in spite of its successes- the hot big bang model also has considerable phenomenological problems. Amusingly, these too can be traced back to the nature of the very early state of the Universe. Let us briefly recall why.

The observable part of our Universe, our present horizon, is about 10^{28} cm large. At earlier times, the horizon was much smaller: a few Planck times after the Big Bang, the horizon was not much bigger that a few Planck lengths, say about 10^{-32} cm. Instead, the portion of space that corresponds to our present horizon was about 1 mm large, i.e. some 30 orders of magnitude larger than the horizon. In other words, at that time, the Universe consisted of $(10^{30})^3 = 10^{90}$ Planckian-size, causally disconnected regions.

The *homogeneity* problem of standard cosmology is the observation that there is no reason to expect that conditions in all those 10^{90} regions were initially the same, since there had never been any causal contact among them. Yet, today, all those regions make up our observable Universe and are homogeneous to one part in 10^5. A related puzzle is the so-called *flatness* problem: How come the primordial Universe had a (spatial) curvature radius 30 orders of magnitude larger that H^{-1}?

In my opinion there are three possible attitudes towards solving the above puzzles, but only two of them are scientifically sound. These are:

- The Universe *was not* particularly homogeneous and flat at the big bang, but a period of accelerated expansion (inflation) *after* the big bang made it that way . This is the standard *post-big bang* inflation idea (see e.g. [4]).

- The Universe *was* already homogeneous and flat at the big bang since a long period of inflation *before* the big bang cooked it up that way. This is basically the *pre-big bang* (PBB) idea.

- There is always, of course, a third answer: some unknown Planckian physics produced an incredibly homogeneous/flat Universe, which is essentially like saying that God decided things that way (see, in this connection, a nice picture in Penrose's book [5]).

How come string theory prefers the second answer in the above list?

3 String Theory hints

The following properties of superstrings strongly speak in favour of the PBB scenario over the one of conventional inflation:

- String theory does *not* automatically give Einstein's General Relativity at low energy/curvature. Rather, it leads to a scalar-tensor theory of the Jordan-Brans-Dicke (JBD) variety, moreover with a dangerously small ω_{JBD} parameter. The extra scalar particle/field, called the dilaton, is unavoidable in string theory.

- The dilaton, which we denote by ϕ, provides the overall strength of all interactions via relations like (here $\lambda_s = \sqrt{\alpha' \hbar}$ is the fundamental string-length parameter):
$$l_P^2/\lambda_s^2 \sim \alpha_{gauge} \sim e^{\phi} \;, \tag{1}$$
from which we deduce that the weak coupling region is $\phi \ll -1$. At present, $e^{\phi} \sim 1/25$, implying $\lambda_s \sim 10 l_P \sim 10^{-32}$ cm.

- Dilaton exchange produces a "fifth force" which threatens precision tests of the equivalence principle [6], hence string theory itself. The simplest way out this problem (see [7] for an amusing alternative) is to assume that the dilaton has a mass of at least 10^{-4} eV, coming from supersymmetry breaking non-perturbative effects. These are negligigle at small coupling and therefore the dilaton, when large and negative, behaves like a massless particle.

- Interesting new symmetries (generically called "dualities") thrive on the existense of ϕ. For cosmology, their most relevant representative is the so-called **Scale-Factor duality** (SFD), whose action is as follows.

A FRW cosmology at $t > 0$, i.e. a *decelerating expansion* of the Universe (apparently) originating *from* a singularity as $t \to 0^+$

gets mapped, via SFD, into

An inflationary cosmology at $t < 0$, i.e. an *accelerated expansion* of the Universe (apparently) going *towards* a singularity as $t \to 0^-$

This second cosmology looks impossible at first since, in order to have inflation, one needs a peculiar equation of state ($3p + \rho < 0$), and a scalar field with no potential energy cannot satisfy that condition. The puzzle is solved, within string (or JBD) gravity, by observing that $G_N^{eff} \sim l_P^2$ is controlled by ϕ through (1). It is indeed perfectly possible for H to grow, even though the energy density decreases, provided ϕ is also growing. We thus come to an important conclusion:

Pre-Big Bang inflation needs a growing dilaton

4 Pre-big bang's postulate/conjecture

If we wish to use a dual cosmology for the prehistory of the Universe, given the positive signs of $\dot{H}$ and $\dot{\phi}$, we have to start from (very?) small initial values for H and e^{ϕ}. Although not strictly necessary, we will also impose, for the sake of simplicity, an almost empty initial Universe. This leads us to the following basic postulate of PBB cosmology:

The Universe started its evolution from the simplest conceivable initial state in string theory, its perturbative vacuum. This corresponds to an (almost)

EMPTY, COLD, FLAT, FREE

Universe as opposed to the standard

DENSE, HOT, HIGHLY-CURVED

initial state of conventional cosmology.

For this assumption to make sense the new initial conditions should be able to provide, at later times, a hot big bang with the desired characteristics. This looks like a very hard task, but, as the following speakers will explain, it is all but impossible.

The above postulate is good news for the theorist. Indeed, if both coupling and curvature are small, physics is adequately described by the tree-level low-energy effective action of string theory, which is something we know all about. Its form reads:

$$\Gamma_{eff} = \frac{1}{2}\int d^4x\sqrt{-g}\; e^{-\phi}\left[\lambda_s^{-2}(\mathcal{R} + \partial_\mu\phi\partial^\mu\phi + H_{\mu\nu\rho}^2) + F_{\mu\nu}^2 + \ldots\right] \quad (2)$$

where we have included the contributions of the Kalb-Ramond antisymmetric tensor field $B_{\mu\nu}$ through its field strength $H_{\mu\nu\rho}$, as well as some typical gauge fields.

The suggestion from string theory now becomes almost a compelling one[8]: Can one combine a standard cosmology at $t > 0$ and a dual PBB cosmology at $t < 0$ to contruct a smooth cosmology containing both? Since, in such a scenario, the Hubble parameter grows for $t < 0$ and decreases for $t > 0$, it should reach its maximum at $t = 0$, instant to be identified with the Big Bang of standard cosmology. The problem is that this maximum is actually infinite if one works in the context of the low-energy effective theory (2). The pre-(post-) big bang phases have singularities in the future(past). However, the low-energy approximation breaks down as soon as the Hubble parameter reaches values $O(\lambda_s^{-1})$ leading us to expect that the maximal value of H, reached at $t = 0$, should be actually finite and of that order. All this leads to the main conjecture of PBB cosmology, the so-called *graceful exit* assumption:

Pre-big bang and FRW phases smoothly join through a high curvature/large coupling (string) phase

5 Pre-big bang as classical gravitational collapse

The PBB postulate can be rephrased in more precise terms as a postulate of **Asymptotic Triviality**(AT), where *asymptotic* means at (timelike and null) past infinity and *triviality* refers both to space-time and to the dilaton. In analogy with QCD's asymptotic freedom AT is very different from just triviality. Actually, while the trivial vacuum of string theory is a very special point in the space of classical solutions, AT vacua are generic, in the technical sense that they depend on as many arbitrary functions as the general solution.

Determining the properties of a generic solution to the field equations implied by (2) is a formidable task, even if one neglects the Kalb-Ramond field. Fortunately, however, we can map our problem into one which has been much investigated, both analytically and numerically, in the literature, the problem of gravitational collapse of a scalar field. This is simply done by "going to the Einstein frame" i.e. by the field redefinition

$$g_{\mu\nu} = g_{\mu\nu}^{(E)} e^{\phi-\phi_0} , \tag{3}$$

in terms of which (2) becomes:

$$\Gamma_{eff} = \frac{1}{2l_P^2} \int d^4x \sqrt{-g^{(E)}} (\mathcal{R}^{(E)} - \tfrac{1}{2} g_{(E)}^{\mu\nu} \partial_\mu \phi \partial_\nu \phi) , \tag{4}$$

where ϕ_0 ($l_P = e^{\frac{1}{2}\phi_0} \lambda_s$) is the present value of the dilaton (of Planck's length).

This problem has been considered precisely in the regime of interest to us, i.e. starting from very "weak" initial data with the aim of finding under which conditions gravitational collapse must later occur. Gravitational collapse basically means that the (Einstein) metric shrinks to zero at a spacelike singularity. However, typically, the dilaton blows up at that same singularity. Given the relation (3) betweew the Einstein and the (physical) string metric, we can easily imagine that the latter blows up near the singularity, thus giving inflation and a big bang. What can we say at the moment about details on all this? Not much, since this part of the game is just starting, but here it is.

Assume that, deeply into the PBB phase, the Universe was not particularly homogeneous, spatial gradients and time derivatives being comparable and, in accordance with the PBB postulate, both tiny in string units. It can be argued [9] that such initial conditions can lead to a chaotic version of PBB inflation since, if a certain patch develops where time-derivatives slightly dominanate over spatial gradients, then inflation turns on in that patch. Thereafter, the evolution of the system can be studied by analytic methods [10], since the approximation of neglecting spatial gradients becomes increasingly accurate within the inflating patch.

In the general case, no analytic methods are known for analyzing the system at even earlier times (i.e. before the inflationary patch forms), while numerical codes are still largely inadequate. Both analytic and numerical methods are instead available for solutions exhibiting some symmetry, e.g. spherical symmetry. In that case, recent numerical work [11] has fully confirmed the occurrence of inflation and the validity of the gradient expansion [10]. However, in the spherically symmetric case, the most promising avenue [12], appears to consist in making use of the powerful analytic results by Christodoulou [13], which provide sufficient criteria for collapse (or for its absence). Those criteria are "scale free": in no way does the collapse theorem of [13] fix the absolute curvature scale at the onset of collapse (i.e. when a closed trapped surface first forms). This curvature scale becomes [12] the scale of inflation (the Hubble parameter) at the onset of inflation in the original PBB problem. But why is this scale so important?

The point is that, in order to solve the homogeneity/flatness problems, dilaton-driven inflation has to last sufficiently long. Its duration is not infinite, since it is limited in the past by the conditions of collapse and, in the future, by the time at which curvatures become so large that we can no-longer trust the low-energy approximation. Thus, as it was actually noticed from the very beginning [8], a successful PBB scenario does require very perturbative initial conditions, so that it takes a long time to reach the BB singularity. Fortunately, both the initial coupling and the initial curvature are free classical parameters, possibly with a random distribution throughout space.

A particular case of this "fine-tuning" was discussed recently by Turner and

E. Weinberg and by Kaloper et al. [14]. These authors consider a homogeneous –but not spatially flat– Universe and notice that the duration of PBB inflation is limited in the past by the initial value of the spatial curvature. This has to be taken very small in string units if sufficient inflation is to be achieved. The issue is whether or not this is fine-tuning. String theory has a single length parameter, λ_s, but, fortunately for us, it also has massless states and low-energy vacua (such as Minkowsky space-time), whose characteristic scale is arbitrarily larger than λ_s. Thus I see no fine-tuning in starting the evolution of the Universe in a state of low-energy, small curvatures, and small coupling. To the contrary, I find it very amusing/exciting to realize that a well-known classical instability pushes the Universe from low energy (curvature) and small coupling towards high energy (curvature) and large coupling.

6 Quantum Mechanical Heating of the Universe and Observable PBB relics

Since there are already several review lectures on this subject (e.g. [15]), I will limit myself to mention the most recent developments simply after recalling the basic physical mechanism underlying particle production in cosmology [16]. A cosmological (i.e. time-dependent) background coupled to a given type of (small) inhomogeneous perturbation Ψ enters the effective low-energy action in the form:

$$I = \frac{1}{2}\int d\eta\; d^3x\; S(\eta)\left[\Psi'^2 - (\nabla\Psi)^2\right]. \tag{5}$$

Here η is the conformal time coordinate, and a prime denotes $\partial/\partial\eta$. The function $S(\eta)$ (sometimes called the "pump" field) is, for any given Ψ, a given function of the scale factor, $a(\eta)$, and of other scalar fields (four-dimensional dilaton $\phi(\eta)$, moduli $b_i(\eta)$, etc.) which may appear non-trivially in the background.

While it is clear that a constant pump field S can be reabsorbed in a rescaling of Ψ, and is thus ineffective, a time-dependent S couples non-trivially to the fluctuation and leads to the producion of pairs of quanta (with opposite momenta). Looking back at Eq. (2), one can easily determine the pump fields for each one of the most interesting perturbations. The result is:

$$\begin{aligned} \text{Gravity waves, dilaton} &\;:\; S = a^2 e^{-\phi} \\ \text{Heterotic gauge bosons} &\;:\; S = e^{-\phi} \\ \text{Kalb}-\text{Ramond, axions} &\;:\; S = a^{-2} e^{-\phi} \end{aligned} \tag{6}$$

A distinctive property of string cosmology is that the dilaton ϕ appears in some very specific way in the pump fields. The consequences of this are very interesting:

- For gravitational waves and dilatons the effect of ϕ is to slow down the behaviour of a (remember that both a and ϕ grow in the pre-big bang phase). This is the reason why those spectra are quite steep [17] and give small contributions at large scales. On the other hand, an interesting stochastic background of gravitational waves in the frequency region where GW decterctors (will) operate, is all but excluded [17].

- For (heterotic) gauge bosons there is no amplification of vacuum fluctuations in standard cosmology, while, in string cosmology, all the "work" is done by the dilaton. In the pre-big bang scenario, the coupling must grow by as large a factor as the one by which the Universe has inflated. This implies a very large amplification of the primordial quantum fluctuation [18], possibly explaining the long-sought origin of seeds for the galactic magnetic fields.

- Finally, for Kalb-Ramond fields and axions, a and ϕ work in the same direction and spectra can be large even at large scales[19]. Note, incidentally, that the power of a in S is determined by the rank of the corresponding tensor. It is well known, however, that the Kalb-Ramond field can be reduced to a (pseudo)scalar field, the axion, through a duality transformation. This turns out to change S into S^{-1}, i.e. the pump field for the axion is actually $a^2 e^{\phi}$. An interesting duality of cosmological perturbations, reminiscent of electric-magnetic (or strong-weak) duality, can be argued [20] to guarantee the equivalence of the Kalb-Ramond and axion spectra.

- Many other spectra, which arise in generic compactifications of superstrings, have also been studied and lead to interesting spectra. For lack of time, I will refer to the existing literature [21, 22].

The possible flatness of axionic spectra in pre-big bang cosmology leads to hopes that, in such a scenario, there is a natural way to generate an interesting spectrum of large-scale fluctuations, one of the much advertised properties of the standard inflationary scenario. This problem has been recently addressed[23] with the conclusion that massless –or extremely light– Kalb-Ramond axions can indeed seed large-scale CMB anisotropy provided they are produced with an primordial narly-flat spectrum. Instead, for axions in the usual mass range (say around $10^{-4}eV$), the primordial axion spectrum has to be nearly-flat at large scales and increasing with frequency at short scales[24]. Both possibilities can be realized in string cosmology but, certainly, a scale-invariant spectrum of large-scale inhomogeneities is all but automatic.

Before closing this section, I wish to recall how one sees the very origin of the hot big bang in this scenario. One can easily estimate the total energy stored in the quantum fluctuations which were amplified by the pre-big bang backgrounds. The result is, roughly,

$$\rho_{quantum} \sim N_{eff}\ H_{max}^4 \ , \tag{7}$$

where N_{eff} is the effective number of species that are amplified and H_{max} is the maximal curvature scale reached around $t = 0$ (this formula has to be modified in case some spectra show negative slopes). We have already argued that $H_{max} \sim \lambda_s^{-1}$ and we know that, in heterotic string theory, N_{eff} is in the hundreds. Yet this rather huge energy density is very far from critical as long as the dilaton is still in the weak-coupling region, justifying our neglect of back-reaction effects. It is very tempting to assume that, precisely when the dilaton reaches a value such that $\rho_{quantum}$ is critical, the Universe will enter the radiation-dominated phase. This too is, at present, the object of active investigation.

7 Conclusions

Pre-big bang cosmology appears to have survived its first 6-7 years of life. Interest in (criticism of) it is growing. It is perhaps time to make a balance sheet.

Conceptual (technical?) and phenomenological problems include:

- Graceful exit from dilaton-driven inflation to FRW cosmology is not fully understood, in spite of recent progress [25]. Possibly, new ideas borrowed from M-theory and D-branes could help in this respect.
- A scale-invariant spectrum of large-scale perturbations is not automatic, although, thanks to possibly flat axion spectra, it does not look inconceivable either.

Attractive features include:

- No need to "invent" an inflaton, or to fine-tune potentials.
- Inflation is "natural" thanks to the duality symmetries of string cosmology.
- The initial conditions problem is decoupled from the singularity problem: a solution to the former is already shaping up and looks exciting.

- A classical gravitational instability (similar to the one giving gravitational collapse and singularities in General Relativity) finds a welcome use in providing inflation.

- A quantum instability (pair creation) is able to heat up an initially cold Universe and generate a hot big bang with the additional virtues of homogeneity and flatness.

- Last but not least: one is dealing with a highly constrained, predictive scheme, which can be tested/falsified by low-energy experiments, as the following talks will explain.

What's my own forecast? It is simple: probably, PBB cosmology will not reach age 10. However, it will not be killed by Hawking, Linde, or Turner, but, more likely, by Planck ... and I do *not* mean Max!

References

1. *Science* (Oct. 31, 1997); Nature (Jan. 8, 1998); The New York Times (Jan. 9, 1998).
2. E. Witten, *The New Republic*, Dec. 29, 1997.
3. S. Hawking, Proceedings of the Texas/ESO-CERN Symposium on Relativistic Astrophysics, Cosmology, and Fundamental Physics, Brighton 1990, (eds. J.D. Barrow, L. Mestel and P.A. Thomas), *Ann. NY Acad. Sci.* **647**, 315 (1991).
4. E. W. Kolb and M. S. Turner, *The Early Universe*, (Addison-Wesley, 1990).
5. R. Penrose, *The Emperor's New Mind*, (Oxford University Press, 1989), Fig. (7.19).
6. T.R. Taylor and G. Veneziano, *Phys. Lett.* B**213**, 459 (1988).
7. T. Damour and A.M. Polyakov, *Nucl. Phys.* B**423**, 532 (1994).
8. G. Veneziano, *Phys. Lett.* B**265**, 287 (1991);
   M. Gasperini and G. Veneziano, *Astropart. Phys.* **1**, 317 (1993); *Mod. Phys. Lett.* **A8**, 3701 (1993); *Phys. Rev.* **D50**, 2519 (1994);
   An updated collection of papers on the PBB scenario is available at *http*://www.to.infn.it/~ gasperin/.
9. G. Veneziano, *Phys. Lett.* B**406** , 297 (1997);
   A. Buonanno, K. A. Meissner, C. Ungarelli and G. Veneziano, *Phys. Rev.* **D57**, 2543 (1998);
   A. Feinstein, R. Lazkoz and M.A. Vazquez-Mozo, *Closed Inhomogeneous String Cosmologies*, hep-th/9704173;

J. D. Barrow and M. P. Dabrowski, *Is there Chaos in Low-energy String Cosmology?*, hep-th/9711049;
J. D. Barrow and K. E. Kunze, *Spherical Curvature Inhomogeneities in String Cosmology*, hep-th/9710018;
K. Saygily, *Hamilton-Jacobi Approach to Pre-Big Bang Cosmology at Long Wavelengths*, hep-th/9710070.

10. V. A. Belinskii and I.M. Khalatnikov, *Sov. Phys. (JETP)* **36**, 591 (1973); N. Deruelle and D. Langlois, *Phys. Rev.* **D52**, 2007 (1995); J. Parry, D. S. Salopek and J. M. Stewart, *Phys. Rev.* **D49**, 2872 (1994).
11. J. Maharana, E. Onofri and G. Veneziano, *A numerical approach to pre-big bang cosmology*, gr-qc/9802001, to appear in JHEP.
12. A. Buonanno, T. Damour and G. Veneziano, to appear.
13. D. Christodoulou, *Comm. P. A. Math.***44**, 339 (1991), and references therein.
14. M. Turner and E. Weinberg, *Phys. Rev.* **D56**, 4604 (1997); N. Kaloper, A.D. Linde and R. Bousso, *Pre-big bang Requires the Universe to be Exponentially Large from the Very Beginning*, hep-th/9801073.
15. G. Veneziano, *Status of String Cosmology: Basic Concepts and Main Consequences*, in *String Gravity and Physics at the Planck Energy Scale*, Erice, 1995, eds. N. Sanchez and A. Zichichi, (Kluver Academic Publishers, Boston, 1996), p. 285.; M. Gasperini, *Status of String Cosmology: Phenomenological Aspects*, ibid., p. 305.
16. See, e.g., V. F. Mukhanov, A. H. Feldman and R. H. Brandenberger, *Phys. Rep.* **215**, 203 (1992).
17. R. Brustein, M. Gasperini, M. Giovannini and G. Veneziano, *Phys. Lett.* B**361**, 45 (1995); R. Brustein et al., *Phys. Rev.* **D51**, 6744 (1995).
18. M. Gasperini, M. Giovannini and G. Veneziano, *Phys. Rev. Lett.* **75**, 3796 (1995); D. Lemoine and M. Lemoine, *Phys. Rev.* **D52**, 1955 (1995).
19. E.J. Copeland, R. Easther and D. Wands, *Phys. Rev.* **D56**, 874 (1997); E.J. Copeland, J.E. Lidsey and D. Wands, *Nucl. Phys.* B**506**, 407 (1997).
20. R. Brustein, M. Gasperini and G. Veneziano, *Duality of Cosmological Perturbations*, CERN-TH/98-53, hep-th/9803018.
21. R. Brustein and M. Hadad, *Phys. Rev.* **D57**, 725 (1998).
22. A. Buonanno, K. A. Meissner, C. Ungarelli and G. Veneziano, *JHEP* **01** (1998) 004.

23. R. Durrer, M. Gasperini, M. Sakellariadou and G. Veneziano, *Seeds of large-scale anisotropy in string cosmology*, CERN-TH/98-69, gr-qc/9804076.
24. M. Gasperini and G. Veneziano, to appear.
25. M. Gasperini, M. Maggiore and G. Veneziano, *Nucl. Phys.* B**494**, 315 (1997);
    R. Brustein and R. Madden,*Phys. Lett.* B**410**, 110 (1997);
    Phys. Rev. **D57**, 712 (1998).

WITH NEUTRINO MASSES REVEALED, PROTON DECAY IS THE MISSING LINK [a]

JOGESH C. PATI
Physics Department, University of Maryland, College Park, MD 20742, USA

By way of paying tribute to Abdus Salam, I recall the ideas of higher unification that he and I initiated. I discuss the current status of those ideas in the light of recent developments, including those of: (a) gauge coupling unification, (b) discovery of neutrino-oscillation at SuperKamiokande, and (c) ongoing searches for proton decay. It is noted that the mass of ν_τ ($\sim 1/20$ eV), suggested by the SuperK result, provides clear support for the route to higher unification based on the ideas of (i) SU(4)-color, (ii) left-right symmetry and (iii) supersymmetry. The change in perspective, pertaining to both gauge coupling unification and proton decay, brought forth by supersymmetry and superstrings, is noted. And, the beneficial roles of string-symmetries in addressing certain naturalness problems of supersymmetry, including that of rapid proton decay, are emphasized. Further, it is noted that with neutrino masses and coupling unification revealed, proton decay is the missing link. Following recent joint work with K. Babu and F. Wilczek, based on supersymmetric unification, it is remarked that the SuperKamiokande result on neutrino oscillation in fact enhances the expected rate of proton decay compared to prior estimates. Thus, assuming supersymmetric unification, one expects that the discovery of proton decay should not be far behind.

I Salam in Perspective

Abdus Salam was a great scientist and a humanitarian. His death was indeed a loss to science and especially to the growth of science in the third world. He will surely be remembered for his contributions to physics, some of which have proven to be of lasting value. These include his pioneering work on electroweak unification for which he shared the Nobel Prize in physics in 1979 with Sheldon Glashow and Steven Weinberg. Contribution of this calibre is rare.

But I believe his most valuable contribution to science and humanity, one that is perhaps unparalleled in the world, is the sacrifice he made of his time, energy and personal comfort in promoting the cause of science in different corners of the globe, in particular the third world. His lifelong efforts in this direction led to the creation of some outstanding research centres, including especially the International Centre for Theoretical Physics (ICTP) at Trieste, Italy, [b] an International Centre for Genetic Engineering and Biotechnology

[a] Based in part on Talk delivered at the Abdus Salam Memorial Meeting, ICTP, Trieste, November 1997.
[b] Now named (at this meeting) the Abdus Salam International Centre for Theoretical Physics.

with components in Trieste and Delhi, and an International Centre for Science and High Technology in Trieste. Salam dreamed of creating twenty international centres like the ICTP, spread throughout the world, emphasizing different areas of science and technology. Approaching developed as well as developing nations, for funding of such institutions, Salam often used the phrase: "science is not cheap, but expenditures on it will repay tenfold" [1]. If only Salam had lived a few more years in good health, many more such institutions would have surely come to fruition.

Salam was also a strong supporter of world peace, and thus of nuclear disarmament and Pugwash. Thus, in addition to his numerous awards for his contributions to physics, including the Nobel prize, he also received some major awards for his contributions to peace and international collaboration, including the Atoms for Peace Award in 1968 and the "Ettore Majorana" - Science for Peace prize in 1989. It is hard to believe that a single individual could accomplish so much both in creating as well as in propagating science. In this sense, Salam was indeed a rare individual—a phenomenon.

I was especially fortunate to have collaborated with Salam closely for over a decade. Of this period, I treasure most the memory of many moments which were marked by the struggle and the joy of research that we both shared. Needless to say, Salam played a central role in the growth of the ideas which we together initiated. To touch upon one aspect of Salam's personality, during the ten year period of our collaboration, there have been many letters, faxes, arguments over the phone and in person and even heated exchanges, about tastes and judgements in physics, but always in a good natured spirit [c]. In our discussions, Salam had some favorite phrases. For example, he would sometimes come up with an idea and get excited. If I expressed that I did not like it for such and such reason, he would get impatient and say to me: "My dear sir, what do you want: Blood?". I would sometimes reply by saying: "No Professor Salam, I would like something better". Whether I was right or wrong, he never took it ill. It is this attitude on his part that led to a healthy collaboration and a strong bond between us. Most important for me, by strongly encouraging from the beginning, and yet often arguing, he had the knack of bring out the best in a collaborator. For this I will remain grateful to him.

By way of paying tribute to Salam therefore, I would first like to recall

[c] A brief account of how our collaboration evolved in the initial phase is given in my article in the Proceedings of the Salamfestschrift [2] which was held here at ICTP in 1993 (that is probably the last scientific meeting that Salam attended), and a shorter version is given in the article written in his honor after he passed away [3]. The first section of this talk is based in part on these two articles.

briefly the ideas on higher unification which we initiated (sec. 2), and then present their current status in the context of subsequent experimental and theoretical developments (secs. 3,4, and 5). The experiments of special relevance are: (a) recent observation of neutrino oscillation at SuperKamiokande, (b) the precision measurements of the gauge couplings at LEP, and (c) ongoing searches for proton decay.

In section 3, I discuss how the recent discovery of atmospheric neutrino-oscillation at SuperKamiokande [d], especially the mass of ν_τ suggested by the SuperK result, agrees well with the gauge coupling unification revealed by the extrapolation of the LEP data on the one hand, and provides clear support for the route to higher unification based on the concepts of SU(4)-color and left-right symmetry on the other hand. On the theoretical side, the major developments of the last two decades are the ideas of supersymmetry, superstrings, and now M-theory. I will briefly remark how these later developments fully retain the basic ideas of higher unification of the 70's, and at the same time, provide a substantially new perspective, because they unify gravity with the other three forces (sec.4). The change in perspective pertains to both gauge coupling unification and proton decay. In discussing the puzzle of proton-longevity in supersymmetry, I remark, following recent work, how string-derived symmetries play an essential role in providing a natural resolution of this puzzle (sec. 5). In the last section, I present a result, recently derived by Babu, Wilczek and me, that exhibits an intimate link between neutrino masses and proton decay in the context of supersymmetric unification (sec. 6). Following the results of a very recent work by the three of us, I remark that the observation of coupling unification as well as the discovery of neutrino-oscillation at SuperK strengthen our expectations for discovery of proton decay in the near future.

II Status of Particle Physics in 1972: The Growth of New Ideas

IIA. The collaborative research of Salam and myself started during my short visit to Trieste in the summer of 1972. At this time, the electroweak SU(2) × U(1)-theory existed[4], but there was no clear idea of the origin of the fundamental strong interaction. The latter was thought to be generated, for example, by the vector bosons (ρ, ω, K^* and ϕ), or even the spin-0 mesons ($\pi, K, \eta, \eta', \sigma$), assumed to be elementary, or a neutral U(1) vector gluon coupled universally to all the quarks[5]. Even the existence of the SU(3)-color degree of freedom[6,7] as a global symmetry was not commonly accepted, because many thought

[d] While the SuperKamiokande discovery occurred some six months after the presentation of this talk, its implications are included here because they are so directly relevant to the unification ideas proposed in the early 70's.

that this would require an undue proliferation of elementary entities. And, of course, asymptotic freedom had not yet been discovered.

In the context of this background, the SU(2) × U(1) theory itself appeared (to us) as grossly incomplete, even in its gauge-sector (not to mention the Higgs sector), because it possessed a set of scattered multiplets, involving quark and lepton fields, with rather peculiar assignment of their weak hypercharge quantum numbers. To remove these shortcomings, we wished: (a) to find a higher symmetry-structure that would organize the scattered multiplets together, and explain the seemingly arbitrary assignment of their weak hypercharge; (b) to provide a rationale for the co-existence of quarks and leptons; further (c) to find a reason for the existence of the weak, electromagnetic as well as strong interactions, by generating the three forces together by a unifying gauge principle; and finally (d) to understand the quantization of electric charge, regardless of the choice of the multiplets, in a way which should also explain [e] why $Q_{\text{electron}} = -Q_{\text{proton}}$.

We realized that in order to meet these *four aesthetic demands*, the following rather unconventional ideas would have to be introduced:

(i) First, one must place quarks and leptons within the same multiplet and gauge the symmetry group of this multiplet to generate simultaneously weak, electromagnetic and strong interactions [8,9].

(ii) Second, the most attractive manner of placing quarks and leptons in the same multiplet, it appeared to us [8], was to assume that quarks do possess the SU(3)-color degree of freedom, and to extend SU(3)-color to the symmetry SU(4)-color, interpreting lepton number as the fourth color. *A dynamical unification of quarks and leptons is thus provided by gauging the full symmetry SU(4)-color.* The spontaneous breaking of SU(4)-color to $SU(3)^c \times U(1)_{B-L}$ at a sufficiently high mass-scale, which makes leptoquark gauge bosons superheavy, was then suggested to explain the apparent distinction between quarks and leptons, as regards their response to strong interactions at low energies. Such a distinction should then disappear at appropiately high energies.

Within this picture, one had no choice but to view *fundamental* strong interactions of quarks as having their origin entirely in the octet of gluons associated with the SU(3)-color gauge symmetry In short, as a *by-product* of our attempts to achieve a higher unification through SU(4)-color, we were led to conclude that low energy electroweak and fundamental strong interactions

[e] We thought that if one could understand why the electron and the proton have equal and opposite charges, one would have an answer to Feynman's question as to why it is that the electron and the proton - rather than the positron and proton - exhibit the same sign of longitudinal polarization in β-decay. The V-A theory of weak interactions did not provide an *a priori* reason for a choice in this regard.

must be generated by the *combined gauge symmetry* $SU(2)_L \times U(1)_Y \times SU(3)^C$, which now constitutes the symmetry of the standard model[8,10,11]. It of course contains the electroweak symmetry $SU(2)_L \times U(1)_Y$[4]. The idea of the SU(3)-color gauge force became even more compelling with the discovery of asymptotic freedom about nine months later[12], which explained approximate scaling in deep inelastic ep-scattering, observed at SLAC.

(iii) Third, it became clear that together with SU(4)-color one must gauge the commuting left-right symmetric gauge structure $SU(2)_L \times SU(2)_R$, rather than $SU(2)_L \times U(1)_{I_{3R}}$, so that electric charge is quantized. In short the route to higher unification should include *minimally* the gauge symmetry[8,9]

$$G(224) = SU(2)_L \times SU(2)_R \times SU(4)^C \tag{1}$$

with respect to which all members of a given family fall into the neat pattern:

$$F^e_{L,R} = \begin{bmatrix} u_r & u_y & u_b & \nu_e \\ d_r & d_y & d_b & e^- \end{bmatrix}_{L,R} . \tag{2}$$

With respect to G(224), the left-right-conjugate multiplets F^e_L and F^e_R transform as (2,1,4) and (1,2,4) respectively; likewise for the mu and the tau families.

Viewed against the background of particle physics of 1972, as mentioned above the symmetry structure G(224) brought some attractive features to particle physics for the first time. They are:

(i) Organization of all members of a family $(8_L + 8_R)$ within one left-right self-conjugate multiplet, with their peculiar hypercharges fully explained.

(ii) Quantization of electric charge, explaining why $Q_{electron} = -Q_{proton}$.

(iii) Quark-lepton unification through SU(4)-color.

(iv) Left-Right and Particle-Antiparticle Symmetries in the Fundamental Laws: With the left-right symmetric gauge structure $SU(2)_L \times SU(2)_R$, as opposed to $SU(2)_L \times U(1)_Y$, it was natural to postulate that at the deepest level nature respects parity and charge conjugation, which are violated only spontaneously[9,13]. Thus, within the symmetry-structure G(224), quark-lepton distinction and parity violation may be viewed as low-energy phenomena which should disappear at sufficiently high energies.

(v) Existence of Right-Handed Neutrinos: Within G(224), there must exist the right-handed (RH) neutrino (ν_R), accompanying the left-handed one (ν_L), for each family, because ν_R is the fourth color - partner of the corresponding RH up-quarks. It is also the $SU(2)_R$-doublet partner of the associated RH charged lepton (see eq. (2)). The RH neutrinos seem to be essential now (see later discussions) for understanding the non-vanishing light masses of the neutrinos, as suggested by the recent observations of neutrino-oscillations.

(vi) B-L as a local Gauge Symmetry: SU(4)-color introduces B-L as a local gauge symmetry. Thus, following the limits from Eötvos experiments, one can argue that B-L must be violated spontaneously. It has been realized, in the light of recent works on electroweak sphaleron effects, that such spontaneous violation of B-L may well be needed to implement baryogenesis via leptogenesis [14].

(vii) Proton Decay: The Hall-Mark of Quark-Lepton Unification:

We recognized that the spontaneous violation of B-L, mentioned above, is a reflection of a more general feature of non-conservations of baryon and lepton numbers in unified gauge theories, including those going beyond G(224), which group quarks and leptons in the same multiplet [9,15]. Depending upon the nature of the gauge symmetry and the multiplet-structure, the violations of B and/or L could be either spontaneous [f], as is the case for the non-conservation of B-L in SU(4) color, and those of B and L (with $\triangle(B-L)=0$ as well as $\triangle$ (B-L) $\neq 0$) in the maximal one-family symmetry like SU(16) [16]; alternatively, the violations could be explicit, which is what happens for the subgroups of SU(16), like SU(5) [17] or SO(10) [18] (see below). One way or another baryon and/or lepton-conservation laws cannot be absolute, in the context of such higher unification. The simplest manifestation of this non-conservation is proton decay ($\triangle B \neq 0, \triangle L \neq 0$); the other is the Majorana mass of the RH neutrinos ($\triangle B = 0, \triangle L \neq 0$), as is encountered in the context of G(224) or SO(10). An unstable proton thus emerges as the crucial prediction of quark-lepton unification [9,17]. Its decay rate would of course depend upon more details including the scale of such higher unification.

IIB. Going Beyond G(224): SO(10) and SU(5)

To realize the idea of a single gauge coupling governing the three forces [8,17], one must embed the standard model symmetry, or G(224), in a simple or effectively simple group (like SU(N) × SU(N)). Several examples of such groups have been proposed. Howard Georgi and Sheldon Glashow proposed the first such group SU(5) [17] which embeds the standard model symmetry, but not G(224). Following the discovery of asymptotic freedom of nonabelian gauge theories [12] and the suggestion of SU(5), Georgi, Helen Quinn and Weinberg showed how renormalization effects, following spontaneous breaking of the unification symmetry, can account for the observed disparity between the three

[f] The case of spontaneous violation arises because the massless gauge particle coupled to any linear combination of B and L (which is gauged) must acquire a mass through SSB in order to conform with the limits from the Eötvos type experiments. The corresponding charge (B and/or L) must then be violated spontaneously.

gauge couplings at low energies [19]. Each of these contributions played a crucial role in strengthening the ideas of higher unification.

To embed G(224) into a simple group, it may be noted that it is isomorphic to SO(4) × SO(6). Thus the smallest simple group to which it can be embedded is SO(10) [18]. By the time SO(10) was proposed, all the advantages of G(224) [(i) to (vi), listed above] and the ideas of higher unification were in place. Since SO(10) contains G(224), the features (i) to (vi) are of course retained by SO(10). In addition, the 16-fold left-right conjugate set $(F_L^e + \bar{F}_R^e)$ of G(224) corresponds to the spinorial 16 of SO(10). Thus, SO(10) preserves even the 16-plet family-structure of G(224), without a need for any extension. By contrast, if one extends G(224) to the still higher symmetry E_6 [20], the advantages (i) to (vi) are retained, but in this case, one must extend the family-structure from a 16 to a 27-plet.

Some distinctions between SU(5) on the one hand versus G(224) or SO(10) on the other hand are worth noting. Historically, SU(5) served an important purpose, being the smallest symmetry that embodies the essential ideas of higher unification. However, it split members of a family into two multiplets: $\bar{5} + 10$. By contrast, SO(10) groups all 16 members of a family into one multiplet. Likewise, G(224), subject to the assumption that parity is a good symmetry at high energies, groups the 16 members into one L-R self-conjugate multiplet. Furthermore, in contrast to G(224) and SO(10), SU(5) violates parity explicitly from the start; it does not contain SU(4)-color, and therefore does not possess B-L as a local symmetry; and the RH neutrino is not an integral feature of SU(5). As I discuss below, these distinctions turn out to be especially relevant to considerations of neutrino masses.

Comparing G(224) with SO(10), as mentioned above, SO(10) possesses *all* the features (i) to (vi) of G(224), but in addition it offers gauge coupling unification. I should, however, mention in passing at this point that the perspective on coupling unification and proton decay changes considerably in the context of supersymmetry and superstrings. In balance, a string-derived G(224) offers some advantages over a string-derived SO(10), while the reverse is true as well. Thus, it seems that a definite choice of one over the other is hard to make at this point. I will return to this discussion in secs. 4 and 5.

III Neutrino Masses: Evidence in Favor of the G(224) Route to Higher Unification

Leaving aside the differences between alternative routes to higher unification, based purely on aesthetic taste, it was of course not clear in the early 70's as to whether the special features of G(224) — i.e. SU(4)-color, left-right symmetry and the RH neutrino — are utilized by nature. The situation has, however, changed dramatically owing to the recent SuperKamiokande (SK) discovery of the oscillation of ν_μ to ν_τ (or ν_X), with a value of $\delta m^2 \approx \frac{1}{2}(10^{-2} - 10^{-3})eV^2$ and an oscillation-angle $sin^2 2\theta > 0.82$ [21]. One can argue (see e.g. [22]) that the SK result, especially the value of δm^2, clearly points to the need for the existence of the RH neutrinos, accompanying the observed LH ones. If one then asks the question: What symmetry on the one hand dictates the existence of the RH neutrinos, and on the other hand also ensures quantization of electric charge, together with quark-lepton unification, one is led to two very beautiful conclusions: (i) quarks and leptons must be unified minimally within the symmetry SU(4)-color, and that, (ii) deep down, the fundamental theory should possess a left-right symmetric gauge structure: $SU(2)_L \times SU(2)_R$. In short, the standard model symmetry must be extended minimally to G(224).

One can now obtain an estimate for the mass of ν_L^τ in the context of G(224) or SO(10) by using the following three steps [22]:

(i) First, assume that B-L and I_{3R}, contained in a string-derived G(224) or SO(10), break near the unification-scale:

$$\mathrm{M_X} \sim 2 \times 10^{16}\mathrm{GeV}, \tag{3}$$

through Higgs multiplets of the type suggested by string-solutions [23] — i.e. $< (1,2,4)_H >$ for G(224) or $< \bar{16}_H >$ for SO(10). (The "empirical" determinations of M_X and the new perspective on unification due to supersymmetry as well as superstrings are discussed in the next section). In the process, the RH neutrinos (ν_R^i), which are singlets of the standard model, can and generically will acquire superheavy Majorana masses of the type $M_R^{ij}\nu_R^{iT}C^{-1}\nu_R^j$, by utilizing the VEV of $< \bar{16}_H >$ and effective couplings of the form:

$$\mathcal{L}_\mathrm{M}(\mathrm{SO(10)}) = \lambda_\mathrm{R}^\mathrm{ij} 16_\mathrm{i} \cdot 16_\mathrm{j} \overline{16}_\mathrm{H} \cdot \overline{16}_\mathrm{H}/\mathrm{M_{P\ell}} + \mathrm{hc} \tag{4}$$

A similar expression holds for G(224). Here i,j=1,2,3, correspond respectively to e, μ and τ families; M_{pl} denotes the reduced Planck mass $\simeq 2 \times 10^{18} GeV$. Such gauge-invariant non-renormalizable couplings might be expected to be induced by Planck-scale physics. (They may well arise - in part or dominantly - by renormalizable interactions through tree-level exchange of superheavy states, such as those in the string tower). Assuming that the Majorana couplings are family-hierarchical, λ_{33} being the leading one, somewhat analogous to those that give the Dirac masses, and ignoring the effects of off-diagonal mixings (for simplicity), one obtains:

$$M_{3R} \approx \frac{\lambda_{33} < \bar{16}_H >^2}{2 \times 10^{18} GeV} \approx \lambda_{33}(4.5 \times 10^{14} GeV)\eta^2 \tag{5}$$

This is the Majorana mass of the RH tau neturino. Guided by the value of M_X, in this estimate, we have substituted $< \bar{16}_H >= (3 \times 10^{16} GeV)\eta$ where $\eta \approx 1/2$ to 2.

(ii) Second, assume that the effective gauge symmetry below the string-scale contains SU(4)-color. Now using SU(4)-color and the Higgs multiplet $(2,2,1)_H$ of G(224) or equivalently 10_H of SO(10), one obtains the relation $m_\tau(M_X) = m_b(M_X)$, which is known to be successful. Thus, there is a good reason to believe that the third family gets its masses primarily from the 10_H or equivalently $(2,2,1)_H$. In turn, this implies:

$$\mathrm{m}(\nu_{\mathrm{D}}^{\tau}) \approx \mathrm{m_{top}}(\mathrm{M_X}) \approx (100 - 120)\mathrm{GeV} \tag{6}$$

(iii) Given the superheavy Majorana masses of the RH neutrinos as well as the Dirac masses, as above, the see-saw mechanism [24] yields naturally light masses for the LH neutrinos. For ν_L^τ (ignoring mixing), one thus obtains, using eqs. (5) and (6),

$$\mathrm{m}(\nu_{\mathrm{L}}^{\tau}) \approx \frac{\mathrm{m}(\nu^{\tau}_{\mathrm{Dirac})^2}}{\mathrm{M_{3R}}} \approx (1/45)\mathrm{eV}(1 \text{ to } 1.44)/\lambda_{33}\eta^2 \tag{7}$$

Considering that on the basis of the see-saw mechanism, we naturally expect that $m(\nu_L^e) \ll m(\nu_L^\mu) \ll m(\nu_L^\tau)$, and assuming that the SuperK observation represents $\nu_L^\mu - \nu_L^\tau$ (rather than $\nu_L^\mu - \nu_X$) oscillation, so that the observed $\delta m^2 \approx 1/2(10^{-2} - 10^{-3})eV^2$ corresponds to $m(\nu_L^\tau)_{obs} \approx (1/15$ to $1/40)$ eV, it seems *truly remarkable* that the expected magnitude of $m(\nu_L^\tau)$, given by eq.(7), is just about what is observed, if $\lambda_{33}\eta^2 \approx 1$ to $1/3$. Such a range of $\lambda_{33}\eta^2$ seems most plausible and natural (see discussion in Ref. 22). It should be stressed that the estimate (7) utilizes the ideas of supersymmetric unification, especially in getting the scale of M_X (eq.(3)), and of SU(4)-color in getting $m(\nu_{Dirac}^\tau)$ (eq.(6)). The agreement between the expected and the SuperK result thus suggests that, *at a deeper level, near the string or the coupling unification scale M_X, the symmetry group G(224) and thus the ideas of SU(4)-color and left-right symmetry are likely to be relevant to nature.*

By providing clear support for G(224), the SK result selects out SO(10) or E_6 as the underlying grand unification symmetry, rather than SU(5). Either SO(10) or E_6 or both of these symmetries ought to be relevant at some scale, and in the string context, as discussed later, that may well be in higher dimensions, above the compactification-scale, below which there need be no more than just the G(224)-symmetry. If, on the other hand, SU(5) were regarded as a fundamental symmetry, first, there would be no compelling reason, based on symmetry alone, to introduce a ν_{R}, because it is a singlet of SU(5). Second, even if one did introduce $\nu_{\mathrm{R}}^{\mathrm{i}}$ by hand, the Dirac masses, arising from the coupling $h^i \bar{5}_i < 5_H > \nu_R^i$, would be unrelated to the up-flavor masses and thus rather arbitrary (contrast with eq. (6)). So also would be the Majorana masses of the ν_R^i's, which are SU(5)-invariant and thus can even be of order Planck scale (contrast with Eq. (5)). This would give $m(\nu_L^\tau)$ in gross conflict with the observed value. In this sense, the SK result appears to disfavor SU(5) as a fundamental symmetry, with or without supersymmetry.

Finally, it is intriguing to note that the SuperK result agrees well with the idea of supersymmetric unification. For this purpose, one could use the mass of $m(\nu_L^\tau)$, suggested by the SuperK data, as an input to obtain the VEV of $< \overline{16}_H >$, that breaks B-L, as an output. By reversing the steps in going from eq. (7) together with eqs. (6) and (5), one obtains, as is to be expected, $< \overline{16}_H > \sim 3 \times 10^{16}$ GeV (if $\lambda_{33} \sim O(1)$). It is rather striking that this is just about the same as the scale of the meeting of the three gauge couplings, which is obtained from extrapolation of their measured value at LEP, in the context of supersymmetry (see next section). *In short, two very different considerations — light neutrino masses on the one hand, and gauge coupling meeting on the other hand — point to one and the same scale for the underlying new physics!* If one assumes supersymmetric unification, one can hardly not notice how beautifully it makes the picture hang together!

In the last section, I will mention briefly how, by adopting familiar ideas of understanding cabibbo-like mixing angles in the quark-sector, one can quite plausibly obtain not only the right magnitude for the mass of ν_τ but also a large $\nu_L^\mu - \nu_L^\tau$ oscillation angle, as observed at SuperK, and simultaneously attribute to the solar neutrino-deficit to $\nu_e - \nu_\mu$ oscillation.

I now discuss the issue of coupling unification.

IV Coupling Unification: A New Perspective Duc To Supersymmetry and Superstrings

It has been recognized from the early 70's, that the concept of higher unification — now commonly called grand unification — has two dramatic consequences: (i) meeting of the gauge couplings at a high scale, and (ii) proton decay[8,9,16,19]. Equally dramatic is the prediction of the light neutrino masses, which is a special feature of only a subclass of grand unification symmetries that contain SU(4)-color, like SO(10) or E_6. As discussed above, this feature seems to be borne out by the SuperKamiokande result on neutrino-oscillations. The status of the first two predictions are discussed in this section and the next.

IVA. Meeting of The Three Gauge Couplings and The Need for Supersymmetry

It has been known for some time that the precision measurements of the standard model coupling constants (in particular $\sin^2\theta_W$) at LEP put severe constraints on the idea of grand unification. Owing to these constraints, the non-supersymmetric minimal $SU(5)$, and for similar reasons, the one-step breaking minimal non-supersymmetric $SO(10)$-model as well, are now excluded [25]. For example, minimal non-SUSY $SU(5)$ predicts: $sin^2\theta_W(m_Z))\,|_{\overline{MS}} = .214 \pm .004$, where as current experimental data show: $sin^2\theta_W(m_Z)_{expt}{}^{LEP} = .2313 \pm .0003$. The disagreement with respect to $sin^2\theta_W$ is reflected most clearly by the fact that the three gauge couplings (g_1, g_2 and g_3), extrapolated from below, fail to meet by a fairly wide margin in the context of minimal non-supersymmetric $SU(5)$ (see fig. 1).

But the situation changes radically if one assumes that the standard model is replaced by the minimal supersymmetric standard model (MSSM), above a threshold of about $1TeV$. In this case, the three gauge couplings are found to meet[26], at least approximately, provided $\alpha_3(m_Z)$ is not too low (see figs. 2a and 2b). Their scale of meeting is given by

$$M_X \approx 2 \times 10^{16} GeV \quad \text{(MSSM or SUSYSU(5))} \tag{8}$$

M_X may be interpreted as the scale where a supersymmetric grand unification symmetry (GUT) (like minimal SUSY SU(5) or SO(10)) — breaks spontaneously into the supersymmetric standard model symmetry $SU(2)_L \times U(1) \times SU(3)^c$.

The dramatic meeting of the three gauge couplings (Fig. 2) thus provides a strong support for both grand unification and supersymmetry.

Considering (a) that a straightforward meeting of the three gauge couplings occurs, only provided supersymmetry is assumed; (b) that supersymmetry provides at least a technical resolution of the gauge hierarchy problem, by preserving the small input value of the ratio of (m_W/M_X), in spite of quantum corrections; and (c) that it is needed for consistency of string theory, it seems apparent that *supersymmetry is an essential ingredient for higher unification.*

IVB. The Issue of Compatibility Between MSSM and String Unifications

The idea of grand unification would be incomplete without incorporating the unity of gravity with the weak, electromagnetic and the strong QCD forces. Superstring theory [27], and now the M theory [28] provide however the only known framework that exhibits the scope for such a unity. It thus becomes imperative that the meeting of the gauge couplings of the three non-gravitational forces, which occur by the extrapolation of the LEP data in the context of MSSM, be compatible with string unification.

Now, string theory does provide gauge coupling unification for the effective gauge symmetry, below the compactification-scale. The new feature is that even if the effective symmety is not simple, like SU(5) or SO(10), but instead is of the form G(213) or G(224) (say), the gauge couplings of G(213) or G(224) should still exhibit familiar unification at the string-scale, for compactification involving appropriate Kac-Moody levels (i.e. $k_2 = k_3 = 1$, $k_Y = \frac{5}{3}$ for G(213)), barring of course string-threshold corrections [29]. And even more, the gauge couplings unify with the gravitational coupling ($\frac{8\pi G_N}{\alpha'}$) as well at the string scale, where G_N is the Newton's constant and α' is the Regge slope.

Thus one can realize coupling unification without having a GUT-like symmetry below the compactification scale. This is the *new perspective* brought forth by string theory. There is, however, an issue to be resolved. Whereas the MSSM-unification scale, obtained by extrapolation of low energy data is given by $M_X \approx 2 \times 10^{16}$ GeV, the expected one-loop level string-unification scale[29] of $M_{st} \approx g_{st} \times (5.2 \times 10^{17} GeV) \approx 3.6 \times 10^{17}$ GeV is about twenty times higher [30,31]. Here, one has used $\alpha_{st} \approx \alpha_{GUT}$ (MSSM) ≈ 0.04.

A few alternative suggestions which have been proposed to remove this mismatch by nearly a factor of 20 between M_X and M_{st}, are as follows:

Matching Through String-Duality: One suggestion in this regard is due to Witten [32]. Using the equivalence of the strongly coupled heterotic $SO(32)$ and the $E_8 \times E_8$ superstring theories in $D = 10$, respectively to the weakly coupled $D = 10$ Type I and an eleven-dimensional M–theory, he observed that the 4-dimensional gauge coupling and $M_{\rm st}$ can both be small, as suggested by MSSM extrapolation of the low energy data, without making the Newton's constant unacceptably large.

Matching Through String GUT: A second way in which the mismatch between M_X and $M_{\rm st}$ could be resolved is if superstrings yield an intact supersymmetric grand unification symmetry like $SU(5)$ or $SO(10)$ with the right spectrum – i.e., three chiral families and a suitable Higgs system $M_{\rm st}$ [33], and if this symmetry would break spontaneously at $M_X \approx (1/20 \text{ to } 1/50) M_{\rm st}$ to the standard model symmetry. However, as yet, there is no realistic, or even close-to realistic, string–derived GUT model [33]. In particular, to date, no string-derived solution exists with a resolution of the doublet-triplet splitting problem, without which one faces the problem of rapid proton decay (see discussions later).

Matching Through Intermediate Scale Matter: A third alternative is based on string–derived standard model–like gauge groups. It attributes the mismatch between M_X and $M_{\rm st}$ to the existence of new matter with intermediate scale masses ($\sim 10^9 - 10^{13}$ GeV), which may emerge from strings[34]. Such a resolution is in principle possible, but it would rely on the delicate balance between the shifts in the three couplings and on the existence of very heavy new matter which in practice cannot be directly tested by experiments.

Matching Through ESSM – A Case for Semi-Perturbative Unification:

Babu and I suggested that a resolution of the mismatch between M_X and M_{st} can come about if there exists two "light" vector-like families ($16 + \bar{16}$) at the TeV scale [35]. Such a spectrum has an apriori motivation in that it provides a simple reason for inter-family mass-hierarchy. It can also be tested at LHC. Including two [35] and even three-loop effects[36], this spectrum leads to a semi-perturbative unification, with $\alpha_{GUT} \approx .2 - .3$, and raises M_X to (1 - 2) $\times 10^{17}$ GeV. Such higher values of α_{GUT} (compared to .04 for MSSM) may provide an additional advantage by helping to stabilize the dilaton.

While each of the solutions mentioned above possesses a certain degree of plausibility (see Ref. 31 for some additional possibilities), it is far from clear which, if any, is utilized by the true string-vacuum. This is of course related to the fact that, as yet, there is no insight as to how the vacuum is selected in the string or in the M-theory.

In summary, string theory, as well as M-theory, fully retain the basic concept of grand unification — i.e. unification of matter and of its gauge forces. But they enrich the scope considerably by (a) unifying all matter of spins 0, 1/2, 1, 3/2, 2 and higher, and (b) unifying gravity with the other forces. As noted above, the perspective on gauge coupling unification however changes in the string context, because such a unification can occur at the string scale, even without having a GUT-like symmetry at that scale. In the next section, I discuss the advantages as well as possible disadvantages of GUT versus non-GUT string solutions, keeping in mind the issues of both coupling unification and rapid proton decay.

I now turn to considerations of proton decay.

V Proton Decay as a Probe to Higher Unification

VA. As mentioned before, one of the hallmarks of grand unification is non-conservation of baryon and lepton numbers, which for most simple models, lead to proton decay[9,17]. The general complexion of baryon and lepton number non-conserving processes, including alternative modes of proton decay, $n - \bar{n}$ oscillation and neutrinoless double beta decay is discussed in my talk at the Oak Ridge Conference[37]. Here I will focus on proton decay.

Almost 25 years have passed since the suggestion of proton decay was first made in the context of unified theories, in 1973. While there was considerable resistance from the theoretical community against such ideas at that point, the psychological barrier against them softened over the years. The growing interest in the prospect of such a decay thus led to the building of proton-decay detectors in different parts of the world, including the most sensitive one of the 80's (IMB) at Cleveland, followed by Kamiokande in Japan. While proton decay is yet to be discovered, it is encouraging that searches for this decay continues at SuperKamiokande with higher sensitivity than ever before and detectors such as ICARUS are planned to come. The dedicated searches for proton decay at IMB (which was operative till a few years ago) and Kamiokande [38] already put severe constraints on grand unification for over a decade. Owing to these constraints, the non-supersymmetric minimal SU(5) and the minimal SO(10) models as well (with one-step breaking) are now excluded. In particular, conservatively, minimal non-SUSY SU(5) predicts: $\Gamma(p \to e^{+}\pi^{0})^{-1} \leq (6-10) \times 10^{31}$ yr, where as current data including those from Superkamiokande [39] yields:

$$\Gamma(p \to e^{+}\pi^{0})^{-1}_{expt} > 1.6 \times 10^{33} yr. \tag{9}$$

VB. The Issue of Proton-Longevity in SUSY Grand Unification

Although non-supersymmetric minimal SU(5) or SO(10) are excluded by proton-decay searches, as well as by precision measurements of $\sin^2\theta_W$, the situation with regard to both issues alters radically, once supersymmetry is combined with the idea of grand unification. First, as mentioned before, SUSY makes it possible for the three gauge couplings to meet at a common scale $M_X \approx 2\times10^{16}$ GeV. If one uses α_3 and α_2 as inputs, it correspondingly leads to the correct prediction for $\sin^2\theta_W$.

As regards proton decay, supersymmetric grand unified theories (GUTS), bring two new features: (i) First, by raising M_X to a higher value compared to the non-supersymmetric case, as above, they strongly suppress the gauge-boson -mediated d=6 proton decay operators, so that one obtains $\Gamma(p \longrightarrow e^{+}\pi^{0})^{-1}_{d=6} \approx 10^{36\pm1.5}$ yr. This is of course compatible with current experimental limits (eq(9)). (ii) Second, they generate d=5 proton decay operators of the form $Q_iQ_jQ_kL_\ell/\mathrm{M}$ and $\overline{UUDE}$ in the superpotential, through the exchange of color triplet Higgsinos, which are the GUT partners of the electroweak Higgsino doublets [40]. These triplets lie, for example, in the $5(\bar{5})$ of SU(5), or in the 10 or SO(10). Since the corresponding amplitudes are damped by just one power of the mass of the color-triplet higgsinos(m_{H_c}), these d=5 operators provide the dominant mechanism for proton decay in supersymmetric GUT.

The d=5 operators have marked effects both on the branching ratios of different decay modes as well as on the rate of proton decay. First, owing to (a) color-antisymmetry, (b) Bose symmetry of the scalar squark and slepton fields, and (c) the family-hierarchical Yukawa couplings, it turns out that these d=5 operators (to be called "standard" d=5) lead to dominant antineutrino modes:

$$p \to \bar{\nu}_\mu K^+, \bar{\nu}_\mu \pi^+ (standard\ d = 5), \tag{10}$$

but highly suppressed $e^+\pi^0$, e^+K^0 and even $\mu^+\pi^0$ and μ^+K^0 modes (at least for small and moderate $\tan\beta \leq 15$). Recall, by contrast, that for non-supersymmetric GUTS, $e^+\pi^0$ is expected to be the dominant mode.

Second, given the Yukawa couplings of the electroweak Higgs doublets (inferred from fermion masses), a typical contribution to the standard d=5 proton decay operator of the form QQQL/M is found to have an effective strength $\approx (m_c m_s \sin\theta_c / v_u v_d)\,(1/M_{H_c}) \approx \frac{10^{-7} tan\beta}{M_{H_c}}$. Now, for plausible values or limits on $m_{\tilde{q}} \leq 1$ TeV, $(m_{\tilde{W}}/m_{\tilde{q}}) \geq 1/6$ and $\tan\beta \geq 3$ (say), the d=5 operator, as noted above, subject to wino-dressing, leads to an inverse decay rate [41]

$$\Gamma^{-1}(p \to \bar{\nu}_\mu K^+) \leq 3 \times 10^{32} yrs (\frac{M_{H_c}}{3 \times 10^{16} GeV})^2 \tag{11}$$

To be conservative, this estimate uses the minimum theoretical value of the hadronic matrix element ($\beta_H = .003 GeV^3$), and assumes a cancellation by a factor of two betwen $\tilde{t}$ and $\tilde{c}$ - contributions, (although, in general, one could gain a factor of 2 to 4 (say) in the rate on each count). Given the current experimental limit of $\Gamma(p \to \bar{\nu} K^+)^{-1} > 5.5 \times 10^{32}$ yrs (90% CL) [42], it follows that the color-triplets must be superheavy. Conservatively [43],

$$M_{H_c} \geq (3 - 5) \times 10^{16} GeV \tag{12}$$

While the color triplets need to be superheavy, their doublet-partners must still be light (≤ 1 TeV). The question arises: How can the color- triplets become superheavy, while the doublet-partners remain naturally light? This is the well-known problem of *doublet-triplet splitting* that faces all SUSY GUTS.

Leaving out the possibility of extreme fine tuning, two of the proposed solutions to this problem are as follows:

(i) The Missing Partner Mechanism [45]: In this case, by introducing suitable large-size Higgs multiplets, such as $50_H + \overline{50}_H + 75_H$, in addition to $5_H + \overline{5}_H$ of $SU(5)$, and introducing couplings of the form $W = C\, 5_H \cdot \overline{50}_H \cdot < 75_H > + D\,\overline{5}_H \cdot 50_H < 75_H >$, one can give superheavy masses to the triplets (anti-triplets) in $5(\overline{5})$ by pairing them with anti-triplets (triplets) in $\overline{50}(50)$. But there do not exist doublets in $50(\overline{50})$ to pair up with the doublets in $5(\overline{5})$, which therefore remain light.

(ii) The Dimopoulos-Wilczek Mechanism [46]: Utilizing the fact that the VEV of 45_H of $SO(10)$ does not have to be traceless (unlike that of 24_H of $SU(5)$), one can give mass to color-triplets and not to doublets in the 10 of $SO(10)$, by arranging the VEV of 45_H to be proportional to $i\tau_2\times$ diag $(x, x, x, 0, 0)$, and introducing a coupling of the form $\lambda 10_{H1} \cdot 45_H \cdot 10_{H2}$ in W. Two $10's$ are needed owing to the anti-symmetry of 45. Because of two $10's$, this coupling would leave two pairs of electroweak doublets massless. One must, however, make one of these pairs superheavy, by introducing a term like $M\, 10_{H2} \cdot 10_{H2}$ in W, so as not to spoil the successful prediction of $\sin^2\theta_W$ of SUSY GUT. In addition, one must also ensure that only 10_{H1} but not 10_{H2} couple to the light quarks and leptons, so as to prevent rapid proton decay. All of these can be achieved by imposing suitable discrete symmetries. There is, however, still some question as to whether the triplets can be sufficiently heavy, or rather $M_{eff} \equiv \lambda < 45_H >^2 /M$ can be of order 10^{18} GeV (that is needed), without conflicting with unification of the gauge couplings.

In summary, solutions to the problem of doublet-triplet splitting needing a suitable *choice* of Higgs multiplets and discrete symmetries are technically feasible. It is however not clear whether any of these mechansims can be consistently derived from an underlying theory, such as the superstring theory. To date, no such mechanism has been realized in a *string-derived* GUT solution [33].

VC. Rapid Proton Decay And The Other Problems of Naturalness in Supersymmetry

In addition to the problem of doublet-triplet splitting that faces SUSY GUT theories, it is important to note that there is a generic problem for all supersymmetric theories, involving either a GUT or a non-GUT symmetry, in the presence of quantum gravity. This is because, in accord with the standard model gauge symmetry $SU(2)_L \times U(1)_Y \times SU(3)^C$, a supersymmetric theory in general permits, in contrast to non-supersymmetric ones, dimension 4 and dimension 5 operators which violate baryon and lepton numbers [40]. Such operators are likely to be induced by Planck-scale physics including especially quantum gravity, unless they are forbidden by symmetries of the theory. Using standard notations, the operators in question are as follows:

$$\begin{aligned} W = & [\eta_1 \overline{U}\,\overline{D}\,\overline{D} + \eta_2 QL\overline{D} + \eta_3 LL\overline{E}] \\ & + [\lambda_1 QQQL + \lambda_2 \overline{U}\,\overline{U}\,\overline{D}\,\overline{E} + \lambda_3 LLH_2H_2]/M. \end{aligned} \tag{13}$$

Here, generation, $SU(2)_L$ and $SU(3)^C$ indices are suppressed. M denotes a characteristic mass scale. The first two terms of $d = 4$, jointly, as well as the $d = 5$ terms of strengths λ_1 and λ_2, individually, induce $\Delta(B - L) = 0$

proton decay with amplitudes $\sim \eta_1\eta_2/m_{\tilde{q}}^2$ and $(\lambda_{1,2}/M)(\delta)$ respectively, where δ represents a loop-factor. Experimental limits on proton lifetime turn out to impose the constraints: $\eta_1\eta_2 \leq 10^{-24}$ and $(\lambda_{1,2}/M) \leq 10^{-23}$ to $10^{-24} GeV^{-1}$. Thus, even if $M \sim M_{string} \sim 10^{18}$ GeV, we must have $\lambda_{1,2} \leq 10^{-5}$ to 10^{-6}, so that proton lifetime will be in accord with experimental limits.

Renormalizable, supersymmetric standard-like and $SU(5)$ [44] models can be constructed so as to avoid, *by choice*, the $d = 4$ operators (i.e. the $\eta_{1,2,3}$-terms) by imposing a discrete or a multiplicative R-parity symmetry: $R \equiv (-1)^{3(B-L)}$, or more naturally, by gauging $B - L$, as in $\mathcal{G}_{224} \equiv SU(2)_L \times SU(2)_R \times SU(4)^C$ or $SO(10)$. Such resolutions, however, do not in general suffice if we permit higher dimensional operators and intermediate or GUT-scale VEVs of fields which violate $(B-L)$ by one unit and thereby R-parity (see below). In string solutions, VEV's of such fields seem to be needed, to generate Majorana masses for the RH neutrinos. Besides, $B - L$ cannot provide any protection against the $d = 5$ operators given by the λ_1 and λ_2 - terms, which conserve $B - L$. As mentioned above these operators are, however, expected to be present in any theory linked with gravity, e.g. a superstring theory, unless they are forbidden by some new symmetries.

These considerations show that, in the context of supersymmetry, the extraordinary stability of the proton is a major puzzle. And, the problem is heightened especially in the context of SUSY GUT theories because of the need for the doublet-triplet splitting in such theories. *The question in fact arises: Why does the proton have a lifetime exceeding* 10^{40} *sec, rather than the apparently natural value, for supersymmetry, of less than 1 sec?* As such, the known longevity of the proton deserves a natural explanation. I believe that it is in fact a major clue to some deeper physics that operates near the Planck-scale.

Apart from the problem of rapid proton decay, supersymmetry in fact generates a few additional problems of similar magnitude. These together constitute the so-called *naturalness problems of supersymmetry.* They include understanding: (i) the extreme smallness of the SUSY-breaking mass-splittings compared to the Planck-scale (i.e. why $(\delta m_s/M_{planck}) \sim 10^{-15}$ rather than order unity), (ii) the smallness of the μ-parameter of MSSM also compared to the Planck-scale, (iii) the strong suppression of the neutrino-Higgsino mixing mass, (that needs to be less than about 1 MeV) in a context where R-parity is violated, and (iv) the smallness of especially the CP-violating part of the $K^o - \bar{K}^o$ amplitude in spite of the potentially large contributions from squark and gluino loops. In addition to this set of problems, which are special to supersymmetry, there is of course the familiar challenge of understanding the hierarchical masses and mixings of quarks and leptons. Resolving these problems would amount to understanding the origins of some extremely small numbers, ranging from 10^{-6} to 10^{-19}, which apriori could be of order unity. As such, I believe that they are a reflection of *new symmetries* which operate near the Planck-scale. In the limit of these symmetries, the respective entities, such as the strengths of the d=4 and d=5 operators and the magnitudes of δm_s and μ, would vanish. Although the symmetries break, quite possibly near the GUT-scale, they need to be powerful enough to provide the needed protection up to sufficiently high order in non-renormalizable terms, scaled by the Planck mass, so as to render the respective numbers as small as they are. Symmetries of this nature simply do not exist in conventional GUTS. They do, however, arise, not so infrequently, in string-solutions, including some which are fairly realistic, possessing three-families and hierarchical Yukawa couplings [47,48,49].

Invariably, these solutions possess non-GUT symmetries such as (i) the (B-L)-preserving standard model-like symmetry G(2113) [47], or (ii) G(224) [48], or (iii) flipped SU(5) × U(1) [49]. Based on some recent work [50], I note below how string symmetries can play an essential role in avoiding the danger of rapid proton decay and also help in resolving some of the other naturalness problems noted above.

VD. The Role of String-Flavor Symmetries in Resolving The Naturalness Problems

To illustrate the usefulness of string-symmetries, I would consider especially a class of three-family string solutions which are based on the free fermionic construction [51] and correspond to a special $Z_2 \times Z_2$ orbifold compactification [47]. They lead, after the applications of all GSO-projections, to a gauge symmetry at the string-scale of the form:

$$\mathcal{G}_{st} = [SU(2)_L \times SU(3)^C \times U(1)_{I_{3R}} \times U(1)_{B-L}] \times [G_M = \prod_{i=1}^{6} U(1)_i] \times G_H. \quad (14)$$

The first factor will be abbreviated as G(2113). Here $U(1)_i$ denote six horizontal symmetries which act non-trivially on the three families (e, μ and τ) and distinguish between them. G_H denotes the hidden-sector symmetry which operate on "hidden" matter. The horizontal symmetries $U(1)_i$ couple to both the observable and the hidden sector matter.

The crucial point is that the pairs (U_1, U_4), (U_2, U_5) and (U_3, U_6), respectively couple to families 1, 2 and 3, in an *identical fashion.* [g] Thus, on the one

[g] While U_1, U_2 and U_3 respectively assign the same charge to all 16 members of families 1,2 and 3, U_4, U_5 and U_6 distinguish between members within a family. Thus U_1, U_2 and U_3 commute with SO(10), but U_4, U_5 and U_6 do not.

hand, these six U(1) symmetries, having their origin in SO(44) [51], distinguish between the three families, unlike a GUT symmetry like SO(10). Thereby they serve as generalized *"flavor" symmetries* and in turn help explain the hierarchical Yukawa couplings of the three families [47]. On the other hand, the coupling of the three pairs (U_1, U_4), (U_2, U_5) and (U_3, U_6) fully preserve the cyclic *permutation symmetry* with respect to the three families.

Turning to the problem of rapid proton decay in the context of these string solutions, there are two features which together help resolve the problem. First, it turns out that for non-GUT solutions of the type obtained in Ref. 47 (this is also true of the G(224)-solution of Ref. 48), in the process of compactification leading to G(2113), the dangerous color triplets are simply projected out of the spectrum altogether. As a result, the problem of doublet-triplet splitting is neatly avoided. *This is an obvious advantage of a non-GUT over a GUT string solution.*

Second, it needs to be said that of the six U(1)'s [Ref. 47], one linear combination — i.e. $U(1)_A = 1/\sqrt{15}[2(U_1 + U_2 + U_3) - (U_4 + U_5 + U_6)]$ — is anomalous, while the other five are anomaly-free (occurrence of such anomalous U(1) is in fact fairly generic in string solutions). Furthermore, the string solutions invariably yield a set of standard model singlet fields $\{\Phi_a\}$ which couple to the flavor symmetries $U(1)_i$. For the solution of Ref. 47, they couple to the six U(1)'s as well as to B-L and I_{3R}. Now a set of these $\{\Phi_i\}$ fields must acquire VEV's of order $(10^{-1} - 10^{-2})\ M_{pl}$ (where $M_{pl} \approx 2 \times 10^{18}$ GeV), in order to cancel the Fayet-Iliopoulos D-term generated by $U(1)_A$, and also all F and D-terms, so that supersymmetry is preserved, barring additional constraints [52].

It turns out that the six flavor symmetries $U(1)_i$, together with certain SUSY-preserving patterns of VEVs of the $\{\Phi_a\}$-fields, suffice to naturally safeguard proton-longevity, to the extent needed, from all potential dangers, including those which may arise through gravity-induced higher dimensional operators (d $\geq$4) and the exchange of color-triplets in the infinite tower of heavy string states [50]. This protection holds in spite of the fact that certain Φ_i's acquiring VEVs carry $\mid B - L \mid = 1$, which help provide superheavy Majorana masses to the RH neutrinos, but, in the process, break R-parity. The protection comes about because the symmetries mentioned above prevent the appearance of the dangerous effective d=4 and d=5 operators, unless one utilizes non-renormalizable operators involving sufficiently high powers of the ratios $< \{\Phi_i\} > /M_{st}$, where each such ratio is naturally $O(1/10)$. *These virtues of the extra flavor symmetries show that, believing in supersymmetry, superstring is suggested just to understand why the proton is so long-lived.*

In above, I have tried to illustrate the beneficial role of string symmetries within one class of fairly realistic string solutions [47]. It still remains to be seen whether such string-symmetries by themselves can account for the desired suppression of the d=4 and the d=5 operators, regardless of the choice of the pattern of VEVs. [For attempts in this direction, see e.g. Refs. 53 and 54].

I should add briefly that the string-flavor symmetries of the type just described are found to play a crucial role in resolving also some of the other problems of naturalness listed above. These include understanding the smallness of SUSY-breaking mass-splittings ($\delta m_s \sim 1$ TeV) on the one hand, and deriving the desired squark-degeneracy that adequately accounts for the suppression of the flavor-changing neutral current processes on the other hand. These two features are realized by implementing supersymmetry-breaking through a non-vanishing D-term of the string-derived anomalous U(1) gauge symmetry, noted above [55,56]. The string-flavor symmetries also help in understanding the strong suppression of the neutrino-higgsino mixing mass [57] and the smallness of the CP violating part of the $K^o - \bar{K}^o$ amplitude [58]. Last but not least, the same flavor symmetries help obtain the qualitatively correct pattern of hierarchical fermion masses and mixings [47]. *Thus the beneficial roles of these string flavor symmetries can hardly be overemphasized.*

One is of course aware that it is premature to take any specific string solution, or even a specific class of solutions, from the vast set of allowed ones, too seriously. Nevertheless it seems feasible that certain features, especially the symmetry properties, may well survive in the final picture that may emerge from the ultimate underlying theory, encompassing string theory, M theory and D-branes. From a purely utilitarian point of view, given the magnitude of the naturalness problems, it seems that such flavor symmetries *should* in fact emerge from the underlying theory, just to help preserve one's belief in supersymmetry.

It needs to be mentioned that while the string-flavor symmetries provide the scope for obtaining a resolution of the problems mentioned above, obtaining a simultaneous resolution of all or most of them in the context of a given string solution is still a challenging task.

VE. A GUT or a Non-GUT String Solution?

In summary, comparing string-GUT with non-GUT solutions, where the latter lead to symmetries like G(2113) or G(224) at the string scale, we see that each has a certain advantage over the other. For a non-GUT solution, the gauge couplings unify only at the string-scale; thus one must assume that somehow a solution of the type discussed in sec. 4 should resolve the mismatch between M_X and M_{st}. This is plausible but not easy to ascertain. In this regard, a string GUT-solution yielding SU(5) or SO(10) appears to have an advantage over a non-GUT solution, because, in the case of the former, the couplings naturally stay together between M_{st} and M_X. Furthermore, a GUT symmetry-breaking scale of $M_{st}/20$ seems to be plausible in the string-context.

On the other hand, as mentioned above, deriving a GUT-solution from strings, while achieving doublet-triplet splitting, is indeed a major burden, and has not been achieved as yet. In this regard, the non-GUT solutions seem to possess a distinct advantage, because the dangerous color-triplets are often naturally projected out [see e.g. Refs. 47 and 48]. Furthermore, these solutions possess new symmetries, which are not available in GUTS, and some of these do not even commute with GUT-symmetries, but they do help in providing the desired protection, even against gravity-induced proton decay, that may otherwise be unacceptably rapid. In addition, as mentioned above, these new symmetries turn out to help in the resolution of the other naturalness-problems of supersymmetry as well (see e.g. Refs. 55, 57 and 58).

Weighing the advantages and possible disadvantages of both, it seems hard to make a clear choice between a GUT versus a non-GUT string-solution. While one may well have a preference for one over the other, it seems reasonable to keep one's options open in this regard and look for other means, based e.g. on certain features of proton decay and the solutions to the naturalness problems, which can help provide a distinction between the two alternatives. Short of making such a choice at this point, one must assume that for a GUT-solution, strings would somehow provide a resolution of the problem of doublet-triplet splitting, while for a non-GUT string-solution, it needs to be assumed that one of the mechanisms mentioned in sec. 4 (for instance, that based on string-duality [32] and/or semi-perturbative unification [35]) is operative so as to remove the mismatch between M_X and M_{st}.

I now discuss how proton decay is influenced by the masses and the mixings of the fermions, especially those of the neutrinos.

VI Link Between Neutrino Masses and Proton Decay

Two important characteristics of supersymmetric unification, based on a gauge symmetry like SO(10) or a string-derived G(224), seem to be borne out by nature. They are: (a) gauge coupling unification at a scale $M_X \sim 2 \times 10^{16}$ GeV, and (b) light neutrino masses ($\ll m_{e,\mu,\tau}$). As discussed in sec. 3, the value of $m(\nu_\tau) \approx 1/20$ eV, suggested by the SuperK (SK) result, is just about what one would expect within the symmetry-structure G(224)/SO(10), if the (B-L)–breaking scale is around M_X. One is thus naturally tempted to ask: *Will proton decay —the other major prediction of grand unification — also reveal in the near future?*

This question acquires a special significance because of the following circumstance. Ordinarily, except for the scale of new physics, involved in the two cases, proton decay, especially its decay modes are considered to be essentially unrelated to the pattern of neutrino masses. However, in a recent paper, Babu, Wilczek and I noted that neutrino masses can have a significant effect on proton decay as regards its rate as well as decay modes [59]. This is because in supersymmetric unified theories, based on SO(10) or G(224), assignment of heavy Majorana masses to the RH neutrinos (as discussed in sec. 3), inevitably introduces *a new set of color-triplets* (unrelated to the electroweak doublets), whose effective couplings to quarks and leptons are related to these Majorana masses (see eqs. (4) and (5)). Exchange of these new color-triplets give rise to a new set of d=5 proton decay operators, which are thus directly related to the neutrino-masses. Assuming that $\nu_e - \nu_\mu$ oscillation is relevant to the MSW explanation of the solar neutrino puzzle, so that $m(\nu_L^\mu) \approx 3 \times 10^{-3}$eV, which corresponds to $M(\nu_R^\mu) \approx 2 \times 10^{12}$ GeV, the strength of the new d=5 operators turns out to be just about right ($\tau_P \approx 10^{31.5 \pm 2.5}$ yrs), for proton decay to be observable at SuperKamiokande [59,60].

The flavor-structure of the new d=5 operators are, however, expected to be distinct from those of the standard d=5 operators, which are governed by the highly hierarchical Dirac masses of quarks and leptons. In contrast to the standard d=5 operators, the new ones can lead to prominent charged lepton decay modes, such as $\ell^+\pi^o, \ell^+K^o$ and $\ell^+\eta$, involving especially μ^+, even for low or moderate values of $tan\beta \leq 10$. The intriguing feature thus is that owing to the underlying SO(10) or just SU(4)-color symmetry, *proton decay operator knows about neutrino masses and vice versa.*

With a *maximal* effective Majorana-coupling for the third family (i.e. $\lambda_{33} \sim \mathcal{O}(1)$), as suggested in sec.3, that corresponds to $M_{3R} \approx (\text{few} \times 10^{14}\text{GeV})$ for the case of no mixing (see eq. (5)), one might worry that proton would decay too fast, because of an enhancement in the new d=5 operators, relative to that considered in Ref. 59. It turns out, however, that because τ^+ is heavier than the proton and also because $\bar{\nu}_\tau K^+$ mode receives a strong suppression-factor from the small mixing angle associated with the third family ($V_{ub} \approx 0.002 - 0.005$), a maximal Majorana-coupling of the third family ($\lambda_{33} \sim \mathcal{O}(1)$), and thus $m(\nu_L^\tau) \approx (1/10 - 1/30)$eV, is still compatible with present limit on proton lifetime [60].

Babu, Wilczek and I have recently attempted to understand the neutrino masses and mixings, as suggested by both the SuperKamiokande result (interpreted as $\nu_\mu - \nu_\tau$ oscillation) and the solar neutrino puzzle, within a predictive *SO(10) or G(224)-based* quark-lepton unified description of the masses and mixings of *all fermions*—i.e. quarks, charged leptons as well as neutrinos [60]. Adopting familiar ideas of generating hierarchical eigenvalues through off-diagonal mixings and correspondingly Cabibbo-like mixing angles, we find that the bizarre pattern of masses and mixings of quarks and charged leptons of all three families can in fact be described adequately (to better than 10% accuracy), within an economical SO(10)-framework, which makes five successful predictions, just for the quark and the charged lepton system.

In the process, the Dirac mass matrices of the neutrinos, as well as of the charged leptons, are fully determined. Taking the Dirac masses of the leptons, thus fixed, together with a simple hierarchical pattern for the Majorana mass matrix of the superheavy right-handed neutrinos, we show that one can obtain quite naturally a large $\nu_L^\mu - \nu_L^\tau$ oscillation angle ($sin^2 2\theta \simeq .85 - .95$), just as observed at SuperK, in spite of highly non-degenerate masses of the three light neutrinos—e.g. with $m(\nu_L^e) \ll m(\nu_L^\mu) \approx (\frac{1}{10} - \frac{1}{20}) m(\nu_L^\tau)$. Such a hierarchical mass-pattern for the light neutrinos is of course natural to see-saw. In this case, $\nu_e - \nu_\mu$ oscillation becomes relevant to the small angle MSW explanation [61] of the solar neutrino puzzle [62]. The distinctive features of this explanation of the neutrino-anomalies are: (a) its origin within an underlying unified theory that relates the masses and mixings of neutrinos to those of quarks and charged leptons, and (b) the emergence of the large oscillation angle without a large mixing in either the $(\nu_\mu - \nu_\tau)$ or the $(\mu - \tau)$ mass-eigenstates.

As an important corollary to this work, owing to the link mentioned above between neutrino masses and proton decay [59], we find that the mass of ν_τ and the large oscillation angle suggested by the SuperKamiokande result in fact imply a net enhancement in the proton decay rate, as well as of the $\mu^+ K^o$ mode [60]. There are of course uncertainties in the prediction for proton-decay rate owing to uncertainties in the SUSY-spectrum, the hadronic matrix element and the relative phases of the many different contributions (see Ref. 41 and in particular Ref. 60 for details). However, given that the individual contributions to the amplitude are enhanced by the neutrino and fermion mass-effects, it seems that it would be hard to reconcile the ideas of supersymmetric unification described here, if the proton life-time exceeds about 10^{34} yrs [h] .

---

[h] This is barring the highly unlikely possibility of a large cancellation (amounting to a net reduction in the amplitude by more than a factor of 5) between the many contributions of unknown relative phases which arise just from the standard as well as those from the new operators, related to neutrino-masses, see Ref. 60 The reduction in the proton decay rate owing to such cancellation in the amplitude is even less likely, when in addition to contributions from the QQQL-operators, one includes those from the $\overline{UUDE}$ operators, using Higgsino-dressing, which become important if the SUSY-parameter μ is moderately large ($\geq$ 500 GeV, say) (See Ref. 42). The expectation of proton lifetime $\leq (1-2) \times 10^{34}$ yrs, (noted above) assumes: $m_{\tilde{q}} \leq (1 - 1.5)$ TeV and $(m_{\tilde{w}}/m_{\tilde{q}}) \geq 1/10$. The prominence of

Assuming such a unification, the prospect for discovery of proton decay at SuperK and/or at ICARUS thus seems strong.

To conclude, with neutrino masses and coupling unification revealed, *proton decay remains as the only missing link.* Its discovery, with dominance of $\bar{\nu}K^+$ and prominence of μ^+K^o modes, would in fact be a double confirmation of both supersymmetric unification through G(224)/SO(10), as well as of the ideas of neutrino masses, described above in this context.

Acknowledgement:

I have greatly benefitted from discussions on topics in this manuscript with Keith Dienes, Alon Faraggi and Edward Witten, and especially from collaborative discussions with Kaladi S. Babu and Frank Wilczek. The research was supported in part by NSF Grant No. Phys-9119745 and in part by the Distinguished Research Fellowhsip awarded by the University of Maryland.

References

1. Source: "Ideals and Realities" - Selected Essays by Abdus Salam, Ed. by C.H. Lai (World Scientific).
2. J.C. Pati in Salam Festschrift, Ed. by A. Ali, J. Ellis and S. Randjbar-Daemi (World Scientific, 1993) pages 368-391.
3. J.C. Pati, "Recollections of Abdus Salam: Scientist and Humanitarian" - Physics Today (abridged version), August (1997); News Letter of the American Chapter of the Indian Physics Association (1997).
4. S. L. Glashow, *Nucl. Phys.* **22**, 57a (1961); S. Weinberg, *Phys. Rev. Lett.* **19**, 1269 (1967); Abdus Salam, in Elementary Particle Theory, Nobel Symposium, ed. by N. Svartholm (Almqvist, Stockholm, 1968), p. 367.
5. For attempts to derive the current algebra framework using the singlet U(1) vector gluon see e.g. H. Fritzsch and M. Gell-Mann, Proc. 15th High Energy Conf., Batavia (1972), pages 135-165, and references there in. The difficulty of SU(9) or SU(12) global symmetry associated with the U(1) vector gluon coupling was one reason for abandoning this possibility: See J.C. Pati and Abdus Salam, ICTP preprint IC/73/81 (unpublished), and L.B. Okun and V.I. Zacharov, *Phys. Lett.* **47B**, 258 (1973).
6. O.W. Greenberg, *Phys. Rev. Lett.* **13**, 598 (1964).
7. M. Han and Y. Nambu, *Phys. Rev.* **139**, B1006 (1965).

μ^+K^o-mode noted above becomes a distinguishing feature of the contributions from the new operators (related to neutrino-masses) especially for low and moderate values of $\tan\beta \leq 20$ (say).

8. J.C. Pati and Abdus Salam; Proc. 15th High Energy Conference, Batavia, reported by J.D. Bjorken, Vol. 2, p. 301 (1972); *Phys. Rev.* **8**, 1240 (1973).
9. J.C. Pati and Abdus Salam, *Phys. Rev. Lett.* **31**, 661 (1973); *Phys. Rev.* **D10**, 275 (1974).
10. The important suggestion of generating strong interactions by gauging SU(3)-color symmetry was first made by Han and Nambu, as early as 1965 [7]. There were two shortcomings in this paper which needed to be removed. First, owing to the coupling of the photon to integer charge-quarks, $SU(3)^c$ was violated explicitly - rather than spontaneously - a feature that spoilt renormalizability. Second, *fundamental* strong interactions were assumed to have two components: the very strong $SU(3)^c$ gauge force and, in addition, a medium strong force, generated by the gauge bosons of flavor symmetry, which were identified with (ρ, ω, K^*, and ϕ) vector bosons. Generating the electroweak forces via the flavor gauge symmetry, as in SU(2) $\times$ U(1), however, precludes (ρ, ω, K^*) being fundamental. This in turn uniquely selects out the SU(3)-color gauge force as the *only source* of fundamental strong interactions, assuming a gauge-origin of all three forces.
11. At the time of our suggestion of the combined symmetry at the 1972 Batavia Conference [Ref. 8], Fritzsch and Gell-Mann also noted the possibility of $SU(3)^c$ gauging at the same conference [Ref. 5]. They, and many others, favored the alternative of fractionally charged quarks (fcq) and unbroken SU(3)-color, with permanent confinement. With little understanding of confinement in those days, Salam and I thought that the case of fcq and permanent confinement should be regarded only as an alternative to the case of spontaneously broken $SU(3)^c$, which would lead to gauge integer charge quarks (icq), with possibly semi-confined but ultimately liberated quarks. The two alternatives arise within the same theory depending upon the nature of SSB. We, (*Phys. Rev. Lett.* **36**, 11 (1976)), as well as others, notably G. Rajasekaran and P. Roy (Pramana **6**, 303 (1975)), therefore proposed key experiments which could distinguish between the two alternatives. Eventually, by the 1980's, on the basis of these tests, especially the two-photon experiments (see e.g. R. Godbole, J. Pati, S. Rindani, T. Jayaraman and G. Rajasekaran, *Phys. Lett.* **142**, 91 (1984)), fcq was clearly favored over gauge icq. The ideas of quark-lepton unification, as also the success of the (combined) standard model symmetry, are of course independent of the nature of quark-charges.

12. D. Gross and F. Wilczek, *Phys. Rev. Lett.* **30**, 1343 (1973); D. Politzer, ibid, **30**, 1346 (1973).
13. R. N. Mohapatra and J. C. Pati, *Phys. Rev.* **D11**, 566 (1974); *Phys. Rev.* **D11**, 2558 (1975); G. Senjanovic and R. N. Mohapatra, *Phys. Rev.* **D12**, 1502 (1975).
14. V. Kuzmin, Va. Rubakov and M. Shaposhnikov, *Phys. Lett.* **BM155**, 36 (1985); M. Fukugita and T. Yanagida, *Phys. Lett.* **B174**, 45 (1986); M. A. Luty, *Phys. Rev.* **D45**, 455 (1992); W. Buchmuller and M. Plumacher, hep-ph/9608308.
15. J.C. Pati, Proc. Seoul Symposium on Elementary Particle Physics in memory of B.W. Lee, page 481 (1978).
16. J.C. Pati, Abdus Salam and J. Strathdee, *Il. Nuov. Com.* **26A**, 77 (1975).
17. H. Georgi and S. L. Glashow, *Phys. Rev. Lett.* **32**, 438 (1974).
18. H. Georgi, in Particles and Fields, ed. by C. Carlson (AIP, NY, 1975), p. 575; H. Fritzsch and P. Minkowski, *Ann. Phys.* **93**, 193 (1975).
19. H. Georgi, H. Quinn and S. Weinberg, *Phys. Rev. Lett.* **33**, 451 (1974).
20. F. Gürsey, P. Ramond and P. Sikivie, *Phys. Lett.* **B60**, 177 (1976).
21. The SuperKamiokande Collaboration: "Evidence for Oscillation of Atmospheric Neutrinos", ICRR-Report-422-98-18 (July, 1998).
22. J.C. Pati "Implications of the SuperKamiokande Result on the Nature of the New Physics", UMD-PP-99-01, hep-ph/9807315; To appear in the Proc. of the Neutrino-98 Conf. held at Takayama, Japan, June, 1998.
23. See e.g. K. R. Dienes and J. March-Russell, hep-th/9604112; K. R. Dienes, hep-ph/9606467.
24. M. Gell-Mann, P. Ramond and R. Slansky, in: *Supergravity*, eds. F. van Nieuwenhuizen and D. Freedman (Amsterdam, North Holland, 1979) p. 315; T. Yanagida, in: *Workshop on the Unified Theory and Baryon Number in the Universe*, eds. O. Sawada and A. Sugamoto (KEK, Tsukuba) 95 (1979); R. N. Mohapatra and G. Senjanovic, *Phys. Rev. Lett.* **44**, 912 (1980).
25. For reviews, see e.g. P. Langacker, talk at Gatlinburg Conference, June 94, hep-ph 9411247, and P. Langacker and N. Polonsky, *Phys. Rev.* **D52**, 3081 (1995), and references there in.
26. For recent work, see P. Langacker and M. Luo, *Phys. Rev.* **D44**, 817 (1991); U. Amaldi, W. de Boer and H. Furstenau, *Phys. Lett.* **B260** (1991); J. Ellis, S. Kelley and D. V. Nanopoulos, *Phys. Lett.* **B260**, 131 (1991); F. Anselmo, L. Cifarelli, A. Peterman and A. Zichichi, *Nuov. Cim.* **A104**, 1817 (1991). The essential features pertaining to coupling unification in SUSY GUTS were noted earlier by S. Dimopoulos, S. Raby

and F. Wilczek, *Phys. Rev.* **D24**, 1681 (1981); W. Marciano and G. Senjanovic, *Phys. Rev.* **D25**, 3092 (1982); M. Einhorn and D.R.T. Jones, *Nucl. Phys.* **B196**, 475 (1982).

27. M. Green and J.H. Schwarz, *Phys. Lett.* **149B**, 117 (1984); D.J. Gross, J.A. Harvey, E. Martinec and R. Rohm, *Phys. Rev. Lett.* **54**, 502 (1985); P. Candelas, G.T. Horowitz, A. Strominger and E. Witten, *Nucl. Phys.* **B258**, 46 (1985). For introductions and reviews, see: M.B. Green, J.H. Schwarz and E. Witten, "Superstring Theory" Vols. 1 and 2 (Cambridge University Press); M. Dine, ed., "String Theory in Four Dimensions" (North Holland, 1988); J. Polchinski, "What is String Theory?", 1994 Les Houches Lectures, hep-th/9411028.
28. The literature on string-dualities and M-theory is large. A few pioneering papers relevant to gauge-coupling unification are: E. Witten, *Nucl. Phys.* **B443**, 85 (1995), P. Horava and E. Witten, *Nucl. Phys.* **B460**, 506 (1996). For reviews, see e.g. J. Polchinski, hep-th/9511157; A. Sen, hep-th/9802051, and references therein. For a layman's guide, see M. Duff (talk at this meeting), hep-ph/9805177 V3.
29. P. Ginsparg, *Phys. Lett.* **B197**, 139 (1987); V.S. Kaplunovsky, *Nucl. Phys.* **B307**, 145 (1988); Erratum: *ibid.* **B382**, 436 (1992).
30. For an early review on this issue, see e.g. S. Weinberg, Summary talk, Proc. 26th Intl. Conf. on High Energy Physics, Dallas, Texas (1992)
31. For a recent discussion, see K. Dienes, *Phys. Reports* **287**, 447 (1997), (hep-th/9602045), and references therein.
32. E. Witten, hep-th/9602070.
33. See e.g. D. Lewellen, *Nucl. Phys.* **B337**, 61 (1990); A. Font, L. Ibanez and F. Quevedo, *Nucl. Phys.* **B345**, 389 (1990); S. Chaudhari, G. Hockney and J. Lykken, *Nucl. Phys.* **B456**, 89 (1995) and hep-th/9510241; G. Aldazabal, A. Font, L. Ibanez and A. Uranga, *Nucl. Phys.* **B452**, 3 (1995); ibid. **B465**, 34 (1996); D. Finnell, *Phys. Rev.* **D53**, 5781 (1996); A.A. Maslikov, I. Naumov and G.G. Volkov, *Int. J. Mod. Phys.* **A11**, 1117 (1996); J. Erler, hep-th/9602032 and G. Cleaver, hep-th/9604183; and Z. Kakushadze and S.H. Tye, hep-th/9605221, and hep-th/9609027; Z. Kakushadze et al, hep-ph/9705202.
34. See e.g. K. Dienes and A. Faraggi, *Nucl. Phys.* **457**, 409 (1995); C. Bachas, C. Fabre and T. Yanagida, *Phys. Lett.* **B370**, 49 (1996).
35. J.C. Pati and K.S. Babu, "*The Problems of Unification-Mismatch and Low α_3: A Solution with Light Vector-Like Matter*", hep-ph/9606215, *Phys. Lett.* **B384**, 140 (1996).
36. C. Kolda and J. March-Russell (hep-ph/9609480).

37. J.C. Pati, "*Baryon Non-Conservation in Unified Theories in the Light of Supersymmetry and Superstrings*", hep-ph/9611371, Proc. Oakridge International Workshop on Baryon Instability, March (1996), pages 7-59.
38. See e.g. J. Stone, Proc. Oakridge International Workshop as in Ref. 37.
39. The SuperKamiokande Collaboration: "Search for Proton Decay Via $p \rightarrow e^+\pi^o$ in a large Water Cherenkov Detector" ICRR Report - 419-98-15, June 1998.
40. S. Weinberg, *Phys. Rev.* **D26**, 287 (1982); N. Sakai and T. Yanagida, *Nucl. Phys.* **B197**, 533 (1982).
41. See e.g. J. Hisano, H. Murayama and T. Yanagida, *Nucl. Phys.* **B402**, 46 (1993); P. Nath, R. Arnowitt and A.H. Chamseddine, *Phys. Rev.* **D32**, 2348 (1985); P. Nath and R. Arnowitt, hep-ph/9708469 and references therin, and K. Babu (private communications).
42. For recent limit on the $\bar{\nu}K^+$ modes, see M. Takita, (SuperK Collaboration), Talk presented in the 29th International Conference on HE physics, Vancouver, July 1998.
43. See also a recent discussion on the importance of the $\overline{UUDE}$ operator, due to Higgsino-dressing (for $\mu \geq 500$ GeV), in T. Goto and T. Nihei, KEK-TH-583 (Aug, 1998), and in V. Lucas and S. Raby, *Phys. Rev.* **D55**, 6986 (1997).
44. S. Dimopoulos and H. Georgi, *Nucl. Phys.* **B193**, 150 (1981); N. Sakai, *Z. Phys.* **C11**, 153 (1982).
45. A. Masiero, D.V. Nanopoulos, K. Tamvakis and T. Yanagida, *Phys. Lett.* **B115**, 380 (1982); B. Grinstein, *Nucl. Phys.* **B206**, 387 (1982).
46. S. Dimopoulos and F. Wilczek, in Proc. Erice Summer School (ed. by A. Zichichi), I.T.P. preprint NSF-ITP-82-07 (1981). For recent discussions see K. S. Babu and S. M. Barr, *Phys. Rev.* **D48**, 5354 (1993); ibid **D51**, 2463 (1995); S. M. Barr and S. Raby, *Phys. Rev. Lett.* **79**, 4748 (1997); C. Albright and s. Barr, hep-ph/9712488; Z. Chacko and R. N. Mohapatra, hep-ph/9810315.
47. A. Faraggi, *Phys. Lett.* **B278**, 131 (1992); *Phys. Lett.* **B274**, 47 (1992); *Nucl. Phys.* **B403**, 101 (1993); A. Faraggi and E. Halyo, *Nucl. Phys.* **B416**, 63 (1994).
48. I. Antoniadis, G. Leontaris and J. Rizos, *Phys. Lett.* **B245**, 161 (1990); G. Leontaris, *Phys. Lett.* **B372**, 212 (1996).
49. I. Antoniadis, J. Ellis, J. Hagelin and D.V. Nanopoulos, *Phys. Lett.* **B231**, 65 (1989). For a review, see J.L. Lopez and D.V. Nanopoulos, hep-ph/9511266)).
50. J.C. Pati, "*The Essential Role of String Derived Symmetries in Ensuring Proton Stability and Light Neutrino Masses*", hep-ph/9607446, *Phys.*

Lett. **B388**, 532 (1996).
51. H. Kawai, D. Lewellen and S.H. Tye, *Phys. Rev. Lett.* **57**, 1832 (1986) and *Nucl. Phys.* **B288**, 1 (1987); I. Antoniadis, C. Bachas and C. Kounnas, *Nucl. Phys.* **B289**, 87 (1987).
52. M. Dine, N. Seiberg and E. Witten, *Nucl. Phys.* **B289**, 585 (1987); J.J. Attick, L. Dixon and A. Sen, *Nucl. Phys.* **B 292**, 109 (1987).
53. See e.g. A. Faraggi, *Nucl. Phys.* **B428**, 111 (1994).
54. J. Ellis, A. Faraggi and D. Nanopoulos, hep-th/9709049.
55. A. Faraggi and J.C. Pati, "*A Family Universal Anomalous U(1) in String Models as the Origin of Supersymmetry Breaking and Squark-degeneracy*", hep-ph/9712516v3, December 97, *Nucl. Phys. B* (to appear).
56. G. Dvali and A. Pomarol, *Phys. Rev. Lett.* **77**, 3728 (1996); P. Binetruy and E. Dudas, *Phys. Lett.* **B389**, 503 (1996).
57. A. Faraggi and J.C. Pati, "*Meeting the Constraint of Neutrino-Higgsino Mixing in Gravity-Unified Theories*", *Phys. Lett.* **B400**, 314 (1997).
58. K.S. Babu and J.C. Pati, "*Towards A Resolution of the Supersymmetric CP Problem Through Flavor and Left-Right Symmetries*", (to appear).
59. K. S. Babu, J. C. Pati and F. Wilczek, *Phys. Lett.* **B423**, 337 (1998), hep-ph/9712307.
60. K.S. Babu, J.C. Pati and F. Wilczek, "*Linking Fermion Masses, Neutrino Oscillations and Proton Decay Via Supersymmetric Unification*" (to appear).
61. S. Mikheyev and A. Smirnov, *Nuovo Cim.* **9C**, 17 (1986), and L. Wolfenstein, *Phys. Rev.* **D17**, 2369 (1978).
62. For a recent comprehensive analysis of the solar neutrino data, see J.N. Bahcall, P.L. Krastev and A. Yu. Smirnov, "Where do we stand with solar neutrino oscillations?", IAS-preprint (July, 1998).

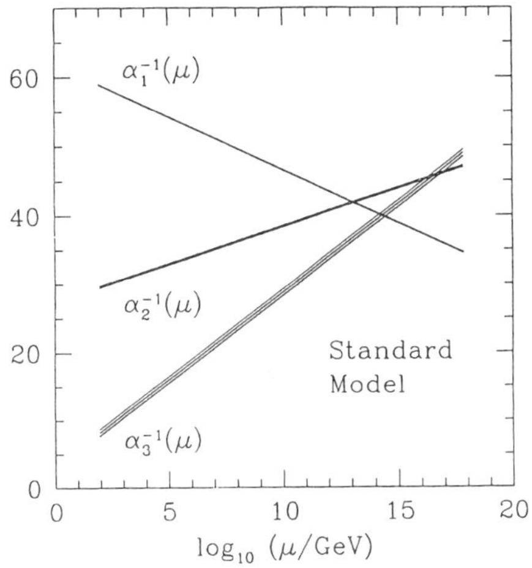

Fig. 1 (From Ref. 31)

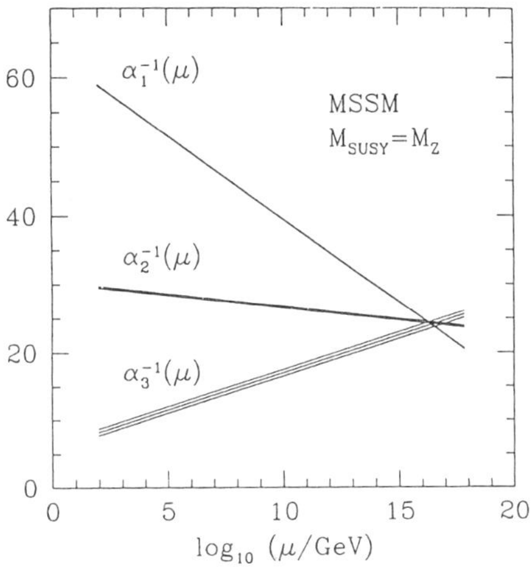

Fig. 2a (From Ref. 31)

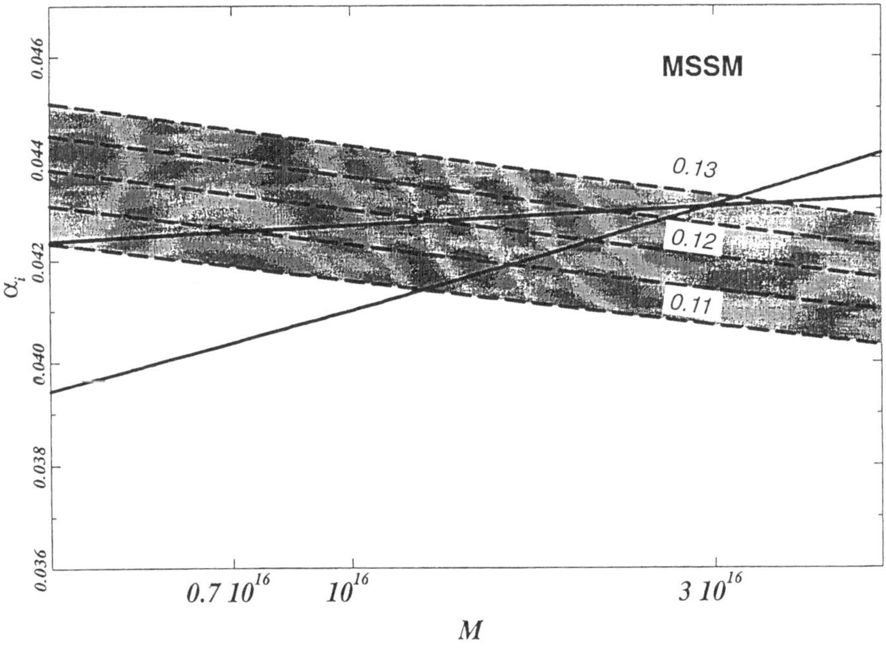

Fig. 2b Unification as a function of α_3 in MSSM (see Ref. 25)

STRING THEORY, BLACK HOLES AND QUANTUM COHERENCE

DANIELE AMATI
SISSA, Trieste, Italy
INFN, Sezione di Trieste
E-mail: amati@sissa.it

It is an honour for me to participate in this tribute to Abdus Salam. On many occasions I had the opportunity of discussions with him on physics, goals, people, ideas, politics and so on. The first time was in 1957 discussing dispersion relations for K-nucleon scattering, a subject on which he and Matthews and Vitale and myself wrote very similar papers. My goodness! How much water has gone under the bridge since then! Hard to imagine today that both of us were hot on the subject.

Abdus had a rich and complex personality, with numerous attributes that are not generally found in a single person. In physics he could alternate rigour with sheer intuition that would lead him to seminal visions or, if I may say so, to reckless models! Society could involve him deeply or flow around him without touching him. A complex personality, who has left us with everlasting legacies that carry his name: a new fundamental understanding of nature on the one side and, on the other, his Centre, as the proof of the role that culture and science play in the struggle for dignity.

Turning to physics, I will show how the information puzzle is solved in string theory based on recently discovered connections between D-branes and black holes. Well defined quantum states do not give rise to black holes. It is the classical limit that provides -on the same token- the loss of quantum coherece, the emergence of a space time description with causal properties and, thus, the formation of a black hole.

Many extraordinary coincidences between string theory and black hole physics have been recently uncovered and different opinions have been advanced on how these coincidences may "explain" or "solve" the well-known information loss puzzle. The idea [1] that very massive string excitations should represent black holes has been better substantiated by analysing massive BPS states that, as known, have properties that are not renormalized, i.e. do not depend on the coupling strength.

In the weak string coupling (g) regime, D-branes in four [2] and five [3] dimensions with a convenient number of charges have been studied. BPS states have been counted as well as nearly BPS states for certain regions of moduli space where perturbative computations are feasible [4]. Decay rates have also been computed [5] - by averaging over the many initial states - and shown to have, for small energy, a typical thermal distribution.

The moduli independence of these results allow the conjecture [6] of their validity beyond the moduli region where they were computed. And their g independence (also suggested by non-renormalization arguments [7]) may imply that they could be continued beyond the weak coupling regime.

An independent treatment - on totally different grounds - of the strong coupling regime substantiates that impression. The large coupling description of the 4 and 5 dimensional systems just discussed is found by solving the 10-d supergravity equations after reduction on the same compact manifold used for the D-brane description. The solution generates a metric [8] that depends on parameters that are related to the charges through the moduli of the compact manifold. The metric shows an event horizon even in the extreme limit; its area in this limit gives the Beckenstein-Hawking entropy of extremal b.h. This entropy and the ADM mass coincide exactly with the mass and entropy (given by the log of the state multiplicity) of the BPS state with the same charges as computed from D-branes in the small coupling regime.

For nearly extremal b.h. the entropy, the rate and the spectrum of evaporation [9] - obtained by solving wave equations in the corresponding metric background - coincide again [5] with those computed for small g/. And, even more remarkably, also deviations from black body spectrum agree [10]. These magic coincidences between such different calculations gave confidence to the g continuation between a unitary D-brane description and the conventional black hole with its information loss.

Different interpretations of this apparent contradiction have been advanced. Hawking questioned the D-brane formalism because the causal properties of space time are not properly taken into account. And even more so having shown [11] that the inconsistency cannot be assigned to the lack of identificability of D-brane states due to their prompt decay. String theorists [12] favour the attitude that black body radiation is only an approximation and that the connection with unitary D-brane formalism guarantees information retrieval in b.h. evaporation. This even more so since duality indicates situations that have flat unitary realizations for $g<<1$ as well as for $g>>1$ with a b.h. intermediate region with quite different space time geometry but with an alledgedly common spectrum.

Let me share the consensus of a smooth g behaviour by considering an S-matrix approach where a continuation has a clear meaning and let me start in the small coupling regime where a perturbative string approach is granted.

If over a well-defined BPS state impinges from far away - where for small g the space is flat - a well-defined quantum state (say a graviton) the S matrix, calculable by perturbative string theory, will be unitary. It describes the absorption of the graviton generating open strings on the D-brane then decaying, through an arbitrary number of steps, to some BPS state plus outgoing closed string excitations.

The final state will be a coherent superposition of the many BPS states (call them different polarizations) and outgoing closed string excitations. S matrix elements on any final state are given by the projection on it of the coherent state produced in the collision. Amplitudes and phases will of course depend on g and will change if one modifies the initial BPS state (initial polarization). These phases, crucial for unitarity, are washed out if one averages over initial polarization as done in order to obtain the (approximate) black body rate as mentioned before.

This averaging, on the other hand, is precisely what is implied by the classical limit. We thus see, even remaining in the small g D-brane regime, that it is the classical treatment that washes away - as expected - the quantum coherence and provides a thermal-like radiation.

To understand better this averaging procedure - always in the perturbative regime - let me return to the well defined initial state whose identity may be recognized from the analysis of the final state it gives rise. Old string theorists will recall however the need to recur to the full set of final state correlations in order to disentangle the high degeneracy of initial states and will also remember the techniques to obtain correlations from multiparticle amplitudes.

Indeed for any initial microstate, each of the many correlation functions may be computed with string perturbation techniques. Results are generally far from trivial and will of course depend on g and on the initial state. The thermalization result discussed before implies the perhaps non surprising fact that, irrespectively of g, each correlation function averages to zero when mediated over the very many initial states, while the spectrum averages to the thermal one. With, eventually, some corrections to black body radiation that may have a classical interpretation.

In going to large g, towards a black hole regime, it is the quantum S matrix that should be analytically continued. The analytic structure may be complex, with new singularities being formed, but unitarity preserved with the concurrence, as before, of the very many phases depending on the initial (black-holish) microstate. And, again, these phases will be washed out in the average implied by the classical limit. The many non trivial correlation functions that averaged to zero in the classical limit for any small g, are expected to continue to do so for large g. In a language more common to this black hole regime, one would say that the very many quantum hairs are washed out by the classical shampoo!

In order to avoid the artificial preparation of a "polarized" initial state (i.e. a microstate) it is convenient to treat the D-brane (or b.h.) formation and evaporation in a consistent quantum S matrix treatment. For small g this is the conventional approach to perturbative string amplitudes. Physically, anyhow, a black hole could be created by imploding "normal" states which are well prepared at infinity in a flat metric: energetic low mass particle collision at zero impact parameter is even sufficient to create a b.h. at a classical or even semiclassical (i.e. with radiation) level [13].

To make a connection with the preceding D-brane discussion, one should prepare different single D-brane initial states so as to provide the charges needed for the expected BPS or near BPS states discussed before. For low coupling, string theory provides a calculable perturbative unitary S matrix in which intermediate bound states appear as poles. Every intermediate state will be a coherent superposition of the different degenerate states with same masses and charges whose (large) multiplicity is the one discussed before. The many S matrix elements will be characterized by phases which are crucial for the unitarity of the approach.

Changing g or modifying the initial colliding states will imply a different intermediate coherent superposition and thus a change of all the many phases in a way which is consistent with unitarity.

These many S-matrix elements are supposed to be analytic in g and thus allow a continuation (allegedly smooth) to a strong coupling regime.

And, as before, it is by averaging incoherently over equivalent initial states that one obtains the thermal spectrum for small or large g.

This means that well defined initial quantum states do not create a black hole even if classically they would be expected to do so. Indeed outgoing spectra will not be thermal and correlation functions far from zero: it is the decoherence procedure that generates the black hole picture. And, at the same time, generates from the string formalism - in which $X\mu$ are operators - a space time description with causal properties and thus event horizons and, consistently, black holes.

Let me recognize that the very large number of degenerate intermediate states, implies that even a very limited average over initial sttes may cause the decoherence that leads to the general relativistic picture. Even hardly unavoidable incoherent soft gravitons around imploding particles (quantum objects) would wash out quantum effects thus reinstalling the conventional black hole picture.

Let me remark that the discussion concerning charged b.h., where extremal configurations are stable, is relevant to the analogous BPS structure of D-branes where it is possible to identify states, count them in small coupling regime and rely on supersymmetric non renormalization theorems to continue some results to large g. In the string S matrix approach then discussed, conditions for black hole formation are met even for small g for sufficient high energies irrispectively of charges. Thus the interplay outlined here between the generation of classical concepts (as space time and causal structure) and the loss of quantum coherence, is expected to hold even for the Schwarzild case where the whole black hole is totally evaporated away.

References

1. M.J. Bowick, L. Smolin and L.C.R. Wijewardhana, *J. of Gen. Rel. and Grav.* (1986); L. Susskind, *Phys. Rev.* **D52**, 6997 (1995), hep-th/9309145.
2. J. Maldacena and A. Strominger, *Phys. Rev. Lett.* **77**, 428 (1996), hep-th9603060; C. Johnson, R. Khuri and R. Myers, *Phys. Lett.* **B378**, 78 (1996), hep-th9603061.
3. A. Strominger and C. Vafa, *Ph. Lett.* **B379**, 99 (1996), hep-th9601029.
4. C. Callan and J. Maldacena, *Nucl. Phys.* **B472**, 591 (1996), hep-th/9602043; G. Horowitz and A. Strominger, *Phys. Rev. Lett.* **77**, 2368 (1996), hep-th9602051.
5. S. Das and S. Mathur, *Nucl. Phys.* **B478**, 561 (1996), hep-th9606185; S. Gubser and I. Klebanov, *Nucl. Phys.* **B482**, 173 (1996), hep-th 9608108.
6. G. Horowitz, J. Maldacena and A. Strominger, *Phys. Lett.* **B383**, 151 (1996), hep-th9603109.
7. J. Maldacena, hep-th961125.
8. M. Cvetic and D. Youm, hep-th 9508058; hep-th 9512127; G. Horowitz, D. Lowe and J. Maldacena, *Phys. Rev. Lett.* **77**, 430 (1996), hep-th9603195.
9. S.W. Hawking, *Nature* **248**, 30 (1974), *Com. Math. Phys.* **43**, 199 (1975).
10. S. Gubser and I. Klebanov, *Phys. Rev. Lett.* **77**, 4491 (1996), hep-th9609076; J. Maldacena and A. Strominger, *Phys. Rev.* **D55**, 861 (1997), hep-th9609026.
11. S.W. Hawking and M.M. Taylor-Robinson, hep-th9702045.
12. G. 't.Hooft, *Int. J. Mod. Phys.* **A11**, 4623 (1996), gr-qc 9607022; L. Susskind, Scientific American, April 1997; G. Horowitz gr-qc 9704072; J.R. David, A. Dar, G. Mandal and S.R. Wadia, hep-th 9610120.
13. P.D. D'Eath and P.N. Payne, *Phys. Rev.* **D46**, 658, 675, 694 (1992).

TOPICS IN M-THEORY

E. SEZGIN

Center for Theoretical Physics, Texas A&M University, College Station, TX 77843, USA

We give a brief history of the passage from strings to branes and we review some aspects of the following topics in M-theory: (a) an extended brane scan, (b) superembedding approach to the dynamics of superbranes and (c) supermembranes in anti de Sitter space, singletons and massless higher spin field theories.

1 Tribute to Abdus Salam

Abdus Salam was a truly unique man with great achievements not only in physics but also in promoting science in developing countries. His place in the annals of science as one of the finest physicists in this century is assured. He is certainly immortalized with his work on the unification of electromagnetic and weak interactions. His achievements in physics extend beyond this monumental work. He contributed many important ideas in particle physics, covering important aspects of renormalization theory, spontaneous symmetry breaking, grand unified theories, superspace, string theory and supermembrane theory.

Abdus Salam's odyssey in physics began in earnest in 1950 when, after having realized that he "sadly lacked the sublime quality of patience" needed for conducting research in experimental particle physics, he started to work under the guidance of Nicholas Kemmer, who advised him to collaborate with Paul Matthews (who was completing his PhD work at Cambridge University at the time) on renormalization of meson theories. This marked the beginning of an amazing journey from the pion-nucleon theory in 1950 to the marvelous discovery of the Standard Model 17 years later. Abdus Salam has given a wonderful account of "the story of the short-lived rise of the pion-nucleon theory as the standard model of 1950-51" [2], and the "story of the rise of chiral symmetry, of spontaneous symmetry breaking and of electroweak unification", including the story of his interactions with Pauli, Peierls, Ward, Weinberg, Glashow and others, in his Nobel Lecture of 1979 [3]. Reading the account of the twists and turns encountered in the remarkable odyssey which lead to the unification of electroweak interactions, one feels the excitement of it and appreciates all the more what the research in our discipline is really all about.

I first met Abdus Salam in 1981 in Trieste, when I joined the Abdus Salam International Centre for Theoretical Physics as a postdoctoral fellow. This marked the beginning of a very enjoyable and fruitful collaboration. I was privileged to have interacted with Abdus Salam for more than a decade. I will always cherish this experience. It was amazing how Abdus Salam treated a young post-doc that I was with so much humility. When we completed our first paper [6] he insisted that I would put my name first. I had to argue vigorously to convince him to put our names in alphabetical order.

During the 80's, we wrote a series of papers [6–19] and we edited a reprint collection with commentaries on supergravities in diverse dimensions [19]. Our work spans topics in higher dimensional Poincaré, anti de Sitter and conformal supergravities, their anomalies and compactifications, string theory and supermembrane theory. Abdus Salam was legendary in being open minded to new ideas. He embraced the developments in the subject of supermembranes when not many others did. He gave his full support for the supermembrane conference which was held in Trieste in 1989. As far as I know, this was the first conference ever to be held on membranes. The last papers we collaborated on [17,18] dealt with connections between membranes, singletons and massless higher spin fields, which are among my favorites. I believe that the full significance of the ideas put forward in those papers will someday be better appreciated, in the process of discovering what M-theory is. It was a joy to speculate about the tantalizing brane-singleton-higher spin gauge theory connections in collaboration with him.

Abdus Salam expressed his motivation in his research very humbly when he said: "I have spent my life working on two problems: first, to discover the basic building blocks of matter; and secondly to discover the basic forces among them" [4]. He was a deeply religous man who realized the limitations of science. He wrote: "my own faith was predicated by the timeless spiritual message of Islam, on matters on which physics is silent, and will remain so" [4]. In the same article, he wrote: " the scientist of today knows when and where he is speculating; he would claim no finality for the associated modes of thought". Such was the humility and wisdom of the man.

Abdus Salam was a man with boundless energy and many creative ideas, not only in physics but also in the process of promoting science in the political domain. He travelled frequently all over the world and in addition to his extremely productive research activities that resulted with over 250 publications, he gave many speeches and he wrote several articles on subjects other than physics. He had an amazing ability to focus on the heart of matters at hand. He was always able to bear in mind the big picture. He was very eloquent in his speeches and his writing. Among many topics, ranging from the importance of transfering science to developing nations to the interaction of science with religion and society, he wrote with passion about the glorious era which some of today's developing countries once had in science. He lamented the decline of science in those countries and he was deeply disappointed with the existing and ever widening gap between the developed and developing countries. He wrote passionately about why most of the developing countries need help in building up scientific infrastructure at all levels and why science transfer must accompany technology transfer if the latter is to take root in those countries. He wrote: "of the two passions of my life, the second has been to stress the importance of "science transfer" for developing countries. After building up the Theoretical Physics Department at Imperial College, London, I have spent 20 years fighting the battle of stressing the necessity of science transfer for developing countries" [5]. He highlighted many important aspects of this and many other related problems in brilliant speeches and essays which have been collected in a book entitled "Ideals and Realities" [1].

Abdus Salam made it a mission to himself to work towards the advancement of science in developing countries. The immensely successful Abdus Salam International Centre for Theoretical Physics in Trieste (AS-ICTP) which he founded in 1964 is a monument to his extraordinary achievement in this sphere. The story of how AS-ICTP came to existence and how it transformed to what it is today is an amazing one which can be glimpsed from various essays that appeared in Ref.1. Abdus Salam was the primary driving force in this process from the very beginning. He always had brilliant ideas for how to expand the functions of the Centre and he saw to it that those ideas were actually put into action [1]. Among many functions of the Centre aimed at promoting scientific activities in the developing countries perhaps the most important one was to make it possible for the physicists from those countries to visit the Centre, and as he put it, to "recharge their intellectual batteries, work on research problems while at the Centre, and then return to their countries carrying a new line of work, refreshed with new ideas and new interactions ".

It should be pointed out that this success did not come easy. Abdus Salam had to fight critical battles at times to ensure the continuation and expansion of the Centre. He worked incredibly hard to this end. He had to make sacrifices, among which was the amount of time he could devote to his beloved research activities. I remember once hopping into a car which was taking him to the Venice airport for one of his frequent trips, so that we could continue our physics discussion en route to the airport. He was constantly at work in trying to ensure the success of the Centre and at the same time trying to carry out his research in physics which he loved so much. With his relentless efforts, he contributed to the advancement of science in the Third World in many ways. Indeed, the legacy of Abdus Salam is not only his epoch-making contribution to physics but also his brilliant contributions in building up the scientific manpower in the developing nations.

Abdus Salam, a great man, brilliant, humanitarian, idealist, visionary, articulate, eloquent and passionate, the man never at rest, is no longer with us but he will always be remembered.

2 From Strings to Branes

A Brief History

The notion of extended objects in the context of elementary particle physics arose several years before the discovery of supersymmetry. The most notable introduction of the idea is due to Dirac[20] who, in 1962, envisaged the possibility of the muon being an excited state of a membrane in four dimension whose ground state corresponds to an electron. With the discovery of supersymmetry in early 70's, the physics of the extended objects took a remarkable turn, though for nearly twenty years the focus was on the simplest extended object, namely the string[21].

It should be pointed out, however, that prior to the proposal of Yoneya, Scherk and Schwarz in 1974 to interpret the dual models as theories of elementary particles rather than hadrons[21], there were attempts[22,23,24,25,26] to generalize the dual models to exhibit the four dimensional conformal $SO(4,2)$ symmetry. Inspired by the connection between the dual resonance models and strings, a further step was taken in Ref.25 (see also Ref.26) to associate the dual models possessing extended conformal symmetry with extended objects (see Ref.27 for a review). In Ref.25, the $(N-1)$ extra spatial dimensions were associated with a globally $SO(N,2)$ invariant theory, and these dimensions were interpreted as the orbital degrees of freedom of $(N-1)$-dimensionally extended objects, related to the "dimension" of the hadronic matter. In Ref.26, the study of the asymptotic behaviour of generalized dual model amplitudes led to the consideration of $2k$ dimensional extended objects. In Ref.27, the intrinsic nonlinearity of Nambu-Goto type actions for 3-dimensional extended objects was recognized and the compactification of a four dimensional worldvolume to a two dimensional worldsheet with continuous internal symmetry was considered.

Turning to the story of super extended objects, to begin with, a manifestly worldsheet supersymmetric Neveu-Schwarz-Ramond formulation of string theory was discovered in 1971 (see Ref.21 for a review). The considerations of anomalies led to the critical target space dimension of $D = 10$. This result, which can be derived in many different ways, is one of the most amazing discoveries in physics. The target space supersymmetry was not manifest in this formulation, but this was remedied by Green and Schwarz [29] who discovered just such a formulation, though they had to sacrifice manifest worldsheet supersymmetry. The Green-Schwarz superstring has a local fermionic symmetry on the worldsheet known as κ-symmetry, which is necessary for theory to make sense for several reasons. The Lorentz covariant quantization of the Green-Schwarz string proved to be a difficult problem, however, and consequently most of the work done in string theory has been based on the Neveu-Schwarz-Ramond formulation.

It turned out that the extension of the superstring construction to higher extended objects heavily favors the Green-Schwarz formalism, where one constructs an action for the map from a bosonic $(p+1)$ dimensional worldvolume to a target superspace. Significant progress towards the generalization of the Green-Schwarz action to higher branes came after a better understanding of its interpretation. A particularly useful such understanding was achieved in a paper by Hughes and Polchinski [31] where the classical Green-Schwarz superstring action in $D = 4$ was understood as the effective low energy action for a Nielsen-Olesen vortex solution of an $N = 1, D = 4$ supersymmetric Abelian Higgs model such that the $N = 1$ supersymmetry is broken down to $(2,0)$ supersymmetry in the $(1+1)$ dimensional worldsheet. Soon after [32], the analog of this phenomenon was shown to arise in the context of a three-brane solution of $(1,0)$ supersymmetric Yang-Mill theory in $D = 6$, such that, this time the $(1,0)$ supersymmetry is broken down to $N = 1$ supersymmetry on the worldvolume of the three-brane. The action was constructed for the collective coordinates which form an $N = 1$ scalar supermultiplet on the three-brane worldvolume.

In 1987, inspired by these results, Bergshoeff, Townsend and the author [33] constructed an eleven dimensional supermembrane action. The target space was taken to be a curved superspace, and the requirement of κ–symmetry was shown to require the equations of motion of the eleven dimensional supergravity! Thus, connection was made between the eleven dimensional supergravity which was invented in its own right nearly a decade before [28] and supersymmetric extended objects. The action was constructed directly without the knowledge of any membrane solution, which was to be found years later [51]. It seemed to be a very natural extension of the Green-Schwarz superstring action in $D = 10$ to a supermembrane action in $D = 11$. Thus, it was very tempting to consider it as a candidate for the description of a fundamental supermembrane theory that went beyond string theory in a natural way. While it was hoped that this passage from string to membrane theory might have welcome consequences in solving some of the outstanding unsolved problems of string theory, the theory was put forward essentially because "it was there". In other words, it was considered as a logical possibility, primarily on the basis of symmetry considerations. It was not invented out of pressing needs in physics based on paradoxes or anomalies, with the possible exception of the desire to "explain" the existence of $D = 11$ supergravity. In fact, once one comes to term with the basic idea of transition from elementary particles to strings, then the passage from strings to membranes is very natural. Indeed, the higher than two dimensional extended objects were also considered and their actions were constructed in Ref.33. A proper classification (with certain assumptions) was made soon afterwards [35] and it was found that the maximum target space dimension allowed was $D = 11$ and the maximum possible extension, p, of the object was $p = 5$. One of the important assumptions made was that the worldvolume fields always form scalar supermultiplets. It was a number of years later that branes which support other supermultiplets, most notably the Maxwell supermultiplets in $(p+1)$ dimensions with $p = 0, 1, .., 9$ and the 2-form supermultiplets in $(5+1)$ dimensions were discovered.

The idea of a fundamental supermembrane in $D = 11$ was pursued for a couple of years after 1987 intensely by a number of authors. Primarily the following issued were addressed: (a) spectrum and stability, (b) anomalies, (c) perturbation theory, (d) covariant quantization, (e) chirality/non-Abelian internal symmetries (f) renormalizability (g) supermembrane in anti de Sitter space and its relation to singleton field theory on the boundary of AdS. We shall come back to these points later.

Considerably intense activities on supermembrane culminated in a Trieste conference in 1989, devoted to the subject [137]. I believe this was the first conference ever on supermembranes. The spectrum problem was emphasized considerably, though other aspects of the supermembranes were also covered. The spectrum issue appeared to be problematic due to the indications that "the supermembrane can grow hair" without cost in energy, which seemed to imply a continuum in the spectrum. No dent could be made in the quantization problem. Despite the lack of covariant quantization scheme, and the lack of any information about the consequences of the κ-symmetry at the quantum level, the theory was widely considered to be nonrenormalizable. Moreover, it appeared to be hopeless that the theory could ever produce a chiral spectrum by any compactification scheme, for it seemed to be intrinsically nonchiral. It also appeared that a realistic internal symmetry gauge groups could not be obtained.

There were, however, some results of promising nature by 1989 (see Ref.138 for an extensive list of references on (super)membranes covering the period by mid 1990). To begin with, the particle[37] and string limits[36] of the supermembrane were obtained. These limits were sensible and they were suggestive of an important role for the supermembrane to play in the larger scheme of things. The study of the zero modes of the supermembrane in the superparticle limit showed that the zero-mode oscillators gave exactly the spectrum of massless states in $D = 11$ supergravity[37]. Even more interesting was the result that a double dimensional reduction of the $D = 11$ supermembrane action yielded the type IIA superstring action[36 a]. The significance of this result was not appreciated at the time, partially due to the fact that the type II strings were considered as academic cases since they could not give rise to any promising physical picture in $D = 4$ in contrast with the heterotic string. It would have been nearly impossible at the time to imagine that one day (in less than a decade, to be more specific) the heterotic string would be related to eleven dimensional supergravity and that all strings would be unified in an eleven dimensional theory! See below.

In a related development, the $D = 11$ supermembrane was quantized in the limit in which the membrane was wrapped around a circle or torus[38]. This procedure brought into focus the study of the Kaluza-Klein modes of the type IIA string as well. Again, the utility of the procedure of wrapping membranes around compact spaces so as to examine the physics of the branes in regions, where they look like particles or strings amenable to perturbation theory, was not fully appreciated until much later.

Another interesting development was the emergence of the area-preserving diffeomorphisms, SDiff, of the supermembrane as a useful tool in the study of the quantum theory[39] (for a generalization to volume preserving diffeomorphisms of super p-branes, see Ref.40). The interesting fact that the supermembrane Hamiltonian in a light-cone gauge turned into a gauge Yang-Mills theory in $(0+1)$ dimensions (i.e. supersymmetric quantum mechanics) with SDiff $\sim SU(\infty)$ as a gauge group was discovered. This story too was to be appreciated later more fully, in the context of the matrix model approach to M-theory[66].

[a]Interestingly, there exists a generalized Virasoro algebra for the type IIA string which can be deduced from the algebra of worldvolume diffeomorphisms and κ-symmetry of the $D = 11$ supermembrane[44].

Last but not least, soon after the eleven dimensional supermembrane was discovered, it was speculated [106] that singletons, which are special representations of the anti de Sitter group [98,100], may play a role in its description. Soon after, it was conjectured [107,108] that a whole class of AdS compactifications of supergravity theories may be closely related to various singleton field theories. The singleton representations of the AdS group are special in that they are ultra short and, strikingly, they admit no Poincaré limit. Indeed, they can be realized in terms of fields that propagate on the boundary of AdS [100]. The kinds of singleton field theories studied back then were free field theories, and that raised the hope that while supermembrane theory may seem to be nonrenormalizable in general, it might miraculously be renormalizable in special backgrounds involving AdS space, where it may be treated as a free singleton field theory. The recent exciting developments in AdS/CFT correspondence [133,134,135] is reminiscent of these hopes, though the exact fashion in which this connection has emerged certainly goes beyond what was known and imagined previously. One thing that was imagined before, and has not materialized yet, is a byproduct of the conjectured brane-singleton connection, namely the possible occurrence of infinitely many massless higher spin fields in AdS_4 as two supersingleton bound states in supermembrane theory! This conjecture was put forward in Refs.17, 18 and 109. We will discuss this topic further in section 3.

Despite these interesting developments, the quantization of the supermembrane and higher superbranes remained an unsolved problem. As early as 1988, however, aspects of branes as solitons or topological defects which break the target space symmetries were revisited [48] and this proved to be a very fruitful move. It was suggested [48] that all p–branes known at the time should correspond to solitonic solutions of certain supersymmetric field theories (just as in the case of 3-brane of Ref.32) or in the case of $10D$ strings and $11D$ supermembrane, they should arise as solutions of appropriate *supergravities*. A few years later, interesting and significant results started to appear in this direction. In 1990, a string solution of $D = 10$ supergravity was found [49]. Soon after, a fivebrane solution of $D = 10$ supergravity coupled to Yang-Mills was discovered [50]. These results meant that string theory contained solitonic objects in its spectrum which were nonperturbative in nature. This implied a simple yet very important fact that supersymmetric extended objects could not possibly be ignored any longer; they were simply there and inevitable! The soliton fever carried over to eleven dimensions as well. In 1991, the supermembrane soliton [51] and in 1992 the superfivebrane soliton of $D = 11$ supergravity were discovered. The discovery of the superfivebrane was somewhat surprising for it was not in the original brane scan. Fascinating aspects of the superfivebranes have been discovered since then and they, of course, occupy an important place in the big scheme of things.

In the early 90's, a whole class of brane solutions in type II string theories were also found. The study of type II branes eventually led to a remarkable discovery by Polchinski in 1995 [61] that the type II p-branes carrying Ramond-Ramond charges can be understood as Dirichlet branes, or D-branes for short. These are p dimensional surfaces on which an open string can end. Thus, it became possible to study the dynamics of at least special kinds of p-branes (at weak coupling limit) by studying the (perturbative) dynamics of an open string! The "D-brane technology" developed rapidly (see Ref.142 for a review) and it provided an important tool for studying the dynamics of intricate brane configurations, leading to discoveries of novel physical phenomena and to breakthroughs in the study of long standing problems in black hole physics.

Concomitant to these developments, and closely related to them, were the important discoveries in the arena of duality symmetries of string theory. Remarkable results on T-duality symmetries relating strings in backgrounds involving a circle of radius R and $1/R$ and S-duality transformations interchanging strong and weak coupling limits started to accumulate with an increasing rate. This is a vast subject, even a brief review of which would take us beyond the scope of this brief introduction. We refer the reader to the excellent reviews of this subject existing in the literature; see for example, Refs.129-138.

Further studies in strong-weak coupling dualities led to major developments in late '94 and early '95 which finally put the string-membrane connection on a firm footing. Firstly, it was observed that the soliton spectrum of the compactified $D = 11$ supergravity agreed with that of compactified type IIA string by the inclusion of the wrapping modes of the supermembrane and superfivebrane and by taking into account the wrapping modes of the type IIA D-branes carrying Ramond-Ramond charges [57]. Next, it was argued that the $D0$–branes of type IIA string correspond to the Kaluza-Klein modes of $D = 11$ supergravity [58]. These were tantalizing new hints for a deep connection between type IIA string and the eleven dimensional supermembrane that went beyond the connection based on double dimensional reduction, because it was not necessary to consider only the zero modes. Finally, Witten in his celebrated paper [59] argued convincingly that the strong coupling limit of type IIA string theory *is* the $D = 11$ supergravity! Furthermore, the ensuing developments showed that all string theories in $D = 10$ were unified via duality relations involving an eleven dimensional origin in one way or another! The view started to develop that there exists an intrinsically nonperturbative and quantum consistent eleven dimensional theory with the properties that (a) it can be approximated by $D = 11$ supergravity at low energies, (b) it contains the supermembrane and superfivebrane in its spectrum and (c) when expanded perturbatively in different coupling constants, it gives different perturbative theories, which can be superstrings or superparticles. This mysterious theory was named (upon a suggestion by Witten) the M-theory [62]. Work on M-theory continues with rapid pace (see, for example, Refs.141, 144 and 145 for technically detailed reviews) and striking new results have emerged from the studies of M-theory in anti de Sitter background [133,134,135].

In the light of these developments, it is interesting now to revisit the questions that arose in the late eighties in the context of the eleven dimensional supermembrane, which were mentioned above. The problem of chirality and non-Abelian internal symmetries found a remarkable answer with the discovery made by Horava and Witten [63] that M-theory compactified on an interval leads to the $E_8 \times E_8$ heterotic string! This phenomenon provides a surprisingly simple and elegant answer to the question of how to obtain a chiral theory starting from eleven dimensions. As for the problems associated with the quantization of the supermembrane, there is no solution in sight (not yet!)

M Theory-Supermembrane Connection

Notwithstanding the presently unsolved problem of how to quantize the supermembrane (perturbatively or nonperturbatively) in eleven dimensions, it is tempting to pose the following question: Is it conceivable that M-theory is nothing but the eleven dimensional supermembrane theory? According to Schwarz [147], "most experts now believe that M-theory cannot be defined as a supermembrane theory". While we are not aware of all the arguments leading to this assertion, some of them go as follows: (a) The supermembrane action is only meant to describe a macroscopic object in M-theory and therefore one should not even attempt to quantize the supermembrane. (b) The fundamental and solitonic supermembranes should be identified [57]. The latter has a finite core due to its horizon [54]. Since the known supermembrane action does not take into account this classical structure of the membrane, it is not an appropriate starting point for quantization [58,60]. (c) As far as the perturbative formulation goes, there is no suitable perturbative expansion parameter (assuming that the theory is not compactified) to order the spacetime amplitudes and moreover the worldvolume perturbation theory is non-renormalizable.

Despite all these considerations, the eleven dimensional supermembrane theory seems to be the only theory that goes beyond $D = 11$ supergravity and which does incorporate supergravity equations of motion. Indeed, it does so already at the classical level, thanks to the power of κ–symmetry. The theory goes beyond $D = 11$ supergravity because we know that the quantization of the supermembrane collapsed to a point yields the $D = 11$ supergravity spectrum, and that the wrapping of the supermembrane around a circle gives type IIA string theory, a perturbative treatment of which yields an infinite set of higher derivative corrections to type IIA supergravity. It is natural to expect a supermembrane origin of these corrections.

In this context, let us recall that while the κ–symmetry of the supermembrane uniquely leads to the $D = 11$ supergravity equations of motion [33], there is one correction to these equations [68] that has been motivated by the considerations of sigma model anomalies in $M5$–brane and one-loop effects in type IIA string [67], and takes the form $C_3 \wedge X_8$, where C_3 is the 3-from potential in $D = 11$ and X_8 is an 8-form made out of the Riemann curvature. Supersymmetrization of this term implies an infinite number of terms in the action, just as in the type IIA theory in $D = 10$. Some aspects of these terms have been deduced from one-loop effects in $D = 11$ supergravity, but this cannot be the full story [b].

What then is the principle which governs the corrections to $D = 11$ supergravity? Since we no longer have the benefit of worldsheet superconformal invariance in curved background that helps in answering a similar question in $D = 10$, we have to find a new principle here. Local supersymmetry is very helpful, but we need more than that, based on the lessons learnt from string theory. We mentioned above that the κ–symmetry of the supermembrane was very restrictive by giving the usual $D = 11$ supergravity equations. Perhaps one should re-examine the issue of κ-symmetry by allowing more general superspace than the standard $D = 11$ Poincaré superspace.

To have control over the higher derivative corrections in $D = 11$, it is very useful to work in superspace. One possible approach is to modify the superspace Bianchi identity as follows

$$dH_7 = G_4 \wedge G_4 + X_8 \ , \tag{1}$$

where the $G_4 = dC_3$ and H_7 is a 7-form whose purely bosonic components are Hodge dual to those of G_4. The $\theta = 0$ component of this equation has been extensively discussed in Ref.68, but not much is known about the superspace consequences of the full equation, essentially due to its complexity. It is possi-

[b] It is nonetheless an amazing fact that these one-loop effects are capable of producing nonperturbative effects in type II string theory [69].

ble that the solution requires the modification of the standard supergeometry by introducing the 2nd and/or 5th rank Γ–matrix terms in the constraint for the dimension zero supertorsion components $T_{\alpha\beta}^{a}$ [70,71]. It would be very interesting to derive the corrections to the minimal superspace geometry from the κ–symmetry considerations, or better yet, from the formalism of superembeddings, which will be summarized in section 4. It is clear that any modification of the standard $D = 11$ supermembrane action, or equations of motion, would be very interesting and it would effect the discussion of just what is the role of the supermembrane in M-theory.

In the above discussion we omitted the superfivebrane. It is natural to expect a sort of duality relationship between supermembranes and superfivebranes (see Ref.143 for a discussion of the membrane/fivebrane duality in $D = 11$). Moreover, it is well known that an open supermembrane can end on a superfivebrane in $D = 11$. Thus, the connection between $M2$–branes and $M5$–branes in $D = 11$ is similar (in some respects) to the connection between open strings and Dp–branes in $D = 10$. In fact, it has been shown [91] that the superfivebrane equations of motion follow from the κ–symmetry of an open supermembrane ending on a superfivebrane!

We conclude this introduction by emphasizing the importance of exploring the consequences of the M-theory unification at the level of interactions. Much of the work done so far understandably has dealt with spectral issues and this has been very beneficial. However, at some point several problems about interactions need to be probed more deeply. Some encouraging results have already emerged. Interesting connections between the string interactions in $D = 10$ and membrane interactions in $D = 11$ have been noted [65]. We already mentioned the one-loop effects in $D = 11$ supergravity giving rise to nonperturbative information about type IIA string amplitudes [69]. Another example is the deduction of the quantum Seiberg-Witten effective action for $N = 2$ supersymmetric Yang-Mills theory [55] from the classical M-fivebrane equations of motion with N three branes moving in its worldvolume [56] [c].

In the next section we will discuss an extended brane scan covering various kinds of branes that have emerged until now. The emphasis will be on the amount of supersymmetry breaking by the branes. The purpose of this section is to give a feel for the panorama of branes in M-theory and their properties. The superembedding theory plays a significant role not only in their classifications but also in the description of their dynamics. With this motivation in

[c] It is rather amusing to see that the Planck constant emerges as a combination of certain integration constants arising in the course of solving the three-brane equations [56].

mind, section 4 is devoted to a summary of the basic ideas behind the superembedding theory. When the target space is taken to be a supercoset involving anti de Sitter space, remarkable things happen, as it has become abundantly clear with recent exciting developments. In section 5, we summarize some aspects of the connections between supermembranes, singletons and higher spin gauge theory.

3 Types of Branes

A Minimal Brane Scan (The Scalar Branes)

Until the discovery of the superfivebrane solution of the $D = 11$ supergravity in 1992 [52], the types of branes that were known were rather limited in number. Some of them were already anticipated in Ref.33 and classified properly in Ref.35. The result is reproduced in Table 1. For uniformity in the nature of the scan, we have left out the type IIB, type I and heterotic strings. The main characteristic of all the branes occurring in this scan is that they all support a scalar multiplet with $1, 2, 4, 8$ scalars and matching spin $1/2$ fermionic degrees of freedom. The branes in this scan fall into four series: The octonionic, quaternionic, complex and real series with co-dimensions 8,4,2,1 embeddings, respectively. All the members of a given series can be obtained from the one that occupies the highest dimension by the process of double dimensional reduction [36]. All branes in this scan for $p > 1$ have minimal possible target space supersymmetry.

The brane scan shown in Figure 1 was originally derived from the requirement of κ-symmetry of their actions [33]. This requirement amounted to the existence of suitable Wess-Zumino terms which was possible whenever a closed super $(p+2)$ in target superspace existed. This in turn meant finding the values of the pairs (p, D) for which the following Γ-matrix identity is satisfied

$$\Gamma_{\mu(\alpha\beta}\Gamma^{\mu\nu_1\cdots\nu_{p-1}}_{\gamma\delta)} = 0 \ , \tag{2}$$

where $\mu = 0, 1, ..., D-1$ and α labels a minimal dimensional spinor in D-dimensions. There is a simple alternative way to deduce the same brane scan. Indeed, using the rule

$$D = (p+1) + n_S \ , \tag{3}$$

where D is the bosonic dimension of the target space, p is the spatial brane dimension and n_S is the number of real scalars in the scalar multiplet, together with the knowledge of which supermultiplets exist in diverse dimensions, the brane scan shown in Table 1 can easily be reproduced.

Of the branes for $p > 1$, only the supermembrane in $D = 11$ attracted the most interest for some time. There was a modest attempt to try to rule out the "other branes" on various grounds [37,41,42,43] though, primarily on the basis of Lorentz anomaly considerations. At the time, intersecting branes [72] were not known and all the scalar branes were considered in their own right. The constraints implied by the κ–symmetry of their actions have been determined long ago [33]. The consequences of these constraints have not been studied so far except for the cases of strings in $D = 3, 4, 6, 10$ [30] and the supermembrane in $D = 11$ [33]. In the case of fivebrane in $D = 10$, one can check that the dual formulation of $(1, 0)$ supergravity is consistent with the κ-symmetry constraints, though it is considerably more difficult to show that it is uniquely implied by these constraints. As for the solution of the κ–symmetry constraints for the remaining branes that occur in the old brane scan (see Figure 1), we expect that the dimensional reduction of the dual $(1, 0)$ supergravity in $D = 10$, which contains a 6-form potential, followed by a truncation of the resulting vector multiplets, provides a solution.

While the old p–branes we have been discussing may emerge in the physics of intersecting branes, it is still interesting to determine if they can describe consistent brane theories formulated in subcritical $D < 10$ dimensional spacetimes. If so, these branes might correspond to an interesting class of M-theory limits in which the $(10 - D)$ or $(11 - D)$ extra dimensions decouple in a rather drastic way. In fact, this is somewhat reminiscent of the situation arising in the so called "little m" theories [81]. However, in the latter case the target space theory typically involves the Yang-Mills supermultiplet but not the supergravity fields. For example, the little m-theory in $D = 7$ involves only the super Yang-Mills theory. This is to be contrasted with the supermembrane in $D = 7$ which arises in the old brane scan (see Figure 1), where the target space theory naturally involves the $N = 1, D = 7$ supergravity but not super Yang-Mills. In fact, this raises the interesting question of whether one can couple supergravity plus Yang-Mills system to the supermembrane in $D = 7$ that arises in the old brane scan. If such a coupling exists, it would be reasonable to expect a limit in which supergravity is decoupled.

Another aspect of the branes listed in Figure 1 which might be worthwhile to study is their anomalies; namely the gravitational anomalies in the target field theories as well as the κ–symmetry and global anomalies on the worldvolume. In doing so, one immediately rules out the fivebrane in $D = 10$ on the basis of

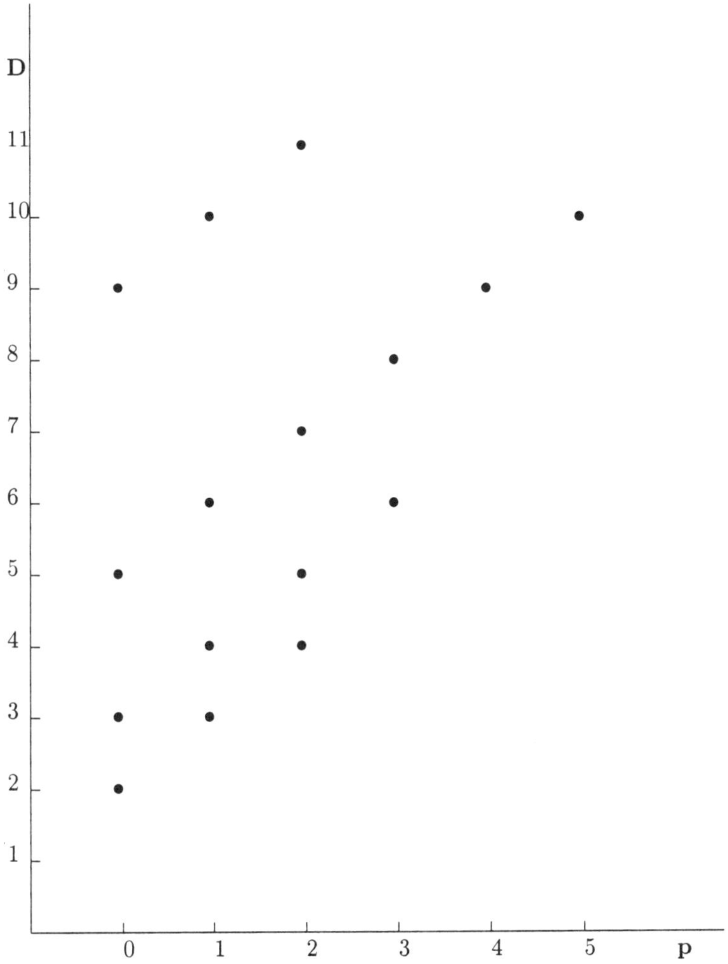

Figure 1: A minimal p-brane scan. These are p-branes in D dimensions for which the collective coordinates form worldvolume scalar supermultiplets.

its incurable gravitational anomaly. However, many of the lower dimensional branes, in particular those for which the target space is odd-dimensional such anomalies do not arise. For example, the supermembrane in $D = 7$ has odd dimensional target as well as odd dimensional worldvolume, and consequently, the pertinent anomalies are the global ones. Indeed, such anomalies have been studied by Witten [82] in the case of the $D = 11$ supermembrane, and it was found that a mild restriction arises on the topology of spacetime and the possible configuration which the membrane Kalb-Ramond field may assume. With similar conditions satisfied, we expect that the supermembrane in $D = 7$ is anomaly free, though we do not know at present how this brane might possibly arise in M theory.

M-Branes and D-Branes

With the discovery of the superfivebrane, the novel situation in which the worldvolume supported a multiplet other than scalar multiplet emerged. Indeed, it was found that the $5 + 1$ dimensional worldvolume theory was that of $(2,0)$ tensor multiplet, containing a two-form potential with self-dual field strength, giving rise to 3 on-shell degrees of freedom, five scalars and 8 on-shell fermi degrees of freedom. The rule (3) still holds. It was speculated later that there should be an analog of this brane with $(1,0)$ tensor multiplet that contains a single scalar, in addition to the a self-dual tensor and 4 on-shell fermions. The single scalar suggests a seven dimensional target space. This model has been analyzed in considerable detail in Ref.97.

In 1985, the D-branes were discovered [61]. These are branes on which open strings can end. The worldvolume dynamics of these branes is described by vector multiplets. Considering the maximally symmetric Maxwell multiplets in various dimensions, i.e. those with 8 bose and 8 fermi on-shell degrees of freedom and using the rule (3), one finds they all imply $D = 10$ target space with type II supersymmetry. Allowing nonmaximal vector supermultiplets on the worldvolume gives rise to a revised brane scan [53] and applying (3) one finds $D = 3, 4, 6$ dimensional target spaces.

At this point, it is tempting to contemplate a classification of all possible branes by classifying all possible globally supersymmetric multiplets in dimensions $p + 1 \leq 10$. There are some complicating factors, however.

Firstly, there is the possibility of dualizing a member of the worldvolume supermultiplet, say a q-form potential, to a $p - q - 1$ form potential. Indeed, one may consider the dualization of a number of forms at the same time. While this may be a simple matter at the free field theory level, it is considerably more difficult for branes where the worldvolume multiplets are self-interacting in a highly nonlinear fashion. In fact, the dualization of forms on the worldvolume is intimately connected with the S, T or more general duality transformations. Consequently, there is the additional question of which branes should be considered as the fundamental ones from which all others can be derived by one duality transformation or by Kaluza-Klein type reductions of the target space and/or the worldvolume.

Secondly, it is possible that the description of the worldvolume theory involves more than one supermultiplet. The simplest example of this is the heterotic string where the worldsheet contains scalar multiplets in the left-moving sector and heterotic fermions or bosons, which are supersymmetry singlets, in the right-moving sector. The fact that worldsheet supersymmetry allows supersymmetric singlets is special to 1 and 2 dimensions, and it cannot occur in higher than two dimensions. Nonetheless, focusing our attention on the fact that the heterotic string is described by two distinct multiplets on the worldsheet, we tentatively refer all branes which support more than one irreducible supermultiplet as heterotic branes.

Given the complications just described, the classification of branes becomes a rather nontrivial task. However, one may consider an alternative scheme in which one focuses on the amount of supersymmetry breaking due to the embedding of superbranes in a given target superspace rather than emphasizing the worldvolume supermultiplet. This approach especially becomes powerful if one considers both the target space as well as the worldvolume to be superspaces. Thus, one considers the embedding theory of supersurfaces, which turns out to be a geometrical and natural framework, not only for classifying the superbranes, but also for providing the manifestly worldvolume and target space supersymmetric dynamical equations. Indeed, the problem of describing the superfivebrane equations of motion was solved for the first time by using this formalism[87,88]. The main criteria in the superembedding theory from the physical point of view is that the basic equations which describe the superembedding lead to sensible equations of motion for the worldvolume fields. This can be typically checked in a reasonably straightforward way at the linearized level, at least for a large class of superbranes.

The classification of all possible branes is still a formidable task despite the universal nature of the superembedding approach. A further complicating factor is that the geometry and topology of the spaces involved can mathematically become rather complicated. For example, the full actions for intersecting branes is yet to be constructed in any approach. Nonetheless the existence of a large class of branes has been deduced from the study of brane solutions to supergravity theories, or from the study of the possible brane charges in supersymmetry algebras, or from the analysis of the linearized embedding equations. Putting together a variety of information available on the nature of existing and predicted types of branes, they can be classified according to the amount of supersymmetry they preserve. For the purposes of the proposed scan, we will assume that the maximum real dimension of supersymmetry is 32 and we shall furthermore consider flat target superspaces with Lorentzian signature.

$32 \to 16$ *Branes*

Assuming Lorentzian signature, the maximum dimension in which real 32 symmetries can occur is $D = 11$. In $D = 11$ the target superspace has (even—odd) dimensions $(11|32)$. Embedding a $(3|16)$ dimensional super worldsurface gives the supermembrane, and embedding a $(6|16,0)$ dimensional super worldsurface gives the superfivebrane. The former is a co-dimension 8 embedding, and the later is a co-dimension 4 embedding. In the latter case, the notation $(16,0)$ means that there are 16 real left-handed spinors and no right-handed spinors. This means $(2,0)$ supersymmetry, since the minimum real dimension of a spinor in $5+1$ dimensions is 8. Such hybrid superspaces can occur in $2 \, mod \, 4$ dimensions (assuming Lorentzian signature).

The supermembrane worldvolume multiplet is an $N = 8$ scalar multiplet, which has 8 real scalars and the superfivebrane worldvolume multiplet is the tensor multiplet of $(2,0)$ supersymmetry which has 5 real scalars and a 2-form potential with self-dual field strength. Thus, the relation (3) holds in both cases.

In $D = 11$ there are two other "superbranes" which preserve half supersymmetry, but for which the relation (3) does not hold. These are the pp-waves, which can be considered in some sense as 1-branes and the Kaluza-Klein monopole, which can be viewed in a certain sense as 6-branes. The existence of 9-branes has also been speculated briefly in Ref.87 and in some more detail in Ref.75. Sometimes the boundaries of the $D = 11$ spacetime arising in the Horava-Witten configuration [63] is also referred to as a $M9$-brane. See Ref.80 for a recent discussion of various aspects of $M9$ branes.

In all these cases, the formula (3) breaks down, and one finds 8 scalars for the pp-wave, 3 scalars for the KK monopole and no scalars for the 9-brane, as opposed to $9, 4, 1$ scalars, respectively. This is essentially due to the fact that the transverse space lacks the isometries of R^9, R^4, R^1, respectively; for example, in the case of KK monopole, the transverse space is a Taub-Nut space. A detailed discussion of these branes can be found, for example, in Ref.75 where they have been called the G-branes (G standing for gravitational).

In classifying the superbranes in $D = 10$, one should take into account the ordinary (vertical) dimensional reduction, or (diagonal) double dimensional reduction of the branes in $D = 11$. Moreover, one may take the eleventh dimension to be S^1 or S^1/Z^2. This would generate a set of branes known to exist in $D = 10$ (See Figure 2).

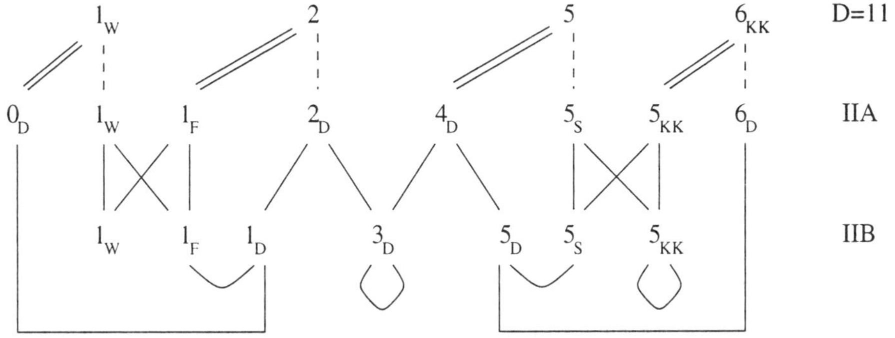

Figure 2: Dimensional reduction and S/T duality maps. The straight arrows between type IIA/B branes denote T dualities, curved lines between the type IIB branes denote S dualities, the dashed denote vertical dimensional reduction and double lines denote double dimensional reduction. The Dp branes for $p = -1, 7, 8, 9$ are special cases which are not shown.

In $D = 10$ we can embed the $(p+1|16)$ dimensional Dp-brane worldsurfaces for $p = 1, 3, 5, 7, 9$ in $(10|16, 16)$ superspaces, where $(16, 16)$ denotes the 16 left-handed and 16 right handed Majorana-Weyl spinor coordinates, associated with type IIA $(1, 1)$ supersymmetry. Similarly, the $(p + 1|16)$ dimensional Dp-brane worldsurfaces for $p = 0, 2, 4, 6, 8$ can be embedded in the $(10|32, 0)$ superspace associated with type IIB $(1, 0)$ supersymmetry. The worldvolume supermultiplets for the Dp-branes are the maximally supersymmetric vector multiplets in $p + 1$ dimensions, which have a single vector and $(9 - p)$ scalars in the bosonic sector. Thus, the relation (3) holds for all these branes.

In $D = 10$, there also exist the fundamental strings, NS5-branes, pp-waves and Kaluza-Klein monopoles (which may be viewed as 5 branes), both in type IIA and type IIB superspaces. These are denoted by 1_F, 5_S, 1_W and 5_{KK}, respectively, in Figure 2. The Dp branes for $p = -1, 7, 8, 9$ are somewhat special cases which are not shown in this Figure. The type IIA/B branes are related to each other by T- and S- dualities as shown in Figure 2. The type IIA theory compactified on a circle of radius R is T-dual to the type IIB theory compactified on a circle of radius $1/R$. The $SL(2, Z)$ S-duality transformation, on the other hand, is a strong-weak coupling symmetry operative in type IIB theory: the $(1_F, 1_D)$ and $(5_S, 5_D)$ form a doublet and 3_D is singlet under this symmetry. The action of the S-duality on the type IIB $D7-$ and $D9$–branes is more involved[76,75,77]. For a more detailed version of Figure 2 which summarizes almost all the known dualities between the type II branes, including the $D7-$ and $D9$–branes, see Ref.77. For the S-duality involving type IIB KK-monopoles, see Ref.78.

The worldvolume field content of all these branes have been obtained and in some cases speculated on in a number of references. We refer to Ref.75 where an extensive discussion and references to earlier work can be found. For reader's convenience and following Refs.75 and 79, the $D = 11$ and type II brane worldvolume contents are listed in Table 1.

Branes	Worldvolume Fields
	D=11 M-Branes
pp-wave	A_1, $8 \times \phi$
M2-brane	$8 \times \phi$
M5-brane	A_2^+, $5 \times \phi$
KK monopole	A_1, $3 \times \phi$
	D=10 Type IIA Branes
Dp-branes	A_1, $(9-p) \times \phi$ $(p = 0, 2, 4, 6, 8)$
Fundamental string	$8 \times \phi$
pp-wave	A_1, $8 \times \phi$
KK monopole	A_1, $3 \times \phi$, $S \sim A_4$
NS 5-brane	A_2^+, $4 \times \phi$, $S \sim A_4$
	D=10 Type IIB Branes
Dp-branes	A_1, $(9-p) \times \phi$ $(p = 1, 3, 5, 7, 9)$
Fundamental string	$8 \times \phi$
pp-wave	$8 \times \phi$
NS 5-brane	A_3, $4 \times \phi$
KK monopole	A_2^+, $3 \times \phi$, $S \sim A_4$, $S' \sim A_4'$

Table 1: Worldvolume bosonic fields. A_q denotes a q-form potential, A_2^+ is 2-form potential with self-dual field strength, S, S' are scalars which can be dualized to appropriate q forms.

$16 \to 8$ *Branes/ Doubly Intersecting Branes*

The maximum dimensional target space admitting 16 real component spinor is $D = 10$. There is only one case to consider here, which is the $(1,0)$ supersymmetric $(10|16,0)$ superspace (or any of its dimensional reductions). We can embed a super 5-brane with $(6|8,0)$ world supersurface. This gives a hypermultiplet of $(1,0)$ supersymmetry on the worldvolume and this is the old super 5-brane which was considered long ago in its own right, prior to the discovery of intersecting branes. The target supergravity theory is the dual formulation of $(1,0)$ supergravity, which has fatal gravitational anomalies (see Ref.19 for a review). There is an alternative way to interpret this embedding, however. It can be viewed as a $D5$-brane within a $D9$-brane of type IIB theory. This is shown as the point $5 \cap 9 = 5$ in Figure 3. It is important to note that both branes are embedded in type IIB superspace which has 32 real supersymmetry, though the 5-brane residing inside the 9-brane possesses only 8 real supersymmetries. This is known as 1/4 supersymmetry preserving brane.

We also emphasize that the supergravity theory in the target of this 5-brane is induced supergravity where all the members of the supergravity multiplet are composites of the type IIB supergravity theory. Since type IIB theory is free from anomalies, the $D5$-brane residing inside the $D9$-brane is presumably anomaly free as well. Starting from the fivebrane inside the ninebrane, one can perform two kinds of T-duality transformation which generates all the intersecting branes shown in Fig. 3. Starting from an intersecting pair $p \cap q$, a vertical move corresponds to a T duality transformation over one of the transverse directions of the q brane. An oblique move corresponds to T-duality over one of the q-brane worldvolume directions. The details of such transformations to obtain one brane solution from the other are treated in a number of papers; see for example, Refs.73 and 74.

Intersecting branes involving NS branes can be obtained by utilizing combined S/T-duality transformations. Not including the cases involving the pp-waves and KK monopoles one thus finds

$$\begin{aligned}
&Dp \cap 1_F = 0\ , && p = 1,2,...,9\ ,\\
&Dp \cap 5_S = (p-1)\ , && p = 1,2,...,6\ ,\\
&1_F \cap 5_S = 1\ , &&\\
&5_S \cap 5_S = 3\ . && \qquad (4)
\end{aligned}$$

For completeness, we also list the by now well known double intersections of

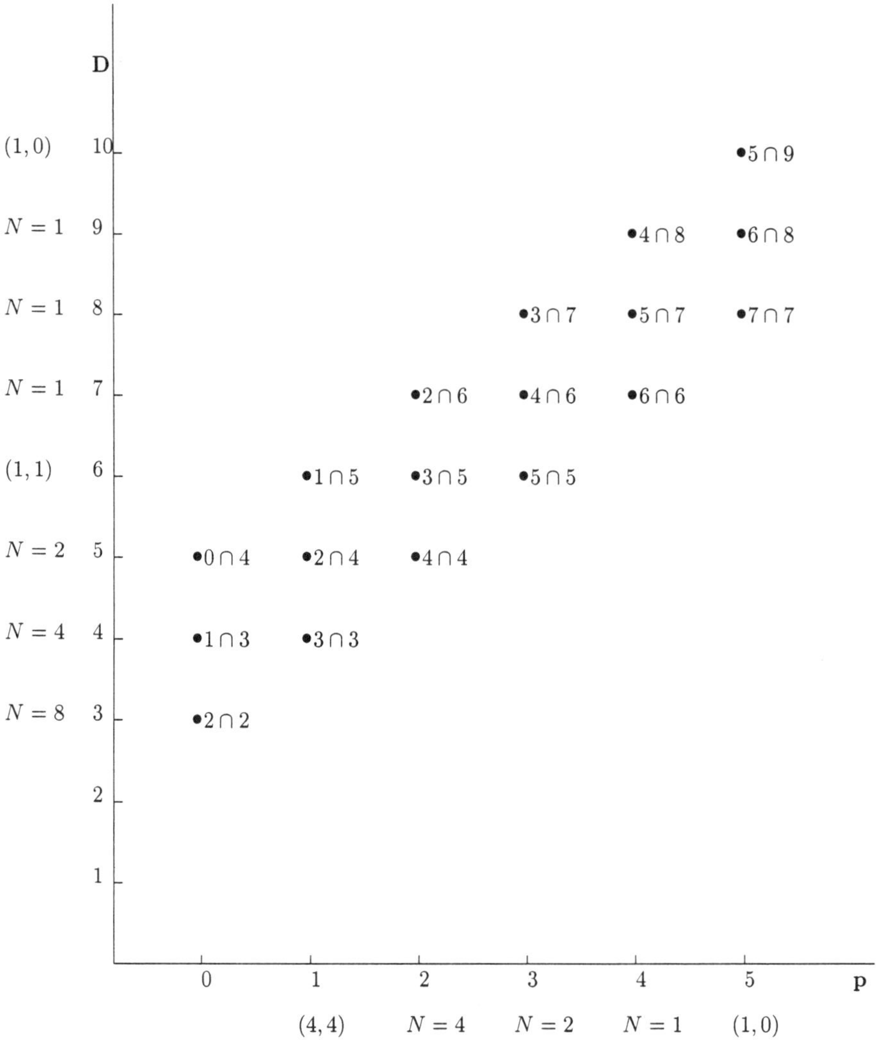

Figure 3: Intersecting D-branes. A Dq_1 brane intersecting a Dq_2 brane over a p–brane is denoted by $q_1 \cap q_2 = p$. The cases $q_1 - q_2 = 4, 2, 0$ correspond to q_1 brane within q_2 brane, ending on q_2 brane or intersecting with q_2 brane, respectively. In all these cases, p–brane is viewed as moving in $D = q_2 + 2$ dimensional target space. Worldvolume and target supersymmetries for these p-branes are shown along the p- and D-axis, respectively. All the points shown in this Figure are related to each other by T–duality transformations described briefly in the text.

M-branes:

$$5 \cap 5 = 3 \,, \qquad 2 \cap 5 = 1 \,, \qquad 2 \cap 2 = 0 \,. \tag{5}$$

The relative and overall transverse dimensions for these intersection are $(4,3)$, $(5,4)$ and $(4,6)$, respectively.

$8 \to 4$ *Branes/ Triply Intersecting Branes*

The maximum dimensional target space admitting 8 real component spinor is the $(1,0)$ superspace in $D = 6$ (or any of its dimensional reductions). We can embed a superstring with $(2|4,0)$ super worldsheet, or a super 3-brane with $(4|4)$ world supersurface in this target superspace. This gives the $(4,0)$ supersymmetric hypermultiplet on the string worldsheet and the $N = 2$ scalar multiplet on the 3-brane worldvolume. Considered as scalar branes, these are shown in Table 1. An alternative way to interpret the above embeddings is to view them as triple intersections of suitable M- or D-branes. The multi-intersections of all known branes have been extensively studied in a number of papers. The most extensive classification known to us can be found in Ref.73. Here, we shall be content with the reproduction of the list for triply intersecting M- and D-branes.

To begin with, the complete list [79] of triple intersections of D-branes is given in Figure 4. The arrows indicate various T-duality transformations whose precise nature is spelled out in Ref.79. The common property of these intersections is that the relative transverse dimension for any pair is always 4. The overall transverse dimension is $(3,2,1)$ for the triple intersections over $(0,1,2)$–branes, respectively.

Finally, the triple intersections of M-branes are given by

$$\begin{aligned} &5 \cap 5 \cap 5 = 3 \,, \qquad 5 \cap 5 \cap 5 = 2 \,, \qquad 5 \cap 5 \cap 5 = 0 \,, \\ &5 \cap 5 \cap 5 = 1 \,, \qquad 2 \cap 5 \cap 5 = 1 \,, \qquad 2 \cap 2 \cap 5 = 0 \,, \\ &2 \cap 2 \cap 2 = 0 \,. \end{aligned} \tag{6}$$

The relative transverse dimension for any pair in all of these intersections is always 4 or 5 (see, for example, Ref.73).

$$\begin{array}{ccccc}
6\cap 2\cap 2=0 & \longrightarrow & 7\cap 3\cap 3=1 & \longrightarrow & 8\cap 4\cap 4=2 \\
\downarrow & & \downarrow & & \downarrow \\
5\cap 3\cap 3=0 & \longrightarrow & 6\cap 4\cap 4=1 & \longrightarrow & 7\cap 5\cap 5=2 \\
\downarrow & & \downarrow & & \downarrow \\
4\cap 4\cap 2=0 & \longrightarrow & 5\cap 5\cap 3=1 & \longrightarrow & 6\cap 6\cap 4=2 \\
\downarrow & & \downarrow & & \downarrow \\
3\cap 3\cap 3=0 & \longrightarrow & 4\cap 4\cap 4=1 & \longrightarrow & 5\cap 5\cap 5=2 \\
\downarrow & & \downarrow & & \downarrow \\
4\cap 2\cap 2=0 & \longrightarrow & 5\cap 3\cap 3=1 & \longrightarrow & 6\cap 4\cap 4=2 \\
\downarrow & & \downarrow & & \downarrow \\
5\cap 3\cap 1=0 & \longrightarrow & 6\cap 4\cap 2=1 & \longrightarrow & 7\cap 5\cap 3=2 \\
\downarrow & & \downarrow & & \downarrow \\
4\cap 4\cap 0=0 & \longrightarrow & 5\cap 5\cap 1=1 & \longrightarrow & 6\cap 6\cap 2=2 \\
\downarrow & & \downarrow & & \downarrow \\
3\cap 3\cap 1=0 & \longrightarrow & 4\cap 4\cap 2=1 & \longrightarrow & 5\cap 5\cap 3=2 \\
\downarrow & & \downarrow & & \downarrow \\
2\cap 2\cap 2=0 & \longrightarrow & 3\cap 3\cap 3=1 & \longrightarrow & 4\cap 4\cap 4=2
\end{array}$$

Figure 4: The triple intersections of D-branes. The arrows indicate various T-duality transformations.

4 Superbrane Dynamics from Superembedding

Background

Superembedding approach to supersymmetric extended objects provides a natural and powerful geometrical framework for describing the dynamics of superbranes. One of the most attractive aspects of this approach is its universality; it seems to apply to any kind of branes, regardless of whether their worldvolume multiplets are scalar, vector, tensor or, indeed, any other supermultiplet. Another virtue of this approach is that the target space *and* worldvolume supersymmetry are both made manifest.

In the superembedding approach to branes, the brane under consideration is described mathematically as a sub-supermanifold (the worldsurface) of the target superspace. The coordinates transverse to the worldsurface are then the Goldstone superfields which encode the information about the worldsurface supermultiplets. The key point is then to impose a natural geometrical constraint on the embedding which can translate into a constraint on the Goldstone superfield. Indeed, a constraint of this nature does exist, and it will be explained below.

The superembedding approach has its origin in what is known as the twistor-like approach to superparticles/strings/branes. This approach was initiated some time ago [83,84] in the context of superparticles. The formalism has been extended to branes and it has been developed by several authors over the years. In particular we refer to Ref.85 and Ref.87 for an extensive list of references.

Starting with Ref.87, in a series of papers [87–96], the superembedding formalism has been developed further. In particular, in Ref.87, the nature of the world-surface supermultiplets emerging from the embedding formalism was spelled out. As applications, the full covariant field equations of the M theory five-branes were obtained for the first time by using this formalism. Moreover, the existence of new types of superbranes were deduced and/or conjectured within this framework. Later, this formalism was used to describe open superbranes. The superembedding formalism yields naturally the field equations rather than an action. However, it is possible to obtain an action as well, and in a recent work the approach of Ref.87 has been extended to cover essentially any superbrane that does not involve worldsurface self-dual field strengths. We will come back to these points briefly later. First, let us begin with the description of the basics of superembedding formalism, followed by some examples.

Basics of the Superembedding Formalism

We consider superembeddings $f : M \to \underline{M}$, where the worldsurface M has (even|odd) dimension $(d|\frac{1}{2}D')$ and the target space has dimension $(D|D')$. In local coordinates M is given as $z^{\underline{M}}(z^M)$, where $z^{\underline{M}} = (x^{\underline{m}}, \theta^{\underline{\mu}})$ and $z^M = (x^m, \theta^\mu)$ (if no indices are used we shall distinguish target space coordinates from worldsurface ones by underlining the former). The embedding matrix $E_A{}^{\underline{A}}$ is defined to be

$$E_A{}^{\underline{A}} = E_A{}^M \left(\partial_M z^{\underline{M}}\right) E_{\underline{M}}{}^{\underline{A}} \,, \tag{7}$$

in other words, the embedding matrix is the differential of the embedding map referred to standard bases on both spaces. Our index conventions are as follows: latin (greek) indices are even (odd) while capital indices run over both types; letters from the beginning of the alphabet are used to refer to a preferred basis while letters from the middle of the alphabet refer to a coordinate basis, the two types of basis being related to each other by means of the vielbein matrix $E_M{}^A$ and its inverse $E_A{}^M$; exactly the same conventions are used for the target space and the worldsurface with the difference that the target space indices are underlined. Primed indices are used to denote directions normal to the worldsurface. We shall also use a two-step notation for worldsurface spinor indices where appropriate: in general discussions, a worldsurface spinor index such as α runs from 1 to $\frac{1}{2}D'$, but it may often be the case that the group acting on this index includes an internal factor as well as the spin group of the worldsurface; in this case we replace the single index α with the pair αi where i refers to the internal symmetry group. A similar convention is used for normal spinor indices.

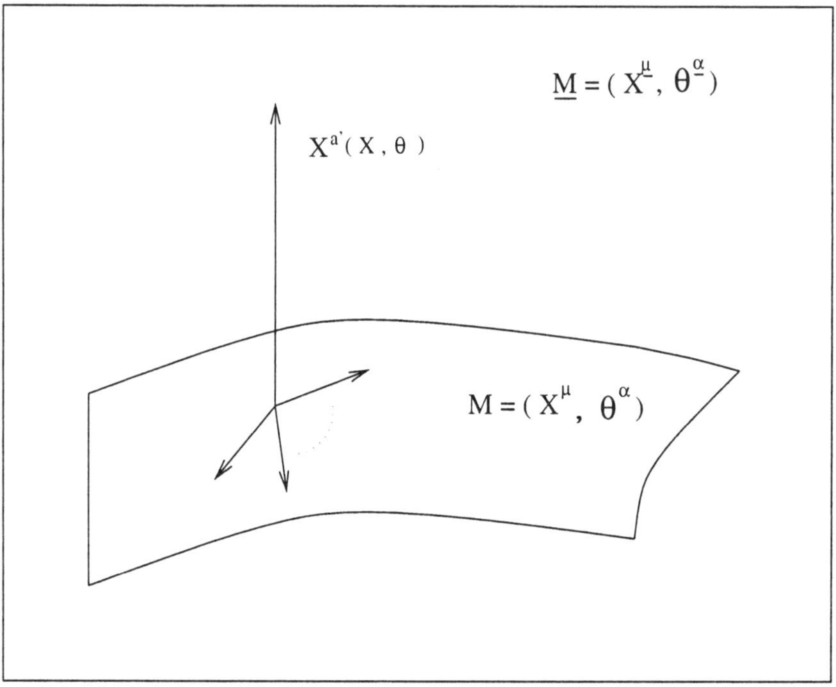

Figure 5: Superembedding of a world supersurface M in a target superspace $\underline{M}$. $X^{a'}$ indicates the transverse coordinates which are the Goldstone superfields associated with the breaking of translations in $\underline{M}$ to translations in M.

The basic embedding condition is

$$E_\alpha{}^{\underline{a}} = 0 \; . \tag{8}$$

It implies that the odd tangent space of the worldsurface is a subspace of the odd tangent space to $\underline{M}$ at each point in $M \subset \underline{M}$. In many cases, equation

(8) determines the equations of motion for the brane under consideration. Moreover, it also determines the geometry induced on the worldsurface and implies constraints on the background geometry which arise as integrability conditions for the existence of such superembeddings. The fact that all this information can be deduced from the simple equation (8) can be intuitively be understood by observing that the curl of the embedding matrix (7) yields the formula

$$\nabla_A E_B{}^{\underline{C}} - (-1)^{AB} \nabla_B E_A{}^{\underline{C}} + T_{AB}{}^C E_C{}^{\underline{C}} = (-1)^{A(B+\underline{B})} E_B{}^{\underline{B}} E_A{}^{\underline{A}} T_{\underline{AB}}{}^{\underline{C}} \quad (9)$$

which involves the super torsion tensors of both, the world and target superspaces. Thus, it is clear that feeding the basic embedding constraint (8) into this integrability equation will have consequences for the worldvolume and target space supergeometries, and hence the dynamics.

Remarkably, the basic embedding equation (8) turns out to be sufficient to determine the full covariant equations of motion for the collective coordinates of the superbrane for most cases. For example, this is the case for both the $M2$ and $M5$ branes. A qualitative aspect of this is that the larger the codimension of the embedding the stronger is the constraint (8). In fact, it was found in Ref.87 that three types of multiplet can arise as a consequence of (8): on-shell, off-shell or underconstrained. In the on-shell case, there can be no superspace actions of the heterotic string type since such actions would necessarily involve the propagation of the Lagrange multipliers that are used in this construction. Nevertheless, on-shell embeddings are useful for deriving equations of motion; for example, the full equations of motion of the M-theory fivebrane were first obtained this way [88]. In the off-shell case, by which it is meant that the worldsurface multiplet is a recognisable off-shell multiplet, it is possible to write down actions of the heterotic string type. The third case that arises, and which we call underconstrained here, typically occurs for branes with low codimension. For example, in codimension one the basic embedding condition gives rise to an unconstrained scalar superfield. In order to get a recognisable multiplet further constraints must be imposed. An example of this is given by IIA D-branes where the basic embedding condition yields an on-shell multiplet for $p = 0, 2, 4$, but an underconstrained one for $p = 6, 8$. Imposing a further constraint which states that there is a worldsurface vector field with the usual modified Bianchi identity whose superspace field strength vanishes unless all indices are bosonic, one recovers on-shell vector multiplets [95]. (For $p = 0, 2, 4$ one can show that the vector Bianchi identity follows from the basic embedding condition.)

The description of the superembedding formalism given above may understandably give the impression that it provides only the superbrane field equations but not an action from which they can be derived. In fact, a rather universal method has been proposed recently [94] by which a superspace action can be obtained for a large class of superbranes. See Ref.94 for a detailed description of the method and comparison with the work of Ref.86.

The power of superembedding formalism has also been put to use in (a) the derivation of the M5–brane equations of motion from those of an open M2–brane ending on the M5–brane [91] and the dynamics of Dp–branes ending on D$(p+2)$–branes [92,93], and (b) the derivation of a Born-Infeld type action for branes involving a higher than 2-form potentials in their worldvolumes [96]. These branes have been called the L-branes [87] because their worldvolume fields form supermultiplets known as the linear multiplets. For example, the L5–brane in $D = 9$ has the linear multiplet on the worldvolume which contains a 4-form potential, and action has been obtained for this object in Ref.96.

Deriving the Field Equations from the Superembedding Constraint

In order to get a feel for how the embedding condition (8) really determines the worldvolume supermultiplet field equations, it suffices to study the linearised version of the constraints resulting from the embedding condition in flat target space limit. The supervielbein for the flat target superspace is,

$$
\begin{aligned}
E^{\underline{a}} &= dx^{\underline{a}} - \frac{i}{2} d\theta^{\underline{\alpha}} (\Gamma^{\underline{a}})_{\underline{\alpha}\underline{\beta}} \theta^{\underline{\beta}} \\
E^{\underline{\alpha}} &= d\theta^{\underline{\alpha}} \ .
\end{aligned}
\tag{10}
$$

Let us choose the physical gauge,

$$
x^{\underline{a}} = \begin{cases} x^a \\ x^{a'}(x,\theta) \end{cases}
\qquad
\theta^{\underline{\alpha}} = \begin{cases} \theta^\alpha \\ \theta^{\alpha'}(x,\theta) \end{cases}
\tag{11}
$$

and take the embedding to be infinitesimal so that $E_A{}^M \partial_M$ can be replaced by $D_A = (\partial_a, D_\alpha)$ where D_α is the flat superspace covariant derivative on the worldsurface, provided that the embedding constraint holds. In this limit the embedding matrix is:

$$E_{\underline{a}}{}^{\underline{b}} = (\delta_a{}^b,\, \partial_a X^{b'})\ , \qquad E_{\underline{a}}{}^{\underline{\beta}} = (0,\, \partial_a \theta^{\beta'})$$
$$E_{\underline{\alpha}}{}^{\underline{b}} = (0,\, D_\alpha X^{a'} - i(\Gamma^{a'})_{\alpha\beta'}\theta^{\beta'})\ , \qquad E_{\underline{\alpha}}{}^{\underline{\beta}} = (\delta_\alpha{}^\beta,\, D_\alpha \theta^{\beta'})\ , \tag{12}$$

where

$$X^{a'} := x^{a'} + \frac{i}{2}\theta^\alpha(\Gamma^{a'})_{\alpha\beta'}\theta^{\beta'}\ . \tag{13}$$

Using the expressions given in (12) in the embedding condition (8), we find, at the linearized level,

$$D_\alpha X^{a'} = i(\Gamma^{a'})_{\alpha\beta'}\theta^{\beta'}\ . \tag{14}$$

This is a general formula. Next, we consider the examples of $M2$ and $M5$ branes. Interestingly, the same embedding constraint (14) yields the on-shell field equations of $N=8$ worldvolume scalar supermultiplet in the M2 brane case, consisting of 8 Bose and 8 Fermi on-shell physical degrees of freedom, and the $(2,0)$ worldvolume tensor multiplet in the case of $M5$ brane, consisting of 5 real scalars, a two-form potential with self-dual field strength describing 3 on-shell degrees of freedom and 8 fermionic on-shell degrees of freedom. Let us show how this works, starting with the case of $M2$ brane.

The M2 Brane

The $M2$ brane worldvolume is an $(3|16)$ dimensional supermanifold embedded in the target superspace of dimension $(11|32)$. The index $\alpha = 1,...,16$ which labels worldsurface fermionic coordinates carry a spinor representation of $SO(2,1)\times SO(8)$, which will be denoted by a pair of indices αA where $\alpha = 1,2$ labels two component Majorana spinor of $SO(2,1)$ and $A = 1,...,8$ labels the chiral spinor of $SO(8)$. The index $\alpha' = 1,...,16$ labels the fermionic directions that are normal to the worldsurface which will be denoted by $\alpha\,\dot{A}$, where $\dot{A}= 1,...,8$ labels the anti-chiral representation of $SO(8)$. The master constraint (8) can then be written as

$$D_{\alpha A}X^{a'} = i(\sigma^{a'})_{A\dot{B}}\,\theta^{\dot{B}}_\alpha\ , \tag{15}$$

where $\sigma^{a'}$ are the chirally projected Dirac Γ-matrices of $SO(8)$ (the van der Wardeen symbols). Differentiating both sides with $D_{\beta B}$ and using the algebra of supercovariant derivatives

$$\{D_\alpha, D_\beta\} = i\,(\gamma^a)_{\alpha\beta}\partial_a\ , \qquad a = 0,1,2\ , \tag{16}$$

after straightforward manipulations one finds the result

$$D_{\alpha A}\theta^{\dot{C}}_\beta = (\sigma^{a'})_A{}^{\dot{C}}\,(\gamma^a)_{\alpha\beta}\,\partial_a X^{a'}\ . \tag{17}$$

The equations of motion now arise as follows. Differentiating (17) with $D_{\gamma B}$, using (14), symmetrizing the equation in $\gamma\alpha$ indices, using the algebra (16) of supercovariant derivatives and multiplying with the $SO(2,1)$ charge conjugation matrix $\epsilon^{\alpha\beta}$, we obtain the Dirac equation

$$(\gamma^a)_\beta{}^\gamma\,\partial_a\theta^{\dot{C}}_\gamma = 0\ . \tag{18}$$

Acting on (17) with $(\gamma^b)^{\alpha\beta}\partial_b$, and using the Dirac equation (18), on the other hand gives the Klein-Gordon equation

$$\partial^a\partial_a\,X^{a'} = 0\ . \tag{19}$$

Continuing in this manner, it can be shows that no new components arise in the Goldstone superfields $X^{a'}$ and $\theta^{\dot{C}}_\beta$. Thus, what we have found is an $N = 8$ on-shell scalar supermultiplet with 8 real scalars obeying the Klein-Gordon equation and 8 two-component Majorana spinors obeying the Dirac equation, altogether representing the 8 fermi and 8 bose on-shell degrees of freedom on the $M2$ brane worldvolume.

The M5 Brane

The procedure for analysing the constraint (14) for the case of $M5$ brane is parallel to the case of $M2$ brane just described in detail. Here, the $M5$ brane worldvolume is an $(6|16)$ dimensional supermanifold embedded in the target superspace of dimension $(11|32)$. The index $\alpha = 1, ..., 16$ which labels worldsurface fermionic coordinates carry a spinor representation of $SO(5,1) \times SO(5)$, which will be denoted by $\theta_{\alpha i}$ where $\alpha = 1, ..., 4$ is the chirally projected spinor index of $SO(5,1)$ and $i = 1,...,4$ labels the spinor of $SO(5)$. The index $\alpha' = 1,...,16$ labels the fermionic coordinates that are normal to the worldsurface which will be denoted by θ_i^α. We are using the well established chiral notation in which the lower α index denotes a chiral spinor, the upper α index denotes and anti-chiral spinor, and these indices are never to be raised and lowered by a charge conjugation matrix. Furthermore, all the spinors in question are symplectic Majorana-Weyl. See Ref.87 for further notation and conventions.

To analyze the constraint (14) for the case of $M5$ brane, we begin by writing it more explicitly as

$$D_{\alpha i} X^{a'} = i\,(\gamma^{a'})_{ij}\theta^j_\alpha\ , \quad a' = 1,...,5\ , \quad \alpha = 1,...,4\ , \tag{20}$$

where $(\gamma^{a'})_{ij}$ are the Dirac γ-matrices of $SO(5)$ (which are antisymmetric). The raising and lowering of the $SO(5)$ spinor index is with the antisymmetric charge conjugation matrix Ω_{ij}.

Starting from (20), manipulations parallel to those described above for the case of $M2$ brane now lead to the result [87]

$$D_{\alpha i}\,\theta_{\beta j} = -\tfrac{1}{2}(\gamma^{a'})_{ij}(\gamma^a)_{\alpha\beta}\partial_a X_{a'} + \Omega_{ij} h_{\alpha\beta}\ , \tag{21}$$

where $(\gamma)_{\alpha\beta}$ are the chirally projected Dirac γ-matrices of $SO(5,1)$, and the symmetric bispinor $h_{\alpha\beta}$ defines a self-dual third-rank antisymmetric tensor

$$h_{\alpha\beta} \equiv (\gamma^{abc})_{\alpha\beta}\, h_{abc}\ , \qquad h_{abc} = \tfrac{1}{3!}\epsilon_{abcdef}h^{def}\ . \tag{22}$$

Comparing the result (21) with (17), reveals that the difference between the $M2$ and $M5$ branes is due to the occurrence of a new worldvolume field h_{abc} in the latter case. Continuing in the manner described for the $M2$ brane case earlier, one finds by applying further spinorial covariant derivatives that the fermion field satisfies the Dirac equation, the scalar fields satisfy the Klein-Gordon equation and the tensor field satisfies the Bianchi identity and field equation for a third-rank antisymmetric field strength tensor. Furthermore, there are no other spacetime components, so that equation (14) defines an on-shell tensor multiplet. It is remarkable that this result follows from the superembedding constraint which takes exactly the same form for both the $M2$ brane as well as the $M5$ brane. This shows the universal nature of the embedding approach; although the worldvolume supermultiplets are rather different in nature, they both arise from one universal superembedding constraint. Unlike the Green-Schwarz type formulation of branes in which one has to search for different kinds of actions depending on the nature of the expected worldvolume supermultiplet, here one starts from a universal and geometrical embedding formula which then determines the worldvolume supermultiplet and provides their equations of motion, if the codimension of the embedding is large enough to make the constraint sufficiently strong to do so. We just saw that this is the case for the $M2$ and $M5$ branes which are described by codimension 8 and 5 embeddings.

The L5 Brane

As a last example to illustrate the universality of the superembedding approach, we examine the $L5$ brane in $D = 9$ [87,96]. The $(6|8)$ dimensional worldvolume superspace is embedded in $(9|16)$ dimensional target superspace. This is a codimension 3 embedding in which the 3 Goldstone superfields give rise to a linear supermultiplet with $(1,0)$ supersymmetry in the $L5$ brane worldvolume. This multiplet consists of 3 real scalars, a 4-form potential describing 5 degrees of freedom and an $Sp(1)$ symplectic Majorana spinor describing 8 real degrees of freedom. This is an example for a superembedding in which the embedding constraint is not sufficient to put the theory of-shell. As a consequence, it is easier to write done an action formula for this theory.

The calculations at the linearized level are again very similar to those explained in detail for the case of $M2$ brane earlier, so it suffices to outline briefly how the worldvolume supermultiplet arises.

The index $\alpha = 1, ..., 8$ now labels worldsurface fermionic coordinates which carry a spinor representation of $SO(5,1) \times SO(3)$, which will be denoted by $\theta_{\alpha i}$ where $\alpha = 1, ..., 4$ is the chirally projected spinor index of $SO(5,1)$ and $i = 1, 2$ labels the spinor of $SO(3)$. The index $\alpha' = 1, ..., 8$ labels the fermionic coordinates that are normal to the worldsurface which will be denoted by θ_i^α. The chiral notation for the spinors is as explained earlier for the case of $M5$ brane. Thus, the master embedding constraint (8) again takes the form (20). The only difference is that the index $i = 1, 2$ now labels an $SO(3)$ spinor. Steps parallel to those described above then lead to the formula

$$D_{\alpha i}\theta_{\beta j} = -(\gamma^{a'})_{ij}(\gamma^a)_{\alpha\beta}\partial_a X^{a'} + \epsilon_{ij}(\gamma^a)_{\alpha\beta}\, h_a \ , \qquad (23)$$

where h_a is the conserved vector in the multiplet, $\partial^a h_a = 0$. This field, together with the 3 scalars $X^{a'}$ and the 8 spinors $\Theta_{\alpha i}$ (evaluated at $\theta = 0$) are the components of the (off-shell) linear multiplet. At the linearized level the field equations are obtained by imposing the free Dirac equation on the spinor field. One then finds the Klein-Gordon equation $\partial_a \partial^a X_{ij} = 0$ for the scalars and the field equation for the antisymmetric tensor gauge field $\partial_{[a} h_{b]} = 0$. The full equations of motion can be obtained [96] either by directly imposing an additional constraint in superspace or by using the recently proposed brane action principle which has the advantage of generating the modified Born-Infeld term for the tensor gauge fields in a systematic way [94].

5 Supermembranes in AdS Background, Singletons and Higher Spin Gauge Theory

String theory has often been studied in Minkowski target spacetime or in a product of Minkowski spacetime with tori, orbifolds or Ricci flat spaces such as K3 or Calabi-Yau manifolds. These are spaces which allow a perturbative formulation of string theory to all orders in α' as a conformal field theory on the string worldsheet. Group manifolds also allow an exact conformal field theoretic treatment and they have been studied in the context of string theory as well, though to somewhat lesser extent. An additional motivation for focusing attention on Ricci flat spaces has been the fact that phenomenologically the most promising string theory is the heterotic string theory which has natural compactifications that require Ricci-flat internal spaces.

The study of duality symmetries in the early 90's and the discovery of D-branes in 1995 brought the type II theories under focus. The most often studied type II backgrounds continued to be Minkowski $\times$ flat or Ricci flat spaces for sometime but that changed drastically with the discovery in 1997 of a remarkable connection between type IIB string on $AdS_5 \times S^5$ and $D = 4, N = 4$ supersymmetric $SU(N)$ Yang-Mills theory [133]. The AdS background has emerged as the near horizon geometry of certain brane solutions, and connections with Yang-Mills have been found by taking particular limits in the parameter space of the theory. The study of branes in AdS space is now in full swing, and it brings together nicely many aspects of brane physics, supersymmetric field theories, Kaluza-Klein supergravities, gauged and conformal supergravities. It has the further dividend of giving new handles on old problems in nonperturbative Yang-Mills gauge theory.

The study of branes in AdS is not altogether a new development though. Already back in late 80's, the $D = 11$ supermembrane was studied in AdS background. Indeed, the solutions of the $D = 11$ supermembrane equations were studied in a series of papers[110,111,112] in $M_4 \times M^7$ background, where M_4 was taken to be AdS_4 or its suitable covering and M_7 to be a suitable seven dimensional Einstein space, such as S^7. Particularly interesting solution was found in which a static spherical membrane resided at the boundary of AdS_4. This was named the Membrane at the End of the Universe. As mentioned earlier, the hope was that a perturbative expansion of the supermembrane around this solution would give a free field theory at the boundary of AdS, thereby having significant consequences for the renormalizability issue. At the time, the properties of the full supermembrane action in AdS background (that is, without expanding around a particular solution) were not investigated. Recent developments, however, have provided abundant motivation to do just that. It is convenient to discuss some general features of the supermembrane in AdS space before we turn to a description of the Membrane at the End of the Universe.

Supermembrane in AdS Background and Singletons

For definiteness, let us consider the supermembrane in $AdS_4 \times S^7$ background, which is a well known $N = 8$ supersymmetric solution of $D = 11$ supergravity[136]. The $D = 11$ supermembrane action in a generic background is given by[33]

$$S = -\int d^3\xi \left(\sqrt{-g} + \epsilon^{ijk} C_{ijk}\right) , \tag{24}$$

where ξ^i $(i = 0, 1, 2)$ are the coordinates on the membrane worldvolume, g_{ij} is the induced metric on Σ and $g = \det g_{ij}$. This metric and the components of the pulled-back 3-form C are defined as

$$g_{ij} = E_i{}^{\underline{a}} E_j{}^{\underline{b}} \eta_{\underline{ab}} , \qquad C_{ijk} = E_i{}^{\underline{A}} E_j{}^{\underline{B}} E_k{}^{\underline{C}} C_{\underline{CBA}} , \tag{25}$$

where $\eta_{\underline{ab}}$ is the Minkowski metric in eleven dimensions, and

$$E_i{}^{\underline{A}} = \partial_i z^{\underline{M}} E_{\underline{M}}{}^{\underline{A}} , \tag{26}$$

and $E_{\underline{M}}{}^{\underline{A}}$ is the target space supervielbein.

Thus, the $OSp(8|4)$ invariant supermembrane action is (25) in a target superspace with isometry group $OSp(8|4)$ which supports a closed 4-form $dH = 0$, which can be locally solved as $H = dC$. The superspace we seek must have $AdS_4 \times S^7$ as a bosonic subspace and consequently it can be chosen to be[115,117]

$$\frac{G}{H} = \frac{OSp(8|4)}{SO(3,1) \times SO(7)} \ . \tag{27}$$

The generators of G and H are

$$\begin{aligned} G : &\quad \overbrace{M_{ab}, P_c}^{SO(3,2)}\ ,\ \overbrace{T_{IJ}, P_J}^{SO(8)}\ ,\ Q_{\alpha A} \\ H : &\quad M_{ab}\ ,\ \ T_{IJ} \end{aligned} \tag{28}$$

where $Q_{\alpha A}$ are the 32 real supergenerators transforming as spinor of $SO(3,2) \times SO(8)$ and the rest of the notation is self explanatory. The supervielbein and the 3-form C on G/H can be calculated straightforwardly from the knowledge of the structure constants of G. See Refs.117, 118 and 119 for further details.

The action (24), with target superspace G/H specified in (27), is manifestly invariant under $OSp(8|4)$ since this is the isometry group of G/H. It is also invariant under the worldvolume local diffeomorphism and local κ–symmetry. Fixing a physical gauge by identifying the worldvolume coordinates with three of the target space coordinates and setting half of the target space fermionic coordinates equal to zero (by means of a suitable projection), breaks the local diffeomorphisms, local κ–symmetry as well as the rigid isometries of G/H. The requirement of maintaining the physical gauge fixes the local symmetry parameters in terms of the rigid parameters, and consequently one arrives at a gauged fixed worldvolume action which is invariant under the rigid G symmetry [116,118].

Thus one obtains an action for the 8 real scalars and 8 Majorana spinors on the worldvolume, which is invariant under the rigid superconformal group $OSp(8|4)$ transformations some of which are linearly realized (and hence manifest) and the rest are nonlinearly realized. All this is perfectly analogous to the discussion of the lightcone gauge fixing in $D = 11$ supermembrane theory in Poincaré superspace [34].

Now, we ask the following question: Is there a vacuum solution of the supermembrane equation such that a perturbative expansion around it yields a free but still $OSp(8|4)$ invariant action? The answer is yes. To see this, it is convenient to use horospherical $\times$ hyperspherical coordinates to parametrize the $AdS_4 \times S^7$ metric as

$$ds^2 = \phi^2 \left(-d\tau^2 + d\sigma^2 + d\rho^2\right) + a^{-2} \left(\frac{d\phi}{\phi}\right)^2 + 4a^{-2}\, d\Omega_7 \;, \qquad (29)$$

where $d\Omega_7$ is the $SO(8)$ invariant metric on S^7. The boundary of this metric is the three dimensional Minkowski space M_3 at $\phi \to \infty$, completed by the point $\phi \to 0$, so that the inversion element of the conformal group action on the boundary is well defined. The nature of this boundary has been discussed in great detail, for example, in Refs.135, 116 and 118.

Denoting the coordinates of the seven sphere by y^I (I=1,...,7), a simple class of solutions to the supermembrane equations is [118]

$$x^i = \xi^i \;, \qquad \phi = \phi_0 \;, \qquad y^I = y_0^I \;, \qquad (30)$$

where $x^i = (\tau, \sigma, \rho)$ and ξ are the worldvolume coordinates, ϕ_0 and y_0^I are arbitrary constants and the fermionic variables are set equal to zero. The singleton action is obtained by expanding around $\phi_0 = 0$ which corresponds to expanding around the boundary of AdS_4 [116,118]. For convenience, one can also set $y_0^I = 0$. In using the normal coordinate expansion formulae [18], it is important to rescale the fluctuation fields appropriately. Defining the fluctuations as [118]

$$\phi = \phi_0 + \sqrt{\phi_0}\, \varphi \;, \qquad y^I = \frac{1}{\sqrt{\phi_0}}\, \varphi^I \;, \qquad \theta = (\phi_0)^{-3/2}\, \lambda \;, \qquad (31)$$

and taking the limit $\phi_0 \to 0$ after substituting to the normal coordinate expansion formulae [18], one finds that the zeroth and first order terms in the normal coordinate expansion vanish and the second order term yields the $N = 8$ singleton action for eight free scalars (φ, φ^I) and eight free fermions λ [116,118]. The action is $OSp(8|4)$ invariant and this does not require a mass term for the bosons, since the boundary of AdS_4 is characterized as a Minkowski space in the coordinate system used here.

Membrane at the End of the Universe

Now, we turn to the description of an interesting set of solutions to the supermembrane equations in $AdS_4 \times S^7$ background which were found sometime ago [110,111] and which are closely related to solution (30). We begin by considering the bosonic field equation for a configuration where the spacetime gravitino and the fermionic co-ordinates are set equal to zero:

$$\partial_i(\sqrt{-h}h^{ij}\partial_j X^N g_{MN}) - \frac{1}{2}\sqrt{-h}h^{ij}\partial_i X^N \partial_j X^P \partial_M g_{NP}$$
$$+\frac{1}{3}\epsilon^{ijk}\partial_i X^N \partial_j X^P \partial_k X^Q H_{MNPQ} = 0 \ , \qquad (32)$$

where h_{ij} is the induced metric

$$h_{ij} = \partial_i X^M \partial\partial_j X^N G_{MN} \ , \qquad (33)$$

$X^M(\tau, \sigma, \rho)$ are the spacetime coordinates of the membrane, g_{MN} is the spacetime metric and $H_{MNPQ} = 4\partial_{[M}B_{NPQ]}$. The κ–symmetry of the supermembrane action requires that g_{MN} and H_{MNPQ} satisfy the usual bosonic equations of $d = 11$ supergravity. Let us consider the solution of these equations is which the spacetime is $AdS_4 \times M_7$, such that AdS_4 has inverse radius a, and M_7 is an Einstein space with Ricci tensor $R_{mn} = \frac{3}{2}a^2 g_{mn}$. In Ref.111 we used the coordinate system in which the AdS_4 metric is given by

$$ds^2 = -(1 + a^2r^2)dt^2 + r^2(d\theta^2 + \sin^2\theta d\phi^2) + (1 + a^2r^2)^{-1}dr^2 \ . \qquad (34)$$

The maximal $SO(3,2)$ symmetry of this metric requires that the period of t is an integer multiple of $2\pi/a$, namely $\Delta t = 2\pi q/a$, where q is an integer [111]. The four-index field strength is taken to be proportional to the Levi-Civita tensor in AdS_4:

$$H_{\mu\nu\rho\sigma} = \tfrac{3}{2}\, a\sqrt{-g}\ \epsilon_{\mu\nu\rho\sigma} \ . \qquad (35)$$

The metric for the seven dimensional internal space will be taken to be an S^7 moded by Z^p, which is a Lens space, $L(1,p)$ that can also be viewed as a $U(1)$ bundle over $CP(3)$ with fibers having the period $2\pi/p$, for some integer p:

$$d\hat{s}^2 = \frac{4}{a^2}[(d\psi + 2A)^2 + ds^2(CP^3)] \ , \qquad (36)$$

where $ds^2(CP^3)$ is the standard Fubini-Study metric on CP^3, with Einstein metric satisfying $R_{ab} = 8g_{ab}$, ψ is the coordinate on the U(1) fibres and A is a one-form potential satisfying $dA = J$, where J is the Kähler two-form on CP^3. The fibre coordinate ψ has the period $\Delta\psi = 2\pi/p$, for some integer p.

The space $L(1,1)$ is the round seven sphere S^7, and $L(1,2)$ is the projective space RP^7.

To solve the supermembrane equations, we make the ansatze [111]

$$\begin{aligned} t &= \tau , & \theta &= \sigma , & \phi &= \rho , \\ r &= r_0 , & y^I &= y_0^I , & \psi &= \alpha\tau/2 , \end{aligned} \qquad (37)$$

where α, r_0 and y_0^I are constants, and y^I are the coordinates on CP^3, when we view the internal space as $U(1)$ bundles over CP^3. Substituting this ansatze into the supermembrane field equations (32), we find that all components of the equation are satisfied identically except in the r direction, which is solved by ($\alpha = 1,\ r_0 = arbitrary$), or by ($\alpha > 1,\ r_0 = \sqrt{4(\alpha^2-1)/3a^2}$). Furthermore we note that the periods of τ and ψ specified above, together with the ansatz (37), imply that $\alpha = 2/pq$. Thus we find the following solutions [111]

$$\begin{aligned} \alpha = 1: \qquad & \widetilde{AdS_4} \times S^7 , & r_0 &= \text{arbitrary} , \\ & AdS_4 \times RP^7 , & r_0 &= \text{arbitrary} , \\ \alpha = 2: \qquad & AdS_4 \times S^7 , & r_0 &= 2/a , \end{aligned} \qquad (38)$$

where $\widetilde{AdS_4}$ denotes the double covering of AdS_4. As was shown in Ref.111, the supersymmetry of the solutions presented above requires that r_0 be taken to infinity (hence the terminology of the "Membrane at the End of the Universe"). This requirement peaks the $\alpha = 1$ solutions above. A normal coordinate expansion of the action around these solutions should yield an $OSp(8|4)$ invariant action formulated on the boundary of AdS_4 (or $\widetilde{AdS_4}$). However, it has been noted [111] that the rescalings of the fluctuation fields needed to extract a nonvanishing quadratic action (which also eliminate all the interaction terms in the limit $r_0 \to \infty$) had the effect of spoiling the extraction of meaningful supersymmetry transformations (which required different rescalings). Thus, a rigorous derivation of the action is still missing. Nonetheless, we expect the standard $OSp(8|4)$ singleton field theory on the boundary of AdS_4 to arise. The action for this theory [107,108] can be described as follows.

The $OSp(8|4)$ singleton supermultiplet consists of 8 real scalars ϕ^I ($I = 1, ..., 8$) in the 8_v representation of $SO(8)$ and 8 four-component spinors $\lambda^{\underline{\alpha}}$ ($\alpha = 1, ..., 8$) in 8_s of $SO(8)$. These fields live on $S_2 \times S^1$ boundary of AdS_4. In addition to the four-dimensional Majorana condition $\bar{\lambda} = \lambda^T C$, the spinor λ satisfies the chirality condition

$$\gamma_0\gamma_1\gamma_2\,\lambda_- = -\lambda_- \ , \tag{39}$$

which, unlike the usual chirality condition, is compatible with the Majorana condition. The $N = 8$ supersingleton action is given by [107,108]

$$\mathcal{L} = -\tfrac{1}{2}\sqrt{-h}\left(h^{ij}\partial_i\phi^I\partial_j\phi^I + \tfrac{1}{4}\,a^2\,\phi^I\phi^I - i\bar{\lambda}^{\underline{\alpha}}_-\gamma^i\nabla_i\lambda^{\underline{\alpha}}_- \right) \ , \tag{40}$$

where ∇_i is the covariant derivative on $S_2 \times S^1$. This action is invariant under the rigid $OSp(8|4)$ transformations, the details of which can be found in Refs.107, 108 and 18. Notice the presence of the mass term for the bosonic singletons in the action. In the case of $S^p \times S^1$ boundary of AdS_{p+2}, this term is given by [114]

$$\frac{p-1}{4p}\,R\,\phi^2 \ , \tag{41}$$

where $R = p(p-1)a^2$ is the curvature scalar of $S^p \times S^1$ and a is the inverse radius of S^p. In fact, the difficulty in obtaining the Lagrangian (40) from the normal coordinate expansion of the supermembrane action on $AdS_4 \times S^7$ around the Membrane at the End of the Universe solutions lies precisely in getting this mass term right, for it seems to vanish if one takes a naive $r_0 \to \infty$ limit [111,116,146].

Singletons and Higher Spin Gauge Theory

The $N = 8$ singleton field theory formulated on $S^2 \times S^1$ boundary of AdS_4 was quantized sometime ago [17,18], with the hope that it might play a role in the quantization of the supermembrane on $AdS_4 \times S^7$ background [107,18]. Furthermore, a remarkable group theoretical property of the singleton representation which states that the direct product of two singletons decomposes into infinitely many massless field with higher spins [99], motivated Bergshoeff, Salam, Tanii and the author [17,18,109] to conjecture that the $D = 11$ quantum supermembrane on $AdS_4 \times S^7$ should give rise to a higher spin gauge theory which contains the usual $N = 8$ AdS supergravity as a subsector. The occurrence of the infinitely many massless higher spin fields implies the existence of infinitely many (local) gauge symmetries analogous to the Yang-Mills, general coordinate and local supersymmetries associated with spin 1, 2 and 3/2, respectively.

Interestingly enough, in a related development Fradkin and Vasiliev [120,121] were in the course of developing a higher spin gauge theory in its own right (see Ref.120 for references to earlier work). These authors succeeded in constructing interacting field theories for higher spin fields. It was observed that the previous difficulties in constructing higher spin theories can be bypassed by formulating the theory in AdS space and to consider an infinite tower of gauge fields controlled by various higher spin algebras based on certain infinite dimensional extensions of super AdS algebras. In particular, the AdS radius could not be taken to infinity since its positive powers occurred in the higher spin interactions and therefore one could not take a naive Poincaré limit.

In a series of papers Vasiliev pursued the program of constructing the *AdS* higher spin gauge theory and simplified the construction considerably. In Ref.122 the spin 0 and 1/2 fields were introduced to the system within the framework of free differential algebras. The theory was furthermore cast into an elegant geometrical form [123] by extending the higher spin algebra to include new auxiliary spinorial variables (see Ref.124 for a review).

Applying the formalism of Vasiliev to a suitable higher spin algebra that contains the maximally extended super AdS algebra $OSp(8|4)$, the resulting spectrum of gauge and matter fields remarkably coincide with the massless states resulting from the symmetric product of two $OSp(8|4$ supersingletons! [17,18,109]. In a recent paper [125], the Vasiliev theory of higher spin fields (which is applicable to a wide class of higher spin superalgebras) was examined in the context of $N = 8$ supersymmetry, and the precise manner in which the $N = 8$ de Wit-Nicolai gauged supergravity [126] can be described within this framework was studied.

In the rest of this section, we will outline some kinematical aspects of the higher spin gauge theory. To begin with, the $N = 8$ singleton representation of the $D = 4$ super AdS group $OSp(8|4)$ [103,101], which are also known as Di's and Rac's, are

$$Rac: \qquad D(\tfrac{1}{2}, 0) \otimes 8_s$$

$$Di: \qquad D(1, \tfrac{1}{2}) \otimes 8_c$$

where $D(E_0, s)$ denotes an UIR of $SO(3,2)$ for which E_0 is the minimal energy eigenvalue of the energy operator M_{04}, and s is the maximum eigenvalue of

$k\backslash s$	0	$\frac{1}{2}$	1	$\frac{3}{2}$	2	$\frac{5}{2}$	3	$\frac{7}{2}$	4	$\cdots$	$2s$	$2s+\frac{1}{2}$	$2s+1$	$2s+\frac{3}{2}$	$2s+2$	$\cdots$
0	$35_{+}+35_{-}$	56	28	8	1											
1	1+1	8	28	56	$35_{+}+35_{-}$	56	28	8	1							
2					1	8	28	56	$35_{+}+35_{-}$	$\cdots$						
$\vdots$																
$s-1$										$\cdots$	1					
s										$\cdots$	$35_{+}+35_{-}$	56	28	8	1	$\cdots$
$s+1$											1	8	28	56	$35_{+}+35_{-}$	$\cdots$
$\vdots$																

Table 2: The $SO(3,2) \times SO(8)$ content of the symmetric tensor product of two $N = 8$ singletons. This product generates the infinite spectrum of an $shs^E(8|4)$ gauge field ($s \geq 1$) and the spectrum of a finite number of additional matter fields ($s < 1$).

the spin operator M_{12} in the lowest energy sector. The decomposition of the symmetric tensor product $[(Rac \oplus Di) \otimes (Rac \oplus Di)]_S$ under the $OSp(8|4)$ leads to the UIR's of $OSp(8|4)$ given in Table 2 [17,18].

The integer $k = 0, 1, 2, ..$ can be considered as level number. Levels $\kappa = 0, 1$ are somewhat special. At level $k = 0$, there is the familiar $N = 8$ supergravity multiplet consisting of 128 Bose and 128 Fermi on-shell degrees of freedom. At level $k = 1$, there is a $256 + 256$ multiplet that has 2 scalars as the lowest member of the supermultiplet and a single spin 4 field as the highest member. For levels $k > 2$ the structure of all the supermultiplets is the same, namely they start with a singlet spin $(2k - 2)$ field and end with a singlet $(2k + 2)$ field, in 8 steps of spin 1/2 increments. The associated $SO(8)$ irreps are: $(1, 8, 28, 56, 35 + 35, 56, 28, 8, 1)$.

Among the interesting and important properties of the spectrum shown in Table 1 is that there are two distinct classes of fields; those with spin $s \geq 1$, which forms an infinite set, and few fields that have spin $s \leq 1/2$. These two classes of fields are separated with a vertical line in Table 2. The fields with spin $s \geq 1$ can be associated with generators of an infinite dimensional algebra, called $shs^E(8|4)$, while the fields with spin $s \leq 1/2$, clearly cannot be associated with any generator. Note however the important fact that all fields shown in Table 2, including those with spin $s \leq 1/2$ are exactly those which arise in the symmetric tensor product of two $OSp(8|4)$ singletons. Physical consistency of gauge field theory based on $shs^E(8|4)$ requires that the complete particle spectrum forms a unitary representation of the full, infinite dimensional algebra

$shs^E(8|4)$ [128]. The product of the $OSp(8|4)$ singletons which give the field content shown in Table 2, indeed does form a unitary representation of $shs^E(8|4)$. Consequently, the matter fields (the left hand side of the vertical line shown in Table 2) must be included, in addition to the gauge fields (the right hand side of the vertical line) in a sensible (consistent and unitary) formulation of higher spin $N = 8$ supergravity theory.

Recently [125], the $N = 8$ higher spin supergravity theory based on $shs^E(8|4)$was investigated in considerable detail, and the precise manner in which it contains the $N = 8$ de Wit-Nicolai gauged supergravity [126] has been shown at the linearized level. In our opinion, this constitutes a positive step towards the understanding of the M-theoretic origin of the massless higher spin gauge theory. To make further progress, one has to compare the interactions of the spin $s \leq 2$ fields in the higher spin theory with those of de Wit-Nicolai gauged supergravity theory at the next order, namely the quadratic order in fields, in the equations of motion. It would be very interesting to find out how the $E_7/SU(8)$ structure of the scalar fields [127] will manifest itself and to determine how the higher spin fields interact with the spin $s \leq 2$ fields. Ultimately, the $N = 8$ higher spin supergravity should emerge from the dynamics of the $N = 8$ singleton field theory defined at the boundary of AdS_4. In this context, we note that the OPE's of the stress energy tensor in the $OSp(8|4)$ singleton theory have been studied [113], but a great deal of work remains to be done to shed more light on the issue of how to extract information about the physics in the bulk of AdS.

In summary, it is worth emphasizing the following points about the $N = 8$ higher spin supergravity whose properties have been outlined above: (a) the existence of the theory is highly nontrivial, (b) the theory is based on an infinite dimensional extension of the $D = 4, N = 8$ super AdS group $OSp(8|4)$, (c) it fuses matter fields with the gauge fields in such a way that the full spectrum of massless states are exactly those which arise from the two $N = 8$ singleton states and (d) the theory contains the equations of motion of the $D = 4, N = 8$ AdS supergravity as a subsector. This last property is very significant in that it is in the spirit of discovering new structures that build upon what we already know. Should this theory survive further scrutiny, then the appropriate question to ask is not if this theory fits into the big picture of M-theory, but rather how it will do so.

The massive Kaluza-Klein states coming from the S^7 compactification of $D = 11$ supergravity must also be taken into account in a suitable extension of the $N = 8$ higher spin supergravity. These states are expected to arise in the product of three or more $N = 8$ singletons. An infinite set of new massive states would appear in the spectrum as a byproduct. A higher spin theory taking into account massive states is yet to be constructed.

An extreme point of view would be to imagine a pure gauge theory formulation of M-theory which contains only the massless fields corresponding to an infinite dimensional symmetry. All the phases of M-theory (the known ones and those yet to be discovered) are then to emerge from the breaking of this master symmetry in various ways.

One can imagine the construction of a higher spin gauge theory directly in $D = 11$ AdS space. However, the anticommutator of two supersymmetries necessarily involves (tensorial) generators in addition to the AdS generators in $D = 11$. One can take the AdS group in $D = 11$ to be the diagonal subgroup of $OSp(1|32) \oplus OSp(1|32)$ [129,130] and study its field theoretic realizations. Apparently, no such realizations are known at present [131,132]. Nonetheless, the singleton representation for this group has been studied recently by Günaydin [130] who found that the product of two such irreps do not contain the $D = 11$ supergravity states, but a further product of the resulting representation does contain the $D = 11$ supergravity fields *and* additional fields as well. This group theoretical result suggests the construction of a $10D$ singleton field theory that lives on the boundary of the $D = 11$ AdS space.

It is clear that much remains to be discovered and that these are exciting times in the quest for an understanding of the mysterious and magic membrane theory.

Acknowledgements

I wish to thank L. Baulieu, E. Bergshoeff, M. Cederwall, C. Chu, M. Duff, E. Eyras, M. Günaydin, J. Lu, B. Nilsson and P. Sundell for useful discussions, and F. Mansouri for drawing my attention to the developments in early 70's pertaining to the relation between generalized dual models and theories of extended objects. I also would like to thank the Abdus Salam International Center for Theoretical Physics, University of Groningen, Universités Pierre et Marie Curie (Paris VI) and Denis Diderot (Paris VII) and the Institute of Theoretical Physics in Göteborg for hospitality. This research has been supported in part by NSF Grant PHY-9722090.

References

1. *Ideals and Realities : Selected Essays of Abdus Salam*, eds. C.H. Lai and A. Kidwai (3rd edition, World Scientific, 1990).
2. Abdus Salam, *Physics and the excellences of the life it brings*, reprinted in Ref.1.
3. Abdus Salam, *Gauge unification of fundamental forces*, reprinted in Ref.1.
4. Abdus Salam, *Islam and science*, reprinted in Ref.1.
5. Abdus Salam, *Particle physics: will Britain kill its own creation?"*, reprinted in Ref.1.
6. A. Salam and E. Sezgin, "Maximal extended supergravity theory in seven dimensions", *Phys. Lett.* **118B**, 359 (1982).
7. A. Salam and E. Sezgin, "SO(4) gauging of N=2 supergravity in seven dimensions", *Phys. Lett.* **126B**, 295 (1983).
8. A. Salam and E. Sezgin, "Chiral compactification on Minkowski $\times S^2$ of $N = 2$ Einstein-Maxwell supergravity", *Phys. Lett.* **147B**, 47 (1984).
9. S. Randjbar-Daemi, A. Salam and E. Sezgin and J. Strathdee, "An anomaly free model in six dimensions", *Phys. Lett.* **151B**, 351 (1985).
10. A. Salam and E. Sezgin, "d=8 supergravity", *Nucl. Phys.* **B258**, 284 (1985).
11. A. Salam and E. Sezgin, "d=8 supergravity: matter couplings, gaugings and Minkowski space compactifications", *Phys. Lett.* **154B**, 37 (1985).
12. A. Salam and E. Sezgin, "Anomaly freedom in chiral supergravities", *Physica Scripta* **32**, 283 (1985).
13. E. Bergshoeff, S. Randjbar-Daemi, A. Salam, H. Sarmadi and E. Sezgin, "Locally supersymmetric sigma model with Wess-Zumino term in two dimensions and critical dimensions for strings", *Nucl. Phys.* **B269**, 77 (1986).
14. E. Bergshoeff, A. Salam and E. Sezgin, "A supersymmetric R^2-action in six dimensions and torsion", *Phys. Lett.* **173B**, 73 (1986).
15. E. Bergshoeff, T.W. Kephardt, A. Salam and E. Sezgin, "Global anomalies in six dimensions", *Mod. Phys. Lett.* **A1**, 267 (1986).
16. E. Bergshoeff, A. Salam and E. Sezgin, "Supersymmetric R^2-actions, conformal invariance and the Lorentz Chern-Simons term in 6 and 10 dimensions", *Nucl. Phys.* **B279**, 659 (1987).
17. E. Bergshoeff, A. Salam, E. Sezgin and Y. Tanii, "Singletons, higher spin massless states and the supermembrane", *Phys. Lett.* **205B**, 237 (1988).
18. E. Bergshoeff, A. Salam, E. Sezgin and Y. Tanii, "N=8 supersingleton quantum field theory", *Nucl. Phys.* **B305**, 497 (1988).

19. A. Salam and E. Sezgin, *Supergravities in Diverse Dimensions*, Vols. 1 & 2 (World Scientific, 1989).
20. P.A.M. Dirac, "An extensible model of electrons", *Proc. Roy. Soc.* **268A**, 57 (1962).
21. J.H. Schwarz, *Superstrings. The First Fifteen Years*, Vols. 1 & 2 (World Scientific, 1985).
22. G. Domokos, "Wave packet realization of lightlike states", *Commun. Math. Phys.* **26**, 15 (1972).
23. F. Gürsey and S. Orfanidis, "Extended hadrons, scaling variables and the Poincaré group", *Nuovo Cimento* **11A**, 225 (1972).
24. E. Del Giudice, P. di Vecchia, S. Fubini and R. Musto, "Lightcone physics and duality", *Nuovo Cimento* **12A**, 813 (1972).
25. F. Mansouri, "Dual models with global SU(2,2) symmetry", *Phys. Rev.* **D8**, 1159 (1973).
26. R.J. Rivers, "Conformal algebraic explanation for the failure of dual theories with satellites", *Phys. Rev.* **D9**, 2920 (1974).
27. F. Mansouri, "Natural extensions of string theories", in *Lewes String Theory Workshop*, July 1985 , eds. L. Clavelli and A. Halpern (World Scientific, 1986).
28. E. Cremmer, B. Julia and J. Scherk, "Supergravity in eleven dimensions", *Phys. Lett.* **B76**, 409 (1978).
29. M.B. Green and J.H. Schwarz, "Covariant description of superstrings", *Phys. Lett.* **B136**, 367 (1984).
30. E. Bergshoeff, E. Sezgin and P.K. Townsend, "Superstring actions in $D = 3, 4, 6, 10$ curved superspace", *Phys. Lett.* **B169**, 191 (1986).
31. J. Hughes and J. Polchinski, "Partially broken global supersymmetry and the superstring", *Nucl. Phys.* **B278**, 147 (1986).
32. J. Hughes, J. Liu and J. Polchinski, "Supermembranes", *Phys. Lett.* **180B**, 370 (1986).
33. E. Bergshoeff, E. Sezgin and P.K. Townsend, "Supermembranes and eleven-dimensional supergravity", *Phys. Lett.* **189B**, 75 (1987).
34. E. Bergshoeff, E. Sezgin and P.K. Townsend, "Properties of the eleven dimensional supermembrane theory", *Ann. Phys.* **185**, 330.
35. A. Achúcarro, J.M. Evans, P.K. Townsend and D.L. Wiltshire, "Super p-branes", *Phys. Lett.* **198B**, 441 (1987).
36. M.J. Duff, P.S. Howe, T. Inami and K.S. Stelle, "Superstrings in D=10 from supermembranes in D=11", *Phys. Lett.* **B191**, 70 (1987).
37. I. Bars, C.N. Pope and E. Sezgin, "Massless spectrum and critical dimension of the supermembrane", *Phys. Lett.* **B198**, 455 (1987).

38. M.J. Duff, T. Inami, C.N. Pope, E. Sezgin and K.S. Stelle, "Semiclassical quantization of the supermembrane", *Nucl. Phys.* **B297**, 515 (1988).
39. B. de Wit, J. Hoppe and H. Nicolai, "On the quantum mechanics of supermembranes", *Nucl. Phys.* **B305**, 545 (1988).
40. E. Bergshoeff, E. Sezgin, Y. Tanii and P.K. Townsend, "Super p-branes as gauge theories of volume preserving diffeomorphisms", *Ann. Phys.* **199**, 340 (1990).
41. U. Marquard and M. Scholl, "Lorentz algebra and critical dimension for the supermembrane", *Phys. Lett.* **227B**, 234 (1989).
42. I. Bars, "First massive level and anomalies in the supermembrane", *Nucl. Phys.* **B308**, 462 (1988).
43. I. Bars and C.N. Pope, "Anomalies in super p-branes", *Class. and Quantum Grav.* **5**, 1157 (1988).
44. C.N. Pope and E. Sezgin, "A generalized Virasoro algebra for the type IIA superstring", *Phys. Lett.* **B233**, 313 (1989).
45. E. Bergshoeff and E. Sezgin, "New spacetime algebras and their Kac-Moody extension", *Phys. Lett.* **B232**, 96 (1989).
46. E. Bergshoeff and E. Sezgin, "Super p-branes and new spacetime superalgebras", *Phys. Lett.* **B354**, 256 (1995), hep-th/9504140.
47. E. Sezgin, "The M-algebra", *Phys. Lett.* **B392**, 323 (1997), hep-th/9609086.
48. P.K. Townsend, "Supersymmetric extended solitons", *Phys. Lett.* **B202**, 53 (1988).
49. A. Dabholkar, G.W. Gibbons, J.A. Harvey and F. Ruiz Ruiz, *Nucl. Phys.* **B340**, 33 (1990).
50. A. Strominger, "Heterotic solitons", *Nucl. Phys.* **B343**, 167 (1990).
51. M.J. Duff and K.S. Stelle, "Multimembrane solutions of D=11 supergravity", *Phys. Lett.* **B253**, 113 (1991).
52. R. Güven, "Black *p*-brane solutions of D=11 supergravity theory", *Phys. Lett.* **B276**, 49 (1992).
53. M.J. Duff and J.X. Lu, *Type II p-branes: the brane scan revisited, Nucl. Phys.* **B390**, 276 (1993), hep-th/9207060.
54. M.J. Duff, G.W. Gibbons and P.K. Townsend, "Macroscopic superstrings as interpolating solitons", *Phys. Lett.* **B332**, 321 (1994), hep-th/9505124.
55. N. Seiberg and E. Witten, "Monopole condensation, and confinement in $N = 2$ supersymmetric Yang-Mills theory", *Nucl. Phys.* **B426**, 19 (1994), hep-th/9407087.
56. P.S. Howe, N.D. Lambert and P.C. West, "Classical M-fivebrane dynamics and quantum $N = 2$ Yang-Mills", hep-th/9710034.

57. C. M. Hull and P. K. Townsend, "Unity of superstring dualities", *Nucl. Phys.* **B438**, 109 (1995), hep-th/9410167.
58. P.K. Townsend, "The eleven dimensional supermembrane revisited", *Phys. Lett.* **B350**, 184 (1995), hep-th/9501068.
59. E. Witten, "String theory dynamics in various dimensions", *Nucl. Phys.* **B443**, 85 (1995), hep-th/9503124.
60. P.K. Townsend, "p-brane democracy", hep-th/9507048.
61. J. Polchinski, "Dirichlet-branes and Ramond-Ramond charges", *Phys. Rev. Lett.* **75**, 4724 (1995), hep-th/9510017.
62. J.H. Schwarz, "The power of M theory", *Phys. Lett.* **B367**, 97 (1996), hep-th/9510086.
63. P. Horava and E. Witten, "Heterotic and type I string dynamics from eleven dimensions", *Nucl. Phys.* **B460**, 506 (1996), hep-th/9510209.
64. C.M. Hull, "String dynamics at strong coupling", *Nucl. Phys.* **B468**, 113 (1996), hep-th/9512181.
65. O. Aharony, J. Sonnenschein and S. Yankielowicz, "Interactions of strings and D-branes from M theory", *Nucl. Phys.* **B474**, 309 (1996), hep-th/9603009.
66. T. Banks, W. Fischler, S.H. Shenker and L. Susskind, "M theory as a matrix model: A conjecture", *Phys. Rev.* **D55**, 5112 (1997), hep-th/9610043.
67. C. Vafa and E. Witten, "One-loop test of string duality", *Nucl. Phys.* **B447**, 261 (1995), hep-th/9505053.
68. M.J. Duff, J.T. Liu and R. Minasian, "Eleven dimensional origin of string-string duality: a one-loop test", *Nucl. Phys.* **452**, 261 (1995), hep-th/9506126.
69. M.B. Green, "Connections between M-theory and superstrings", hep-th/9712195.
70. P.S. Howe, "Weyl superspace", *Phys. Lett.* **B415**, 149 (1997), hep-th/9707184.
71. B.E.W. Nilsson, "A superspace approach to branes and supergravity", in *Theory of Elementary Particles*, eds. H. Dorn, D. Lüst and G. Weight (Wiley-Vch, 1998).
72. G. Papadopoulos and P.K. Townsend, "Intersecting M-branes", *Phys. Lett.* **B380**, 273 (1996), hep-th/9603087.
73. E. Bergshoeff, M. de Roo, E. Eyras, B. Janssen and J. P. van der Schaar, "Multiple intersections of D-branes and M-branes", *Nucl. Phys.* **B494**, 119 (1997), hep-th/9612095.
74. J.P. Gauntlett, "Intersecting branes", hep-th/9705011.

75. C.M. Hull, "Gravitational duality, branes and charges", *Nucl. Phys.* **B509**, 216 (1998), hep-th/9705162.
76. G.W. Gibbons, M.B. Green and M.J. Perry, "Instantons and seven-branes in type IIB superstring theory", *Phys. Lett.* **B370**, 37 (1996), hep-th/9511080.
77. P. Meessen and T. Ortin, "An Sl(2,Z) multiplet of nine-dimensional type II supergravity theories", hep-th/9806120.
78. E. Eyras, B. Janssen and Y. Lozano, "5-branes, KK-monopoles and T-duality", hep-th/9806169.
79. E. Bergshoeff, B. Janssen and T. Ortin, "Kaluza-Klein monopoles and gauged sigma models", *Phys. Lett.* **B410**, 131 (1997), hep-th/9706117.
80. E. Bergshoeff, J.P. van der Schaar, "On M9 branes", hep-th/9806069.
81. A. Losev, G. Moore and S.L. Shatashvili, "M & m's", *Nucl. Phys.* **B522**, 105 (1998), hep-th/9707250.
82. E. Witten, "On flux quantization in M-theory and the effective action", *J. Geom. Phys.* **22**, 1 (1997), hep-th/9609122.
83. D. Sorokin, V. Tkach and D.V. Volkov , "Superparticles, twistors and Siegel symmetry", *Mod. Phys. Lett.* **A4**, 901 (1989).
84. D. Sorokin, V. Tkach, D.V. Volkov and A. Zheltukhin, "From superparticle Siegel supersymmetry to the spinning particle proper-time supersymmetry", *Phys. Lett.* **B216**, 302 (1989).
85. I. Bandos, P. Pasti, D. Sorokin, M. Tonin and D. Volkov, "Superstrings and supermembranes in the doubly supersymmetric geometrical approach", *Nucl. Phys.* **B446**, 79 (1995), hep-th/9501113.
86. I. Bandos, D.Sorokin and D. Volkov, "On the generalized *Phys. Lett.* **B352**, 269 (1995), hep-th/9502141.
87. P.S. Howe and E. Sezgin, "Superbranes", *Phys. Lett.* **B390**, 133 (1997), hep-th/9607227.
88. P.S. Howe and E. Sezgin, "D=11,p=5", *Phys. Lett.* **B394**, 62 (1997), hep-th/9611008.
89. P.S. Howe, E. Sezgin and P.C. West, "Covariant field equations of the M-theory five-brane", *Phys. Lett.* **399B**, 49 (1997), hep-th/9702008.
90. P.S. Howe, E. Sezgin and P.C. West, "Aspects of superembeddings", hep-th/9705093.
91. C.S. Chu and E. Sezgin, "M-Fivebrane from the open supermembrane", *JHEP* **12**, 001 (1997), hep-th/9710223.
92. C.S. Chu, P.S. Howe and E. Sezgin, "Strings and D-branes with boundaries", hep-th/9801202.
93. C.S. Chu, P.S. Howe, E. Sezgin and P.C. West, "Open superbranes", hep-th/9803041.

94. P.S. Howe, O. Raetzel and E. Sezgin, "On brane actions and superembeddings", hep-th/9804051.
95. P.S. Howe, O. Raetzel, E. Sezgin and P. Sundell, "D-brane superembeddings", in preparation.
96. P.S. Howe, O. Raetzel, I. Rudychev and E. Sezgin, "L-branes", in preparation.
97. T. Adawi, M. Cederwall, U. Gran, M. Holm and B.E.W. Nilsson, "Superembeddings, nonlinear supersymmetry and 5-branes", hep-th/9711203.
98. P.A.M. Dirac, "A remarkable representation of the 3+2 de Sitter group", *J. Math. Phys.* **4**, 901 (1963).
99. M. Flato and C. Fronsdal, "One massless particle equals two Dirac singletons", *Lett. Math. Phys.* **2**, 421 (1978).
100. M. Flato and C. Fronsdal, "Quantum field theory of singletons. The Rac", *J. Math. Phys.* **22**, 1100 (1981).
101. H. Nicolai and E. Sezgin, "Singleton representations of $OSp(N|4)$", *Phys. Lett.* **143B**, 389 (1984).
102. E. Sezgin, "The spectrum of the eleven dimensional supergravity compactified on the round seven sphere", Trieste preprint IC-83-220, Nov 1983, in *Phys. Lett.* **B138**, 57 (1984), and in *Supergravities in Diverse Dimensions*, eds. A. Salam and E. Sezgin (World Scientific, 1988).
103. M. Günaydin, *Oscillator-like unitary representations of noncompact groups and supergroups and extended supergravity theories*, in *Lecture Notes in Physics, Vol. 180* (1983), eds. E. Inönü and M. Serdaroğlu.
104. E. Sezgin, "The spectrum of D=11 supergravity via harmonic expansions on $S^4 \times S^7$", *Fortsch. Phys.* **34**, 217 (1986).
105. M. Günaydin, L.J. Romans and N.P. Warner, "Spectrum generating algebras in Kaluza-Klein theories", *Phys. Lett.* **B146**, 401 (1984).
106. M.J. Duff, "Supermembranes: the first fifteen weeks", *Class. Quantum Grav.* **5**, 189 (1988).
107. M.P. Blencowe and M.J. Duff, "Supersingletons", *Phys. Lett.* **B203**, 229 (1988).
108. H. Nicolai, E. Sezgin and Y. Tanii, "Conformally invariant supersymmetric field theories on $S^p \times S^1$ and super p-branes", *Nucl. Phys.* **B305**, 403 (1988).
109. E. Bergshoeff, E. Sezgin and Y. Tanii, "A quantum consistent supermembrane theory", Trieste preprint, IC/88/5. Jan 1988.
110. E. Bergshoeff, M.J. Duff, C.N. Pope and E. Sezgin "Supersymmetric supermembrane vacua and singletons", *Phys. Lett.* **199B**, 69 (1987).

111. E. Bergshoeff, M.J. Duff, C.N. Pope and E. Sezgin, "Compactifications of the eleven-dimensional supermembrane", *Phys. Lett.* **B224**, 71 (1989).
112. M.J. Duff, C.N. Pope and E. Sezgin, "A stable supermembrane vacuum with a discrete spectrum", *Phys. Lett.* **225B**, 319 (1989).
113. E. Bergshoeff, E. Sezgin and Y. Tanii, "Stress tensor commutators and Schwinger terms in singleton theories", *Int. J. Mod. Phys.* **A5**, 3599 (1990).
114. E. Sezgin and Y. Tanii, "Superconformal sigma models in higher than two dimensions", *Nucl. Phys.* **B443**, 70 (1995), hep-th/9412163.
115. P.S. Howe, unpublished.
116. P. Claus, R. Kallosh, J. Kumar, P.K. Townsend and A. Van Proeyen, "Conformal theory of M2, D3, M5 and 'D1+D5' branes", hep-th/9801206.
117. R. Kallosh, J. Rahmfeld and A. Rajaraman, "Near horizon superspace", hep-th/9805217.
118. G. Dall'Agata, D. Fabbri, C. Fraser, P. Fré, P. Termonia and M. Trigiante, "The $Osp(8|4)$ singleton action from the supermembrane", hep-th/9807115.
119. B. de Wit, K. Peeters, J. Plefka and A. Sevrin, "The M-theory two-brane in $AdS_4 \times S^7$ and $AdS_7 \times S^4$", hep-th/9808052.
120. E.S. Fradkin and M.A. Vasiliev, "On the gravitational interaction of massless higher spin fields", *Phys. Lett.* **B189**, 89 (1987).
121. E.S. Fradkin and M.A. Vasiliev, "Cubic interaction in extended theories of massless higher-spin fields", *Nucl. Phys.* **B291**, 141 (1987).
122. M.A. Vasiliev, "Consistent equations for interacting massless fields of all spins in the first order curvatures", *Ann. Phys.* **190**, 59 (1989).
123. M.A. Vasiliev, "More on equations of motion for interacting massless fields of all spins in $3+1$ dimensions", *Phys. Lett.* **B285**, 225 (1992).
124. M.A. Vasiliev, "Higher-spin gauge theories in four, three and two dimensions", hep-th/9611024.
125. E. Sezgin and P. Sundell, "Higher spin $N = 8$ supergravity", to appear in Nuclear Phys. B, hep-th/9805125.
126. B. de Wit and H. Nicolai, "$N = 8$ supergravity", *Nucl. Phys.* **B208**, 323 (1982).
127. E. Cremmer and B. Julia, "The SO(8) supergravity", *Nucl. Phys.* **B159**, 141 (1979).
128. S.E. Konstein and M.A. Vasiliev, "Massless representations and admissibility condition for higher spin superalgebras", *Nucl. Phys.* **B312**, 402 (1989).

129. J.W. van Holten and A. van Proeyen, "N=1 supersymmetry algebras in d=2,3,4 mod 8", *J. Phys.* **A15**, 763 (1982).
130. M. Günaydin, "Unitary supermultiplets of OSp(1/32,R) and M-theory", hep-th/9803138.
131. R. D'Auria and P. Fré, "Geometric supergravity in D=11 and its hidden supergroup", *Nucl. Phys.* **B201**, 101 (1982).
132. L. Castellani, P. Fré, F. Giani, K. Pilch and P. van Nieuwenhuizen, "Gauging of d=11 supergravity?", *Ann. Phys.* **146**, 35 (1983).
133. J.M. Maldacena, "The large N limit of superconformal field theories and supergravity", hep-th/9711200.
134. S.S. Gubser, I.R. Klebanov and A.M. Polyakov, "Gauge theory correlators from non-critical string theory", hep-th/9802109.
135. E. Witten, "Anti-de Sitter space and holography", hep-th/9802150.
136. M.J. Duff, B.E.W. Nilsson and C.N. Pope, "Kaluza-Klein supergravity", *Phys. Rep.* **130**, 1 (1986).
137. M.J. Duff, C.N. Pope and E. Sezgin (eds), *Supermembranes and Physics in (2+1)-Dimensions* (World Scientific, 1990), Proceedings of Trieste Conference, Trieste, July 1989.
138. E. Sezgin, *Comments on supermembrane theories*, in *Strings, Membranes and QED*, eds. C. Aragone, A. Restuccia and S. Salamó (Universidad Simón Bolívar, 1992).
139. A. Giveon, M. Porrati and E. Rabinovici, "Target space duality in string theory", *Phys. Rep.* **244**, 77 (1994), hep-th/9401139.
140. M.J. Duff, R.R. Khuri and J.X. Lu, "String solitons", *Phys. Rep.* **259**, 213 (1995), hep-th/9412184.
141. J.H. Schwarz, *Lectures on superstrings and M theory dualities*, hep-th/9607201.
142. J. Polchinski, *TASI lectures on D-branes*, hep-th/9611050.
143. M.J. Duff, "Supermembranes", hep-th/9611203.
144. P.K. Townsend, *Four lectures on M theory*, hep-th/9612121.
145. A. Sen, "An introduction to non-perturbative string theory", hep-th/9802051
146. M.J. Duff, "A layman's guide to M theory", hep-th/9805177.
147. J.H. Schwarz, "From superstrings to M theory", hep-th/9807135.
148. M.J. Duff, "Anti-de Sitter space, branes, singletons, superconformal field theories and all that", hep-th/9808100.

A LAYMAN'S GUIDE TO M-THEORY

M. J. DUFF
Center for Theoretical Physics, Texas A&M University,
College Station, Texas 77843, USA

The best candidate for a fundamental unified theory of all physical phenomena is no longer ten-dimensional superstring theory but rather eleven-dimensional *M-theory*. In the words of Fields medalist Edward Witten, "M stands for 'Magical', 'Mystery' or 'Membrane', according to taste". New evidence in favor of this theory is appearing daily on the internet and represents the most exciting development in the subject since 1984 when the superstring revolution first burst on the scene.

1 Abdus Salam

The death of Abdus Salam was a great loss not only to his family and to the physics community; it was a loss to all mankind. For he was not only one of the finest physicists of the twentieth century, having unified two of the four fundamental forces in Nature, but he also dedicated his life to the betterment of science and education in the Third World and to the cause of world peace. Although he won the Nobel Prize for physics, a Nobel Peace Prize would have been entirely appropriate.

At the behest of Patrick Blackett, Salam moved to Imperial College, London, in 1957 where he founded the Theoretical Physics Group. He remained at Imperial as Profesor of Physics for the rest of his carreer. I was fortunate enough to be his PhD student at Imperial College from 1969 to 1972, and then his postdoc at the ICTP from 1972 to 1973.

Among Salam's earlier achievements was the role played by renormalization in quantum field theory when, in particular, he amazed his Cambridge contempories with the resolution of the notoriously thorny problem of overlapping divergences. His brilliance then burst on the scene once more when he proposed the famous hypothesis that *All neutrinos are left-handed*, a hypothesis which inevitably called for a violation of parity in the weak interactions. He was always fond of recalling his visit to Switzerland where he submitted (or should I say "humbly" submitted [a]) his two-component neutrino idea to the formidable Wolfgang Pauli. Pauli responded with a note urging Salam to "think of something better"! So Salam delayed publication until after Lee and Yang had conferred the mantle of respectibility on parity violation. That

[a] "I am a humble man" was something of a catchphrase for Salam and used whenever anyone tried to make physics explanations more complicated than necessary.

taught Salam a valuable lesson and he would constantly advise his students never to listen to grand old men.[b] It also taught him to adopt a policy of publish or perish, and his scientific output was prodigious with over 300 publications.

Of course, the work that won him the 1979 Nobel Prize that he shared with Glashow and Weinberg was for the electroweak unification which combined several of his abiding interests: renormalizability, non-abelian gauge theories and chirality. His earlier work in 1960 with Goldstone and Weinberg on spontaneous symmetry breaking and his work with John Ward in the mid 1960s on the weak interactions was no doubt also influential. Looking back on my time as a student in the Theory Group at Imperial from 1969 to 1972, a group that included not only Abdus Salam but also Tom Kibble, one might think that I would have been uniquely poised to take advantage of the new ideas in spontaneously broken gauge theories. Alas, it was not to be since no-one suggested that electroweak unification would be an interesting topic of research. In fact I did not learn about spontaneous symmetry breaking until after I got my PhD! The reason, of course, is that neither Weinberg nor Salam (nor anybody else) fully realized the importance of their model until 't Hooft proved its renormalizability in 1972 and until the discovery of neutral currents at CERN. Indeed, the Nobel Committee was uncharacteristically prescient in awarding the Prize to Glashow, Weiberg and Salam in 1979 because the W and Z bosons were not discovered experimentally at CERN until 1982. Together with Pati, Salam went on to propose that the strong nuclear force might also be included in this unification. Among the predictions of this *Grand Unified Theory* are magnetic monopoles and proton decay: phenomena which are still under intense theoretical and experimental investigation. More recently, it was Salam, together with his lifelong collaborator John Strathdee who first proposed the idea of superspace, a space with both commuting and anticommuting coordinates, which underlies all of present day research on supersymmetry.

However, it is to Abdus Salam that I owe a tremendous debt as the man who first kindled my interest in the Quantum Theory of Gravity: a subject which at the time was pursued only by mad dogs and Englishmen. (My thesis title: *Problems in the Classical and Quantum Theories of Gravitation* was greeted with hoots of derision when I announced it at the Cargese Summer School en route to my first postdoc in Trieste. The work originated with a bet between Abdus Salam and Hermann Bondi about whether you could generate the Schwarzschild solution using Feynman diagrams. You can (and I did) but I never found out if Bondi ever paid up.) It was inevitable that Salam would not rest until the fourth and most enigmatic force of gravity was unified with the other three. Such a unification was always Einstein's dream and it remains the

[b]I hope this student, at least, has lived up to that advice!

most challenging tasks of modern theoretical physics and one which attracts the most able and active researchers, such as those here today.

I should mention that being a student of someone so bursting with new ideas as Salam was something of a mixed blessing: he would assign a research problem and then disappear on his travels for weeks at a time.[c] On his return he would ask what you were working on. When you began to explain your meagre progess he would usually say "No, no, no. That's all old hat. What you should be working on is this", and he would then allocate a completely new problem!

I think it was Hans Bethe who said that there are two kinds of genius. The first group (to which I would say Steven Weinberg, for example, belongs) produce results of such devastating logic and clarity that they leave you feeling that you could have done that too (if only you were smart enough!). The second kind are the "magicians" whose sources of inspiration are completely baffling. Salam, I believe, belonged to this magic circle and there was always an element of eastern mysticism in his ideas that left you wondering how to fathom his genius.

Of course, these scientific achievements reflect only one side of Salam's character. He also devoted his life to the goal of international peace and cooperation, especially to the gap between the developed and developing nations. He firmly believed that this disparity will never be remedied until the Third World countries become the arbiters of their own scientific and technological destinies. Thus this means going beyond mere financial aid and the exportation of technology; it means the training of a scientific elite who are capable of discrimination in all matters scientific. He would thus vigorously defend the teaching of esoteric subjects such as theoretical elementary particle physics against critics who complained that the time and effort would be better spent on agriculture. His establishment of the ICTP in Trieste was an important first step in this direction.

It is indeed a tragedy that someone so vigorous and full of life as Abdus Salam should have been struck down with such a debilitating disease. He had such a wonderful *joie de vivre* and his laughter, which most resembled a barking sea-lion, would reverberate throughout the corridors of the Imperial College Theory Group. When the deeds of great men are recalled, one often hears the cliché "He did not suffer fools gladly", but my memories of Salam at Imperial College were quite the reverse. People from all over the world would arrive and knock on his door to expound their latest theories, some of them quite bizarre. Yet Salam would treat them all with the same courtesy and respect.

[c]Consequently, it was to Chris Isham that I would turn for practical help with my PhD thesis.

Perhaps it was because his own ideas always bordered on the outlandish that he was so tolerant of eccentricity in others; he could recognize pearls of wisdom where the rest of us saw only irritating grains of sand. A previous example was provided by the military attache from the Israeli embassy in London who showed up one day with his ideas on particle physics. Salam was impressed enough to take him under his wing. The man was Yuval Ne'eman and the result was flavor $SU(3)$.

Let me recall just one example of a crazy Salam idea. In that period 1969-72, one of the hottest topics was the Veneziano Model and I distinctly remember Salam remarking on the apparent similarity between the mass and angular momentum relation of a Regge trajectory and that of an extreme black hole. Nowadays, of course, string theorists will juxtapose black holes and Regge slopes without batting an eyelid but to suggest that black holes could behave as elementary particles back in the late 1960's was considered preposterous by minds lesser than Salam's. (A comparison of the gyromagnetic ratios of spinning black holes and elementary string states is the subject of some of my recent research, so in this respect Salam was 25 years ahead of his time!) As an interesting historical footnote let us recall that at the time Salam had to change the gravitational constant to match the hadronic scale, an idea which spawned his *strong gravity*; today the fashion is the reverse and we change the Regge slope to match the Planck scale!

Theoretical physicists are, by and large, an honest bunch: occasions when scientific facts are actually deliberately falsified are almost unheard of. Nevertheless, we are still human and consequently want to present our results in the best possible light when writing them up for publication. I recall a young student approaching Abdus Salam for advice on this ethical dilemma: "Professor Salam, these calculations confirm most of the arguments I have been making so far. Unfortunately, there are also these other calculations which do not quite seem to fit the picture. Should I also draw the reader's attention to these at the risk of spoiling the effect or should I wait? After all, they will probably turn out to be irrelevant." In a response which should be immortalized in *The Oxford Dictionary of Quotations*, Salam replied: "When all else fails, you can always tell the truth".

Amen.

2 Magical Mystery Membranes

Up until 1995, hopes for a final theory [1] that would reconcile gravity and quantum mechanics, and describe all physical phenomena, were pinned on *superstrings*: one-dimensional objects whose vibrational modes represent the elementary particles and which live in a ten-dimensional universe.[2] All that has now changed. In the last two years ten-dimensional superstrings have been subsumed by a deeper, more profound, new theory: eleven-dimensional *M-theory*.[3] The purpose of the present paper is to convey to the layman some of this excitement.

According to the standard model of the strong nuclear, weak nuclear and electromagnetic forces, all matter is made up of certain building block particles called *fermions* which are held together by force-carrying particles called *bosons*. This standard model does not incorporate the gravitational force, however. A vital ingredient in the quest to go beyond this standard model and to find a unified theory embracing all physical phenomena is *supersymmetry*, a symmetry which (a) unites the bosons and fermions, (b) requires the existence of gravity and (c) places an upper limit of *eleven* on the dimension of spacetime. For these reasons, in the early 1980s, many physicists looked to eleven-dimensional *supergravity* in the hope that it might provide that elusive superunified theory.[5] Then in 1984 superunification underwent a major paradigm-shift: eleven-dimensional supergravity was knocked off its pedestal by ten-dimensional superstrings. Unlike eleven-dimensional supergravity, superstrings appeared to provide a quantum consistent theory of gravity which also seemed capable, in principle, of explaining the standard model.[d]

Despite these startling successes, however, nagging doubts persisted about superstrings. First, many of the most important questions in string theory, in particular how to confront it with experiment and how to accommodate quantum black holes, seemed incapable of being answered within the traditional framework called *perturbation theory*, according to which all quantities of interest are approximated by the first few terms in a power series expansion in some small parameter. They seemed to call for some new, *non-perturbative*, physics. Secondly, why did there appear to be *five* different mathematically consistent superstring theories: the $E_8 \times E_8$ heterotic string, the $SO(32)$ heterotic string, the $SO(32)$ Type I string, the Type IIA and Type IIB strings? If one is looking for a unique *Theory of Everything*, this seems like an embarrassment of riches! Thirdly, if supersymmetry permits eleven dimensions, why do superstrings stop at ten? This question became more acute with the discoveries of the *supermembrane* in 1987 the *superfivebrane* in 1992. These are bubble-like supersymmetric extended objects with respectively two and five dimensions moving in an eleven-dimensional spacetime, which are related to one another by a *duality* reminiscent of the electric/magnetic duality that relates an electric monopole (a particle carrying electric charge) to a mag-

[d]For an up-to-date non-technical account of string theory, the reader is referred to the forthcoming popular book by Brian Greene.[4]

$$\left.\begin{array}{l} E_8 \times E_8 \;\; \textit{heterotic string} \\ SO(32) \;\; \textit{heterotic string} \\ SO(32) \;\; \textit{Type I string} \\ \textit{Type IIA string} \\ \textit{Type } \mathrm{IIB} \textit{ string} \end{array}\right\} \textit{M theory}$$

Table 1: The five apparently different string theories are really just different corners of M-theory.

netic monopole (a hypothetical particle carrying magnetic charge). Finally, therefore, if we are going to generalize zero-dimensional point particles to one-dimensional strings, why stop there? Why not two-dimensional membranes or more generally p-dimensional objects (inevitably dubbed *p-branes*)? In the last decade, this latter possibility of spacetime bubbles was actively pursued by a small but dedicated group of theorists,[6] largely ignored by the orthodox superstring community.

Although it is still too early to claim that all the problems of string theory have now been resolved, M-theory seems a big step in the right direction. First, it is intrinsically non-perturbative and already suggests new avenues both for particle physics and black hole physics. Secondly, it is is an eleven-dimensional theory which, at sufficently low energies, looks, ironically enough, like eleven-dimensional supergravity. Thirdly, it subsumes all five consistent string theories and shows that the distinction we used to draw between them is just an artifact of perturbation theory. See Table 2. Finally, it incorporates supermembranes and that is why M stands for Membrane. However, it may well be that we are only just beginning to scratch the surface of the ultimate meaning of M-theory, and for the time being therefore, M stands for Magic and Mystery too.

3 Symmetry and Supersymmetry

Central to the understanding of modern theories of the fundamental forces is the idea of *symmetry*: under certain changes in the way we describe the basic quantities, the laws of physics are nevertheless seen to remain unchanged. For example, the result of an experiment should be the same whether we perform it today or tomorrow; this symmetry is called *time translation invariance*. It should also be the same before and after rotating our experimental apparatus; this symmetry is called *rotational invariance*. Both of these are examples of *spacetime symmetries*. Indeed, Einstein's general theory of relativity is based on the requirement that the laws of physics should be invariant under *any* change in the way we describe the positions of events in spacetime. In the standard model of the strong, weak and electromagnetic forces there are other kinds of *internal* symmetries that allow us to change the roles played by different elementary particles such as electrons and neutrinos, for example. These statements are made precise using the branch of mathematics known as *Group Theory*. The standard model is based on the group $SU(3) \times SU(2) \times U(1)$, where $U(n)$ refers to *unitary* $n \times n$ matrices and S means unit determinant. Grand Unified Theories, which have not yet received the same empirical support as the standard model, are even more ambitious and use bigger groups, such as $SU(5)$, which contain $SU(3) \times SU(2) \times U(1)$ as a subgroup. In this case, the laws remain unchanged even when we exchange the roles of the *quarks* and electrons. Thus it is that the greater the unification, the greater the symmetry required. The standard model symmetry replaces the three fundamental forces: strong, weak and electromagnetic, with just two: the strong and electroweak. Grand unified symmetries replace these two with just one strong-electroweak force. In fact, it is not much of an exaggeration to say that the search for the ultimate unified theory is really a search for the right symmetry.

At this stage, however, one might protest that some of these internal symmetries fly in the face of experience. After all, the electron is very different from a neutrino: the electron has a non-zero mass whereas the neutrino is massless.[e] Similarly, the electrons which orbit the atomic nucleus are very different from the quarks out of which the protons and neutrons of the nucleus are built. Quarks feel the strong nuclear force which holds the nucleus together, whereas electrons do not. These feelings are, in a certain sense, justified: the world we live in does not exhibit the $SU(2) \times U(1)$ of the standard model nor the $SU(5)$ of the grand unified theory. They are what physicists call "broken symmetries". The idea is that these theories may exist in several different *phases*, just as water can exist in solid, liquid and gaseous phases. In some of these phases the symmetries are broken but in other phases, they are exact. The world we inhabit today happens to correspond to the broken-symmetric phase, but in conditions of extremely high energies or extremely high temperatures, these symmetries may be restored to their pristine form. The early stages of our universe, shortly after the Big Bang, provide just such an environment. Looking back further into the history of the universe, therefore, is also a search for greater and greater symmetry. The ultimate symmetry we are looking for may well be the symmetry with which the Universe began.

M-theory, like string theory before it, relies crucially on the idea, first put

[e] Or at least has a very tiny mass.

forward in the early 1970s, of a spacetime *supersymmetry* which exchanges bosons and fermions. Just as the earth rotates on its own axis as it orbits the sun, so electrons carry an intrinsic angular momentum called "spin" as they orbit the nucleus in an atom. Indeed all elementary particles carry a spin s which obeys a quantization rule $s = nh/4\pi$ where $n = 0, 1, 2, 3, ...$ and h is Planck's constant. Thus particles may be divided into *bosons* or *fermions* according as the spin, measured in units of $h/2\pi$, is integer $0, 1, 2, ...$ or half-odd-integer $1/2, 3/2, 5/2...$ Fermions obey the *Pauli exclusion principle*, which states that no two fermions can occupy the same quantum state, whereas bosons do not. They are said to obey opposite *statistics*. According to the standard model, the quarks and leptons which are the building blocks of all matter, are spin 1/2 fermions; the gluons, W and Z particles and photons which are the mediators of the strong, weak and electromagnetic forces, are spin 1 bosons and the Higgs particle which is responsible for the breaking of symmetries and for giving masses to the other particles, is a spin 0 boson. Unbroken supersymmetry would require that every elementary particle we know of would have an unknown super-partner with the same mass but obeying the opposite statistics: for each boson there is a fermion; for each fermion a boson. Spin 1/2 quarks partner spin 0 *squarks*, spin 1 photons partner spin 1/2 *photinos*, and so on. In the world we inhabit, of course, there are no such equal mass partners and bosons and fermions seem very different. Supersymmetry, if it exists at all, is clearly a broken symmetry and the new supersymmetric particles are so heavy that they have so far escaped detection. At sufficiently high energies, however, supersymmetry may be restored. Supersymmetry may also solve the so-called *gauge hierarchy problem*: the energy scale at which the grand unified symmetries are broken is vastly higher from those at which the electroweak symmetries are broken. This raises the puzzle of why the electrons, quarks, and W-bosons have their relatively small masses and the extra particles required by grand unification have their enormous masses. Why do they not all slide to some common scale? In the absence of supersymmetry, there is no satisfactory answer to this question, but in a supersymmetric world this is all perfectly natural. The greatest challege currently facing high-energy experimentalists at Fermi National Laboratory (Fermilab) in Chicago and the European Centre for Particle Physics (CERN) in Geneva is the search for these new supersymmetric particles. The discovery of supersymmetry would be one of the greatest experimental achievements of the century and would completely revolutionize the way we view the physical world.[f]

---

[f] This will present an interesting dilemma for those pundits who are predicting the *End of Science* on the grounds that all the important discoveries have already been made. Presumably, they will say "I told you so" if supersymmetry is not discovered, and "See, there's

Symmetries are said to be *global* if the changes are the same throughout spacetime, and *local* if they differ from one point to another. The consequences of *local* supersymmetry are even more far-reaching: it predicts gravity. Thus if Einstein had not already discovered General Relativity, local supersymmetry would have forced us to invent it. In fact, we are forced to a *supergravity* in which the graviton, a spin 2 boson that mediates the gravitational interactions, is partnered with a spin 3/2 *gravitino*. This is a theorist's dream because it confronts the problem from which both general relativity and grand unified theories shy away: neither takes the other's symmetries into account. Consequently, neither is able to achieve the ultimate unification and roll all four forces into one. But local supersymmetry offers just such a possiblity, and it is this feature above all others which has fuelled the theorist's belief in supersymmetry in spite of twenty-five years without experimental support.

4 Eleven-Dimensional Supergravity

Supergravity has an even more bizarre feature, however, it places an upper limit on the dimension of spacetime! We are used to the idea that space has three dimensions: height, length and breadth; with time providing the fourth dimension of *spacetime*. Indeed this is the picture that Einstein had in mind in 1916 when he proposed general relativity. But in the early 1920's, in their attempts to unify Einstein's gravity and Maxwell's electromagnetism, Theodore Kaluza and Oskar Klein suggested that spacetime may have a hidden fifth dimesion. This idea was quite succesful: Einstein's equations in five dimensions not only yield the right equations for gravity in four dimensions but Maxwell's equations come for free. Conservation of electric charge is just conservation of momentum in the fifth direction. In order to explain why this extra dimension is not apparent in our everyday lives, however, it would have to have a different *topology* from the other four and be very small. Whereas the usual four coordinates stretch from minus infinity to plus infinity, the fifth cordinate would lie between 0 and $2\pi R$. In other words, it describes a circle of radius R. To get the right value for the charge on the electron, moreover, the circle would have to be tiny, $R \sim 10^{-35}$ meters, which satisfactorily explains why we are unaware of its existence. It is difficult to envisage a spacetime with such a topology but a nice analogy is provide by a garden hose: at large distances it looks like a line but closer inspection reveals that at each point of the line, there is a little circle. So it was that Kaluza and Klein suggested there is a little circle at each point of four-dimensional spacetime. Moreover, this explained for the first time the empirical fact that all particles come with

one thing less left to discover" if it is.

an electric charge which is an integer multiple of the charge on the electron, in other words, why electric charge is *quantized.*

The Kaluza-Klein idea was forgotten for many years but was revived in the early 1980s when it was realized by Eugene Cremmer, Bernard Julia and Joel Scherk from the Ecole Normale in Paris that supergravity not only permits up to seven extra dimensions, but in fact takes its simplest and most elegant form when written in its full eleven-dimensional glory. Moreover, the kind of four-dimensional picture we end up with depends on how we *compactify* these extra dimensions: maybe seven of them would allow us to derive, a la Kaluza-Klein, the strong and weak forces as well as the electromagnetic. In the end, however, eleven dimensional supergravity fell out of favor for several reasons.

First, despite its extra dimensions and despite its supersymmetry, eleven-dimensional supergravity is still a *quantum field theory* and runs into the problem from which all such theories suffer: the quantum mechanical probability for certain processes yields the answer *infinity.* By itself, this is not necessarily a disaster. This problem was resolved in the late 1940s in the context of Quantum Electrodynamics (QED), the study of the electromagnetic interactions of photons and electrons, by showing that these infinities could be absorbed in to a redefinition or *renormalization* of the parameters in the theory such as the mass and charge of the electron. This *renormalization* resulted in predictions for physical observables which were not only finite but in spectacular agreement with experiment. Spurred on by the success of QED, physicists looked for renormalizable quantum field theories of the weak and strong nuclear interactions which in the 1970s culminated in the enormously successful standard model that we know today. One might be tempted, therefore, to conclude that *renormalizability,* namely the ability to absorb all infinities into a redefinition of the parameters in the theory, is a prerequisite for any sensible quantum field theory. However, the central quandary of all attempts to marry quantum theory and gravity, such as eleven-dimensional supergravity, is that Einstein's general theory of relativity turns out *non-renormalizable*! Does this mean that Einstein's theory should be thrown on the scrapheap? Actually, the modern view of renormalizability is a little more forgiving. Suppose we have a renormalizable quantum field theory describing both light particles and heavy particles of mass m. Even such a renormalizable theory can be made to look non-renormalizable if we eliminate the heavy particles by using their equations of motion. The resulting equations for the light particles are then non-renormalizable but perfectly adequate for describing processes at energies less than mc^2, where c is the velocity of light. We run into trouble only if we try to extrapolate them beyond this range of validity, at which point we should instead resort to the original version of the theory with the massive particles put back in. In this light, therefore, the modern view of Einstein's theory is that it is perfectly adequate to explain gravitational phenomena at low energies but that at high energies it must be replaced by some more fundamental theory containing massive particles. But what is this energy, what are these massive particles and what is this more fundamental theory?

There is a natural energy scale associated with any quantum theory of gravity. Such a theory combines three ingredients each with their own fundamental constants: Planck's constant h (quantum mechanics), the velocity of light c (special relativity) and Newton's gravitational constant G (gravity). From these we can form the so-called Planck mass $m_P = \sqrt{hc/G}$, equal to about 10^{-8} kilograms, and the Planck energy $M_P c^2$, equal to about 10^{19} GeV. (GeV is short for giga-electron-volts=10^9 electron-volts, and an electron-volt is the energy required to accelerate an electron through a potential difference of one volt.) From this we conclude that the energy at which Einstein's theory, and hence eleven-dimensional supergravity, breaks down is the Planck energy. On the scale of elementary particle physics, this energy is enormous [g]: the world's most powerful particle acclerators can currently reach energies of only 10^4 GeV. So it seemed in the early 1980s that we were looking for a fundamental theory which reduces to Einstein's gravity at low energies, which describes Planck mass particles and which is supersymmetric. Whatever it is, it cannot be a quantum field theory because we already know all the supersymmetric ones and they do not fit the bill.

Equally puzzling was that an important feature of the real world which is incorporated into both the standard model and grand unified theories is that Nature is *chiral*: the weak nuclear force distinguishes between right and left. (As Salam had noted with his left-handed neutrino hypothesis). However, as emphasized by Witten among others, it is impossible via conventional Kaluza-Klein techniques to generate a chiral theory from a non-chiral one and unfortunately, eleven-dimensional supergravity, in common with any *odd*-dimensional theory, is itself non-chiral.

5 Ten-Dimensional Superstrings

For both these reasons, attention turned to ten-dimensional superstring theory. The idea that the fundamental stuff of the universe might not be pointlike elementary particles, but rather one-dimensional strings had been around from the early 1970s. Just like violin strings, these relativistic strings can vibrate

[g] For this reason, incidentally, the *End of Science* brigade like to claim that, even if we find the right theory of quantum gravity, we will never be able to test it experimentally! As I will argue shortly, however, this view is erroneous.

and each elementary particle: graviton, gluon, quark and so on, is identified with a different mode of vibration. However, this means that there are *infinitely many* elementary particles. Fortunately, this does contradict experiment because most of them, corresponding to the higher modes of vibration, will have masses of the order of the Planck mass and above and will be unobservable in the direct sense that we observe the lighter ones. Indeed, an infinite tower of Planck mass states is just what the doctor ordered for curing the non-renormalizability disease. In fact, because strings are *extended*, rather than pointlike, objects, the quantum mechanical probabilities involved in string processes are actually *finite*. Moreover, when we take the *low-energy limit* by eliminating these massive particles through their equations of motion, we recover a ten-dimensional version of supergravity which incorporates Einstein's gravity. Now ten-dimensional quantum field theories, as opposed to eleven-dimensional ones, also admit the possibility of *chirality*. The reason that everyone had still not abandoned eleven-dimensional supergravity in favor of string theory, however, was that the realistic-looking Type I string, which incorporated internal symmetry groups containing the $SU(3) \times SU(2) \times U(1)$ of the standard model, seemed to suffer from inconsistencies or *anomalies*, whereas the consistent non-chiral Type IIA and chiral Type IIB strings did not seem realistic.

Then came the September 1984 superstring revolution. First, Michael Green from QueenMary and Westfield College, London, and John Schwarz from the California Institute of Technology showed that the Type I string was free of anomalies provide the group was uniquely $SO(32)$ where O(n) stands for *orthogonal* $n \times n$ matrices. They suggested that a string theory based on the exceptional group $E_8 \times E_8$ would also have this property. Next, David Gross, Jeffrey Harvey, Emil Martinec and Ryan Rohm from Princeton University discovered a new kind of heterotic (hybrid) string theory based on just these two groups: the $E_8 \times E_8$ heterotic string and the $SO(32)$ heterotic string, thus bringing to *five* the number of consistent string theories. Thirdly, Philip Candelas from the University of Texas, Austin, Gary Horowitz and Andrew Strominger from the University of California, Santa Barbara and Witten showed that these heterotic string theories admitted a Kaluza-Klein compactification from ten dimensions down to four. The six-dimensional compact spaces belonged to a class of spaces known to the pure mathematicians as *Calabi-Yau manifolds*. The resulting four-dimensional theories resembled quasi-realistic grand unfied theories with chiral representations for the quarks and leptons! Everyone dropped eleven-dimensional supergravity like a hot brick. The mood of the times was encapsulated by Nobel Laureate Murray Gell-Mann in his closing address at the 1984 Santa Fe Meeting, when he said: "Eleven Dimensional

Supergravity (Ugh!)".

6 Ten to Eleven: It is not too late

After the initial euphoria, however, nagging doubts about string theory began to creep in.

Theorists love *uniqueness*; they like to think that the ultimate *Theory of Everything*[1] will one day be singled out, not merely because all rival theories are in disagreement with experiment, but because they are mathematically inconsistent. In other words, that the universe is the way it is because it is the only possible universe. But string theories are far from unique. Already in ten dimensions there are five mathematically consistent theories: the Type I $SO(32)$, the heterotic $SO(32)$, the heterotic $E_8 \times E_8$, the Type IIA and the Type IIB. (Type I is an *open* string in that its ends are allowed to move freely in spacetime; the remaining four are *closed strings* which form a closed loop.) Thus the first problem is the *uniqueness problem*.

The situation becomes even worse when we consider compactifying the extra six dimensions. There seem to be billions of different ways of compactifying the string from ten dimensions to four (billions of different Calabi-Yau manifolds) and hence billions of competing predictions of the real world (which is like having no predictions at all). This aspect of the uniqueness problem is called the *vacuum-degeneracy problem*. One can associate with each different phase of a physical system a *vacuum state*, so called because it is the quantum state corresponding to no real elementary particles at all. However, according to quantum field theory, this vacuum is actually buzzing with virtual particle-antiparticle pairs that are continually being created and destroyed and consequently such vacuum states carry energy. The more energetic vacua, however, should be unstable and eventually decay into a (possibly unique) stable vacuum with the least energy, and this should describe the world in which we live. Unfortunately, all these Calabi-Yau vacua have the same energy and the string seems to have no way of preferring one to the other. By focussing on the fact that strings are formulated in ten spacetime dimensions and that they unify the forces at the Planck scale, many critics of string theory fail to grasp this essential point. The problem is not so much that strings are unable to produce four-dimensional models like the standard model with quarks and leptons held together by gluons, W-bosons, Z bosons and photons and of the kind that can be tested experimentally in current or forseeable accelerators. On the contrary, string theorists can dream up literally billions of them! The problem is that they have no way of discriminating between them. What is lacking is some dynamical mechanism that would explain why the theory singles out one particular Calabi-Yau manifold and hence why we live in one particular vacuum; in other words, why the world is the way it is. Either this problem will not be solved, in which case string theory will fall by the wayside like a hundred other failed theories, or else it will be solved and string theory will be put to the test experimentally. Neither string theory nor M-theory is relying for its credibility on building thousand-light-year accelerators capable of reaching the Planck energy, as some *End-of-Science* Jeremiahs have suggested.

Part and parcel of the vacuum degeneracy problem is the *supersymmetry-breaking problem.* If superstrings are to describe our world then supersymmetry must be broken, but the way in which strings achieve this, and at what energy scales, is still a great mystery.

A third aspect of vacuum degeneracy is the *cosmological constant* problem. Shortly after writing down the equations of general relativity, Einstein realized that nothing prevented him from adding an extra term, called the *cosmological term* because it affects the rate at which the universe as a whole is expanding. Current astrophysical data indicates that the coefficent of this term, called the *cosmological constant*, is zero or at least very small. Whenever an *a priori* allowed term in an equation seems to be absent, however, theorists always want to know the reason why. At first sight supersymmetry seems to provide the answer. The cosmological constant measures the energy of the vacuum, and in supersymmetric vacua the energy coming from virtual bosons is exactly cancelled by the energy coming from virtual fermions! Unfortunately, as we have already seen, the vacuum in which our universe currently finds itself can at best have broken supersymmetry and so all bets are off. As with cake, we can't have our cosmological constant and eat it too! In common with all other theories one can think of, superstrings as yet provide no resolution of this paradox.

On the subject of gravity, let us not forget *black holes.* According to Cambridge University's Stephen Hawking, they are not as black as they are painted: quantum black holes radiate energy and hence grow smaller. Moreover, they radiate energy in the same way irrespective of what kind of matter went to make up the black hole in the first place. The rate of radiation increases with diminishing size and the black hole eventualy explodes leaving nothing behind, not even the grin on the Cheshire Cat. All the information about the original constituents of the black hole has been lost and this leads to the *information loss paradox* because such a scenario flies in the face of traditional quantum mechanics. On a more pragmatic level, another unsolved problem was that the thermodynamic entropy formula of the black hole radiation, first written down by Jacob Bekenstein (Hebrew University), had never received a *microscopic* explanation. The entropy of a system is a measure of its disorder, and is related to the number of quantum states that the system is allowed to occupy. For a black hole, this number seems incredibly high but what microscopic forces are at work to explain this? Not even strings, with their infinite number of vibrational modes seemed to have this capability.

Given all the good news about string theory, though, string enthusiasts were reluctant to abandon the theory notwithstanding all these problems. Might the faults lie not with the theory itself but rather with the way the calculations are carried out? In common with the standard model and grand unified theories, the equations of string theory are just too complicated to solve exactly. We have to resort to an approximation scheme and the time-honored way of doing this in physics is *perturbation theory*. Let us recall quantum electrodynamics, for example, and denote by e the electric charge on an electron. The ratio $\alpha = 2\pi e^2/hc$ is a dimensionless number called, for historical reasons, the fine structure constant. Fortunately for physicists, α is about $1/137$: much less than 1. Consequently, if we can express processes (such as the probability of one electron scattering off another) in a power series in this coupling constant α, then we can be confident that keeping just the first few terms in the series will be a good approximation to the exact result. As a simple example of approximating a mathematical function $f(x)$ by a power series, consider

$$f(x) = (1-x)^{-1} = 1 + x + x^2 + x^3 \ldots \tag{1}$$

Provided x is very much less than unity, the first few terms provide a good approximation. This is precisely what Richard Feynman was doing when he devised his *Feynman diagram* technique. The same perturbative techniques work well in the weak interactions where the corresponding dimensionless coupling constant is about 10^{-5}. Indeed, this is how the weak interactions justify their name. When we come to the strong interactions, however, we are not so lucky. Now the strong fine structure constant which governs the strength of low-energy nuclear processes, for example, is of order unity and perturbation theory can no longer be trusted: each term in the power series expansion is just as big as the others. The whole industry of *lattice gauge theory* is devoted to an attempt to avoid perturbation theory in the strong interactions by doing numerical simulations on supercomputers. It has proved enormously difficult.

The point to bear in mind, however, (and one that even string theorists sometimes forget) is that "God does not do perturbation theory"; it is merely a technique dreamed up by poor physicists because it is the best they can do. Furthermore, although theories such as quantum electrodynamics manage to avoid it, there is a possible fatal flaw with perturbation theory. What happens if the process we are interested in depends on the coupling constant in an intrinsically non-perturbative way which does not even admit a power series expansion? Such mathematical functions are not difficult to come by: the function

$$f(x) = e^{1/x^2}, \tag{2}$$

for example, cannot be approximated by a power series in x no matter how small x happens to be. The equations of string theory are sufficiently complicated that such non-perturbative behaviour cannot be ruled out. If so, might our failure to answer the really difficult problems be more the fault of string *theorists* than string *theory*?

An apparently different reason for having mixed feelings about superstrings, of course, especially for those who had been pursuing Kaluza-Klein supergravity prior to the 1984 superstring revolution, was the dimensionality of spacetime. If supersymmetry permits eleven spacetime dimensions, why should the theory of everything stop at ten?

This problem rose to the surface again in 1987 when Eric Bergshoeff of the University of Groningen, Ergin Sezgin, now at Texas A&M University, and Paul Townsend from the University of Cambridge discovered *The eleven-dimensional supermembrane*. This membrane is a bubble-like extended object with two spatial dimensions which moves in a spacetime dictated by our old friend: eleven-dimensional supergravity! Moreover, Paul Howe (King's College, London University), Takeo Inami (Kyoto University), Kellogg Stelle (Imperial College) and I were then able to show that if one of the eleven dimensions is a circle, then we can wrap one of the membrane dimensions around it so that, if the radius of the circle is sufficiently small, it looks like a string in ten dimensions. In fact, it yields precisely the Type IIA superstring. This suggested to us that maybe the eleven-dimensional theory was the more fundamental after all.

7 Supermembranes

Membrane theory has a strange history which goes back even further than strings. The idea that the elementary particles might correspond to modes of a vibrating membrane was put forward originally in 1960 by the British Nobel Prize winning physicist Paul Dirac, a giant of twentieth century science who was also responsible for two other daring postulates: the existence of *anti-matter* and the existence of *magnetic monopoles*. Anti-particles carry the same mass but opposite charge from particles and were discovered experimentally in the 1930s. Magnetic monopoles carry a single magnetic charge and to this day have not yet been observed. As we shall see, however, they do feature prominently in M-theory. When string theory came along in the 1970s, there were some attempts to revive Dirac's membrane idea but without much success.

The breakthrough did not come until 1986 when James Hughes, James Liu and Joseph Polchinski of the University of Texas showed that, contrary to the expectations of certain string theorists, it was possible to combine the membrane idea with supersymmetry: the *supermembrane* was born.

Consequently, while all the progress in superstring theory was being made a small but enthusiastic group of theorists were posing a seemingly very different question: Once you have given up 0-dimensional particles in favor of 1-dimensional strings, why not 2-dimensional membranes or in general p-dimensional objects (inevitably dubbed p-*branes*)? Just as a 0-dimensional particle sweeps out a 1-dimensional *worldline* as it evolves in time, so a 1-dimensional string sweeps out a 2-dimensional *worldsheet* and a p-brane sweeps out a d-dimensional *worldvolume*, where $d = p + 1$. Of course, there must be enough room for the p-brane to move about in spacetime, so d must be less than the number of spacetime dimensions D. In fact supersymmetry places further severe restrictions both on the dimension of the extended object and the dimension of spacetime in which it lives. One can represent these as points on a graph where we plot spacetime dimension D vertically and the p-brane dimension $d = p + 1$ horizontally. This graph is called the *brane-scan*. See Table 2. Curiously enough, the maximum spacetime dimension permitted is eleven, where Bergshoeff, Sezgin and Townsend found their 2-brane. In the early 80s Green and Schwarz had showed that spacetime supersymmetry allows classical superstrings moving in spacetime dimensions $3, 4, 6$ and 10. (Quantum considerations rule out all but the ten-dimensional case as being truly fundamental. Of course some of these ten dimensions could be curled up to a very tiny size in the way suggested by Kaluza and Klein. Ideally six would be compactified in this way so as to yield the four spacetime dimensions with which we are familiar.) It was now realized, however, that there were twelve points on the scan which fall into four sequences ending with the superstrings or 1-branes in $D = 3, 4, 6$ and 10, which were now viewed as but special cases of this more general class of supersymmetric extended object. These twelve points are the ones with $d \geq 2$ and denoted by S in Table 2. For completeness, we have also included the superparticles with $d = 1$ in $D = 2, 3, 5$ and 9.

The letters S, V and T refer to *scalar*, *vector* and *tensor* respectively and describe the different kinds of particles that live on the worldvolume of the p-brane. Spin 0 bosons and their spin 1/2 fermionic partners are said to form a scalar supermultiplet. An example is provided by the eleven-dimensional supermembrane that occupies the $(D = 11, d = 3)$ slot on the branescan. It has 8 spin 0 and 8 spin 1/2 particles living on the three-dimensional (one time, two space) worldvolume of the membrane. But as we shall see, it was subsequently realized that there also exist branes which have higher spin bosons

$D \uparrow$	0	1	2	3	4	5	6	7	8	9	10	11
11	.			S			T					
10	.	V	S/V	V	V	V	S/V	V	V	V	V	
9	.	S				S						
8	.				S							
7	.			S			T					
6	.	V	S/V	V	S/V	V	V					
5	.	S		S								
4	.	V	S/V	S/V	V							
3	.	S/V	S/V	V								
2	.	S										
1	.											
0	.	.	.	.	.	.	.	.	.	.	.	.
	0	1	2	3	4	5	6	7	8	9	10	11 $d \rightarrow$

Table 2: The brane scan, where S, V and T denote scalar, vector and tensor multiplets.

on their worldvolume and belong to vector and tensor supermultiplets.

A particularly interesting solution of eleven-dimensional supergravity, found by Bergshoeff, Sezgin and myself in collaboration with Chris Pope of Texas A&M University, was called "The membrane at the end of the universe." It described a four-dimensional spacetime with the extra seven dimensions curled up into a seven-dimensional sphere and in which the supermembrane occupied the three-dimensional boundary of the four-dimensional spacetime (rather as the two-dimensional surface of a soap bubble encloses a three-dimensional volume). This spacetime is of the kind first discussed earlier this century by the Dutch physicist Willem de Sitter and has a non-zero cosmological constant. It fact it is called anti-de Sitter space because the cosmological constant is negative. Now shortly after he first wrote down the equations of the membrane, Dirac pointed out in a (at the time unrelated) paper that anti de Sitter space admits some strange kinds of fields that he called *singletons* which have no analogue in ordinary flat spacetime. These were much studied by Christian Fronsdal and collaborators at the University of California, Los Angeles, who pointed out that they reside not in the bulk of the anti-de Sitter space but on the three dimensional boundary. In 1988, the present author noted that, in the case of the seven-sphere compactification of eleven-dimensional supergravity, the singletons correspond to the same 8 spin 0 plus 8 spin 1/2 scalar supermultiplet that lives on the worldvolume of the supermembrane, and it was natural to suggest that the membrane and the singletons should be identified. In this way, via the *the membrane at the end of the universe*, the physics in the bulk of the four-dimensional spacetime was really being determined by the physics on the three-dimensional boundary.

Notwithstanding these and subsequent results, the supermembrane enterprise (Type *II* A&M Theory?) was ignored by most adherents of conventional superstring theory. Those who had worked on eleven-dimensional supergravity and then on supermembranes spent the early eighties arguing for *spacetime* dimensions greater than four, and the late eighties and early nineties arguing for *worldvolume* dimensions greater than two. The latter struggle was by far the more bitter.[h]

8 Solitons, Topology and Duality

Another curious twist in the history of supermembranes concerns their interpretation as *solitons*. In their broadest definition, solitons are classical solutions of a field theory corresponding to lumps of field energy which are prevented from dissipating by a *topological* conservation law, and hence display particle-like properties. The classic example of a such soliton is provided by magnetic monopole solutions of four-dimensional grand unified theories found by Gerard ‘t Hooft of the the University of Utrecht in the Netherlands and Alexander Polyakov, now at Princeton. Solitons play a ubiquitous role in theoretical physics appearing in such diverse phenomena as condensed matter physics and cosmology, where they are frequently known as *topological defects*.

To understand the meaning of a *topological conservation law*, we begin by recalling that in 1917 the German mathematician Emmy Noether had shown that to every global symmetry, there corresponds a quantity that is conserved in time. For example, invariance under time translations, space translations and rotations give rise to the laws of conservation of energy, momentum and angular momentum, respectively. Similarly, conservation of electric charge corresponds to a change in the phase of the quantum mechanical wave functions that describe the elementary particles. One might naively expect that conservation of magnetic charge would admit a similar explanation but, in fact, it has a completely different *topological* origin, and is but one example of what are now termed topological conservation laws. Topology is that branch of mathematics which concerns itself just with the shape of things. Topologically speaking, therefore, a teacup is equivalent to a doughnut because each two-

[h]One string theorist I know would literally cover up his ears whenever the word “membrane” was mentioned within his earshot! Indeed, I used to chide my more conservative string theory colleagues by accusing them of being unable to utter the M-word. That the current theory ended up being called M-theory rather than Membrane theory was thus something of a Pyrrhic victory.

dimensional surface has just one hole: one can continuosly deform one into the other. The surface of an orange, on the other hand, is topologically distinct having no holes: you cannot turn an orange into a doughnut. So it is with the intricate field configurations describing magnetic monopoles: you cannot turn a particle carrying n units of magnetic charge into one with n' units of magnetic charge, if $n \neq n'$. Hence the charge is conserved but for topological reasons; not for any reasons of symmetry.

In 1977, however, Claus Montonen of the University of Helsinki and David Olive, now at the University of Wales at Swansea, made a bold conjecture. Might there exist a *dual* formulation of fundamental physics in which the roles of Noether charges and topological charges are reversed? In such a dual picture, the magnetic monopoles would be the fundamental objects and the quarks, W-bosons and Higgs particles would be the solitons! They were inspired by the observation that in certain *supersymmetric* grand unified theories, the masses M of all the particles whether elementary (carrying purely electric charge Q), solitonic (carrying purely magnetic charge P) or *dyonic* (carrying both) are described by a universal formula

$$M^2 = v^2(Q^2 + P^2) \tag{3}$$

where v is a constant. Note that the mass formula remains unchanged if we exchange the roles of P and Q! The Montonen-Olive conjecture was that this electric/magnetic symmetry is a symmetry not merely of the mass formula but is an exact symmetry of the entire quantum theory! The reason why this idea remained merely a conjecture rather than a proof has to do with the whole question of perturbative versus non-perturbative effects. According to Dirac, the electric charge Q is quantized in units of e, the charge on the electron, whereas the magnetic charge is quantized in units of $1/e$. In other words, $Q = me$ and $P = n/e$, where m and n are integers. The symmetry suggested by Olive and Montonen thus demanded that in the dual world, we not only exchange the integers m and n but we also replace e by $1/e$ and go from a regime of weak coupling to a regime of strong coupling! This was very exciting because it promised a whole new window on non-perturbative effects. On the other hand, it also made a proof very difficult and the idea was largely forgotten for the next few years.

Although the original paper by Hughes, Liu and Polchinski made use of the soliton idea, the subsequent impetus in supermembrane theory was to mimic superstrings and treat the p-branes as fundamental objects in their own right (analagous to particles carrying an electric Noether charge). Even within this framework, however, it was possible to postulate a certain kind of duality between one p-brane and another by relating them to the geometrical concept of *p-forms*. (Indeed, this is how p-branes originally got their name.) In their classic text on general relativity, *Gravitation*, Misner, Thorne and Wheeler [7] provide a way to visualize *p-forms* as describing the way in which surfaces are stacked. Open a cardbord carton containing a dozen bottles, and observe the honeycomb structure of intersecting north-south and east-west cardboard separators between the bottles. That honeycomb structure of tubes is an example of a 2-form in the context of ordinary 3-dimensional space. It yields a number (number of tubes cut) for each choice of 2-dimensional surface slicing through the 3-dimensional space. Thus a 2-form is a device to produce a number out of a surface. All of electromagnetism can be summarized in the language of 2-forms, honeycomb-like structures filling all of 4-dimensional spacetime. There are two such structures, Faraday= F and Maxwell= $*F$ each *dual*, or perpendicular, to the other. The amount of electric charge or magnetic charge in an elementary volume is equal respectively to the number of tubes of the Maxwell 2-form $*F$ or Faraday 2-form F that end in that volume. (In a world with no magnetic monopoles, no tubes of F would ever end.) To summarize, in 4-dimensional spacetime, an electric 0-brane is dual to a magnetic 0-brane.

An equivalent way to understand why 0-dimensional point particles produce electric fields which are described by 2-forms is to note that in 4-dimensional spacetime a pointlike electric charge can be surrounded by a two-dimensional sphere. Similarly, a string (1-brane) in 4-dimensional spacetime can be "surrounded" by a 1-dimensional circle, and so the electric charge per unit length of a string is described by a 1-form Maxwell field but its magnetic dual perpendicular to it is described by a 3-form Faraday field. By contrast, in 5 spacetime dimensions, although the Faraday field of a 0-brane is still a 2-form, the dual Maxwell field is a now a 3-form, consistent with the fact that you now need a 3-dimensional sphere to surround the pointlike electric charge. But a 3-form is just the Faraday field produced by a string. Consequently, in 5 spacetime dimensions, the magnetic dual of an electric 0-brane is a string. Though in practice it is harder to visualize, it is straightforward in principle to generalize this duality idea to any p-brane in any spacetime dimension D. The rule is that the Faraday field is a $(p+2)$ form and the dual Maxwell field perpendicular to it is $(D-p-2)$-form. Consequently, the magnetic dual of an electric p-brane is a $\tilde{p}$-brane where $\tilde{p} = D - p - 4$. In particular, in the critical $D = 10$ spacetime dimensions of superstring theory, a string ($p = 1$) is dual to a fivebrane ($\tilde{p} = 5$). (If you have trouble imagining that in 10 dimensions you need a 3-dimensional sphere to surround a 5-brane, don't worry, you are not alone!)

Now the low energy limit of 10-dimensional string theory is a 10-dimensional supergravity theory with a 3-form Faraday field and dual 7-form Maxwell field, just as one would expect if the fundamental object is a string. However, 10-dimensional supergravity had one puzzling feature that had long been an enigma from the point of view of string theory. In addition to the above version there existed a *dual* version in which the roles of the Faraday and Maxwell fields were interchanged: the Faraday field was a 7-form and the Maxwell field was a 3-form! This suggested to the present author in 1987, in analogy with the Olive-Montonen conjecture, that perhaps this was indicative of a dual version of string theory in which the fundamental objects are fivebranes! This became known as the *string/fivebrane duality conjecture*. The analogy was still a bit incomplete, however, because at that time the fivebrane was not regarded as a soliton.

The next development came in 1988 when Paul Townsend of Cambridge University revived the Hughes-Liu-Polchinski idea and showed that many of the super p-branes also admit an interpretation as topological defects (analogous to particles carrying a magnetic topological charge). Of course, this involved generalizing the usual notion of a soliton: it need not be restricted just to a 0-brane in four dimensions but might be an extended object such as a p-brane in D-dimensions. Just like the monopoles studied by Montonen and Olive, these solitons preserve half of the spacetime supersymmetry and hence obey a relation which states that their mass per unit p-volume is given by their topological charge. Then in 1990, a major breakthrough for the string/fivebrane duality conjecture came along when Strominger found that the equations of the 10-dimensional heterotic string admit a fivebrane as a soliton solution which also preserves half the spacetime supersymmetry and whose mass per unit 5-volume is given by the topological charge associated with the Faraday 3-form of the string. Moreover, this mass became larger, the smaller the strength of the string coupling, exactly as one would expect for a soliton. He went on to suggest a complete strong/weak coupling duality with the strongly coupled string corresponding to the weakly coupled fivebrane. By generalizing some earlier work of Rafael Nepomechie (University of Florida, Gainesville) and Claudio Teitelboim (University of Santiago), moreover, it was possible to to show that the electric charge of the fundamental string and the magnetic charge of the solitonic fivebrane obeyed a Dirac quantization rule. In this form, string/fivebrane duality was now much more closely mimicking the electric/magnetic duality of Montonen and Olive. Then Curtis Callan (Princeton University), Harvey and Strominger showed that similar results also appear in both Type IIA and Type IIB string theories; they also admit fivebrane solitons. However, since most physicists were already sceptical of electric/magnetic duality in four dimensions, they did not immediately embrace string/fivebrane duality in ten dimensions!

Furthermore, there was one major problem with treating the fivebrane as a fundamental object in its own right; a problem that has bedevilled supermembrane theory right from the beginning: no-one knows how to quantize fundamental p-branes with $p > 2$. All the techniques that worked so well for fundamental strings and which allow us, for example, to calculate how one string scatters off another, simply do not go through. Problems arise both at the level of the worldvolume equations where our old *bete noir* of non-renormalizability comes back to haunt us and also at the level of the spacetime equations. Each term in string perturbation theory corresponds to a two-dimensional worldsheet with more and more holes: we must sum over all topologies of the worldsheet. But for surfaces with more than two dimensions we do not know how to do this. Indeed, there are powerful theorems in pure mathematics which tell you that it is not merely hard but impossible. Of course, one could always invoke the dictum that *God does not do perturbation theory*, but that does not cut much ice unless you can say what He does do! So there were two major impediments to string/fivebrane duality in 10 dimensions. First, the electric/magnetic duality analogy was ineffective so long as most physicists were sceptical of this duality. Secondly, treating fivebranes as fundamental raised all the unresolved issues of non-perturbative quantization.

The first of these impediments was removed, however, when Ashoke Sen (Tata Institute) revitalized the Olive-Montonen conjecture by establishing that certain dyonic states, which their conjecture demanded, were indeed present in the theory. Many duality sceptics were thus converted. Indeed this inspired Nathan Seiberg (Rutger's University) and Witten to look for duality in more realistic (though still supersymmetric) approximations to the standard model. The subsequent industry, known as Seiberg-Witten theory, provided a wealth of new information on non-perturbative effects in four-dimensional quantum field theories, such as quark-confinement and symmetry-breaking, which would have been unthinkable just a few years ago.

The Olive-Montonen conjecture was originally intended to apply to four-dimensional grand unified field theories. In 1990, however, Anamarie Font, Luis Ibanez, Dieter Lust and Fernando Quevedo at CERN and, independently, Soo Yong Rey (University of Seoul) generalized the idea to four-dimensional superstrings, where in fact the idea becomes even more natural and goes by the name of S-duality.

In fact, superstring theorists had already become used to a totally different kind of duality called T-duality. Unlike, S-duality which was a non-perturbative symmetry and hence still speculative, T-duality was a perturbative symmetry and rigorously established. If we compactify a string theory on a circle then, in addition to the Kaluza-Klein particles we would expect in an ordinary field theory, there are also extra *winding* particles that arise because a string can wind around the circle. T-duality states that nothing changes if we exchange the roles of the Kaluza-Klein and winding particles provided we also exchange the radius of the circle R by its inverse $1/R$. In short, a string cannnot tell the difference between a big circle and a small one!

9 String/String Duality in Six Dimensions

Recall that when wrapped around a circle, an 11-dimensional membrane behaves as if it were a 10-dimensional string. In a series of papers between 1991 and 1995, a team at Texas A&M University involving Ramzi Khuri, James T. Liu, Jianxin Lu, Ruben Minasian, Joachim Rahmfeld, and myself argued that this may also be the way out of the problems of 10-dimensional string/fivebrane duality. If we allow four of the ten dimensions to be curled up and allow the solitonic fivebrane to wrap around them, it will behave as if it were a 6-dimensional solitonic string! The fundamental string will remain a fundamental string but now also in 6-dimensions. So the 10-dimensional string/fivebrane duality conjecture gets replaced by a 6-dimensional string/string duality conjecture. The obvious advantage is that, in contrast to the fivebrane, we do know how to quantize the string and hence we can put the predictions of string/string duality to the test. For example, one can show that the coupling constant of the solitonic string is indeed given by the inverse of the fundamental string's coupling constant, in complete agreement with the conjecture.

When we spoke of string/string duality, we originally had in mind a duality between one heterotic string and another, but the next major development in the subject came in 1994 when Christopher Hull (Queen Mary and Westfield College, London University) and Townsend suggested that, if the four-dimensional compact space is chosen suitably, a six-dimensional heterotic string can be dual to a six-dimensional Type IIA string! These authors also added futher support to the idea that the Type IIA string originates in eleven dimensions.

It occurred to the present author that string/string duality has another unexpected pay-off. If we compactify the six-dimensional spacetime on two circles down to four dimensions, the fundamental string and the solitonic string will each acquire a T-duality. But here is the miracle: the T-duality of the solitonic string is just the S-duality of the fundamental string, and vice-versa! This phenomenon, in which the non-perturbative replacement of e by $1/e$ in one picture is just the perturbative replacement of R by $1/R$ in the dual picture, goes by the name of *Duality of Dualities*. See Table 3. Thus four-dimensional electric/magnetic duality, which was previously only a conjecture, now emerges

	Fundamental string	Dual string
T − duality :	$Radius \leftrightarrow 1/(Radius)$	$charge \leftrightarrow 1/(charge)$
	$Kaluza - Klein \leftrightarrow Winding$	$Electric \leftrightarrow Magnetic$
S − duality :	$charge \leftrightarrow 1/(charge)$	$Radius \leftrightarrow 1/(Radius)$
	$Electric \leftrightarrow Magnetic$	$Kaluza - Klein \leftrightarrow Winding$

Table 3: Duality of dualities

automatically if we make the more primitive conjecture of six-dimensional string/string duality.

10 M-Theory

All this previous work on T-duality, S-duality, and string/string duality was suddenly pulled together under the umbrella of M-theory by Witten in his, by now famous, talk at the University of Southern California in February 1995. Curiously enough, however, Witten still played down the importance of supermembranes. But it was only a matter of time before he too succumbed to the conclusion that we weren't doing just string theory any more! In the coming months, literally hundreds of papers appeared in the internet confirming that, whatever M-theory may be, it certainly involves supermembranes in an important way. For example, in 1992 R. Gueven (University of Ankara) had shown that eleven-dimensional supergravity admits a solitonic fivebrane solution dual to the fundamental membrane solution found the year before by Stelle and myself. See the $(D = 11, d = 6)$ point marked by a T in Table 2. It did not take long to realize that 6-dimensional string/string duality (and hence 4-dimensional electric/magnetic duality) follows from 11-dimensional membrane/fivebrane duality. The fundamental string is obtained by wrapping the membrane around a one-dimensional space and then compactifying on a four-dimensional space; whereas the solitonic string is obtained by wrapping the fivebrane around the four-dimensional space and then compactifying on the one-dimensional space. Nor did it take long before the more realistic kinds of electric/magnetic duality envisioned by Seiberg and Witten were also given an explanation in terms of string/string duality and hence M-theory.

Even the chiral $E_8 \times E_8$ string, which according to Witten's earlier theorem could never come from eleven-dimensions, was given an eleven-dimensional explanation by Petr Horava (Princeton University) and Witten. The no-go theorem is evaded by compactifying not on a circle (which has no ends), but on a line-segment (which has two ends). It is ironic that having driven the nail into the coffin of eleven-dimensions (and having driven Gell-Mann to utter "Ugh!"), Witten was the one to pull the nail out again! He went on to argue that if the size of this one-dimensional space is large compared to the six-dimensional Calabi-Yau manifold, then our world is approximately five-dimensional. This may have important consequences for confronting M-theory with experiment. For example, it is known that the strengths of the four forces change with energy. In supersymmetric extensions of the standard model, one finds that the fine structure constants $\alpha_3, \alpha_2, \alpha_1$ associated with the $SU(3) \times SU(2) \times U(1)$ all meet at about 10^{16} GeV, entirely consistent with the idea of grand unification. The strength of the dimensionless number $\alpha_G = GE^2$, where G is Newton's contant and E is the energy, also almost meets the other three, but not quite. This near miss has been a source of great interest, but also frustration. However, in a universe of the kind envisioned by Witten, spacetime is approximately a narrow five dimensional layer bounded by four-dimensional walls. The particles of the standard model live on the walls but gravity lives in the five-dimensional bulk. As a result, it is possible to choose the size of this fifth dimension so that all four forces meet at this common scale. Note that this is much less than the Planck scale of 10^{19} GeV, so gravitational effects may be much closer in energy than we previously thought; a result that would have all kinds of cosmological consequences.

Thus this eleven-dimensional framework now provides the starting point for understanding a wealth of new non-perturbative phenomena, including string/string duality, Seiberg-Witten theory, quark confinement, particle physics phenomenology and cosmology.

11 Black Holes and D-Branes

Type II string theories differ from heterotic theories in one important respect: in addition to the usual Faraday 3-form charge, called the Neveu-Schwarz charge after Andre Neveu (University of Montpelier) and Schwarz, they also carry so-called Ramond charges, named after Pierre Ramond of the University of Florida, Gainesville. These are associated with Faraday 2-forms and 4-forms in the case of Type IIA and Faraday 3-forms and 5-forms in the case of Type IIB. Accordingly in 1993, Jiaxin Lu and I were able to find new solutions of the Type IIA string equations describing super p-branes with $p = 0, 2$ and their duals with $p = 6, 4$ and new solutions of Type IIB string equations with $p = 1, 3$ and their duals with $p = 5, 3$. Interestingly enough, the Type IIB superthreebrane is *self-dual*, carrying a magnetic charge equal to its electric charge. This meant that there were more points on the brane-scan than had previously been appreciated. These occupy the V slots in Table 2. For all these solutions, the mass per unit p-volume was given by the charge, as a consequence of the preservation half of the spacetime supersymmetry. However, we recognized that they were in fact just the extremal mass=charge limit of more general non-supersymmetric solutions found previously by Horowitz and Strominger. These solutions, whose mass was greater than their charge, exhibit *event horizons*: surfaces from which nothing, not even light, can escape. They were *black* branes!

Thus another by-product of these membrane breakthroughs has been an appreciation of the role played by black holes in particle physics and string theory. In fact they can be regarded as black branes wrapped around the compactified dimensions. These black holes are tiny (10^{-35} meters) objects; not the multi-million solar mass objects that are gobbling up galaxies. However, the same physics applies to both and there are strong hints by Lenny Susskind (Stanford University) and others that M-theory may even clear up many of the apparent paradoxes of quantum black holes raised by Hawking.

As we have already discussed, one of the biggest unsolved mysteries in string theory is why there seem to be billions of different ways of compactifying the string from ten dimensions to four and hence billions of competing predictions of the real world. Remarkably, Brian Greene of Cornell University, David Morrison of Duke University and Strominger have shown that these wrappped around black branes actually connect one Calabi-Yau vacuum to another. This holds promise of a dynamical mechanism that would explain why the world is as it is, in other words, why we live in one particular vacuum. A fuller discussion may be found in Greene's book.[4]

Another interconnection was recently uncovered by Polchinski who realized that the Type II super p-branes carrying Ramond charges may be identified with the so-called Dirichlet-branes (or D-branes, for short) that he had studied some years ago by looking at strings with unusual boundary conditions. Dirichlet was a French mathematician who first introduced such boundary conditions. These D-branes are just the surfaces on which open strings can end. In the process, he discovered an 8-brane in Type IIA theory and a 7-brane and 9-brane in Type IIB which had previously been overlooked. See Table 2. This D-brane technology has opened up a whole new chapter in the history of supermembranes. In particular, it has enabled Strominger and Cumrun Vafa from Harvard to make a comparison of the black hole entropy calculated from the degeneracy of wrapped-around black brane states with the Bekenstein-Hawking entropy of an extreme black hole. Their agreement provided the first microscopic explanation of black hole entropy. Moreover, as Townsend had shown earlier, the extreme black hole solutions of the ten-dimensional Type IIA string (in other words, the Dirichlet 0-branes) were just the Kaluza-Klein particles associated with wrapping the eleven-dimensional membrane around a circle. Moreover, four-dimensional black holes also admit the interpretation of intersecting membranes and fivebranes in eleven-dimensions. All this holds promise of a deeper understanding of black hole physics via supermembranes.

12 Eleven to Twelve: Is it still too early?

We have remarked that eleven spacetime dimensions are the maximum allowed by super p-branes. This is certainly true if we believe that the Universe has only one time dimension. Worlds with more than one time present all kind of headaches for theoretical physicists and they prefer not to think about them. For example, there would be no "before" and "after" in the conventional sense. Just for fun, however, in 1987 Miles Blencowe (Imperial College, University of London) and I imagined what would happen if one relaxed this one-time requirement. We found that we could not rule out the possibility of a supersymmetric extended object with a (2 space, 2 time) worldvolume living in a (10 space, 2 time) spacetime. We even suggested that the Type IIB string with its (1 space, 1 time) worldsheet living in a (9 space, 1 time) spacetime might be descended from this object in much the same way that the Type IIA string with its (1 space, 1 time) worldsheet living in a (9 space, 1 time) spacetime is descended from the (2 space, 1 time) worldvolume of the supermembrane living in a (10 space, 1 time) spacetime. This idea lay dormant for almost a decade but has recently been revived by Vafa and others in the context of *F-theory*. The utility of F-theory is certainly beyond dispute: it has yielded a wealth of new information on string/string duality. But should the twelve dimensions of F-theory be taken seriously? And if so, should F-theory be regarded as more fundamental than M-theory? (If M stands for Mother, maybe F stands for Father.) To make sense of F-theory, however, it seems necessary to somehow freeze out the twelfth timelike dimension where there appears to be no dynamics. Moreover, Einstein's requirement that the laws of physics be invariant under changes in the spacetime coordinates seems to apply only to ten or eleven of the dimensions and not to twelve. So the symmetry of the theory, as far as we can tell, is only that of ten or eleven dimensions. The more conservative interpretation of F-theory, therefore, is that the twelfth dimension is just a mathematical artifact with no profound significance. Time (or perhaps I should say "Both times") will tell.

13 So what is M-Theory?

Is M-theory to be regarded literally as membrane theory? In other words should we attempt to "quantize" the eleven dimensional membrane in some, as yet unknown, non-perturbative way? Personally, I think the jury is still out on whether this is the right thing to do. Witten, for example, strongly believes that this is not the correct approach. He would say, in physicist's jargon, that we do not even know what the right degrees of freedom are. So although M-theory admits 2-branes and 5-branes, it is probably much more besides.

Recently, Tom Banks and Stephen Shenker at Rutgers together with Willy Fischler from the Univerity of Texas and Susskind have even proposed a rigorous definiton of M-theory known as M(atrix) theory which is based on an infinite number of Dirichlet 0-branes. In this picture spacetime is a fuzzy concept in which the spacetime coordinates $x, y, z, \ldots$ are matrices that do not commute e.g. $xy \neq yx$. This approch has generated great excitement but does yet seem to be the last word. It works well in high dimensions but as we descend in dimension it seems to break down before we reach the real four-dimensional world.

Another interesting development has recently been provided by Juan Maldacena at Harvard, who has suggested that M-theory on anti-de Sitter space, including all its gravitational interactions, may be completely described by a non-gravitational theory on the boundary of anti-de Sitter space. This holds promise not only of a deeper understanding of M-theory, but may also throw light on non-perturbative aspects of the theories that live on the boundary, which in some circumstances can include the kinds of quark theories that govern the strong nuclear interactions. Models of this kind, where a bulk theory with gravity is equivalent to a boundary theory without gravity, have also been advocated by 't Hooft and independently by Susskind who call them *holographic* theories. The reader may notice a striking similarity to the earlier idea of "The membrane at the end of the universe" [6] and interconnections between the two are currently being explored.

M-theory has sometimes been called the *Second Superstring Revolution*, but we feel this is really a misnomer. It certainly involves new ideas every bit as significant as those of the 1984 string revolution, but its reliance upon supermembranes makes it is sufficently different from traditional string theory to warrant its own name. One cannot deny the tremendous historical influence of the last decade of superstrings on our current perspectives. Indeed, it is the pillar upon which our belief in a quantum consistent M-theory rests. In my opinion, however, the focus on the perturbative aspects of one-dimensional objects moving in a ten-dimensional spacetime that prevailed during this period will ultimately be seen to be a small (and perhaps physically insignificant) corner of M-theory. The overriding problem in superunification in the coming years will be to take the Mystery out of M-theory, while keeping the Magic and the Membranes.

References

1. Steven Weinberg, *Dreams of a Final Theory*, Pantheon, 1992.
2. Michael Green, *Superstrings*, Scientific American 255 (1986) 44-56 (September).
3. Michael Duff, *The theory formerly known as strings*, Scientific American 278 (1998) 54-59 (February).
4. Brian Greene, *The Elegant Universe*, W. W. Norton, 1998.
5. Daniel Z. Freedman and Peter van Nieuwenhuizen, *The Hidden Dimensions of Spacetime*, Scientific American 252 (1985) 62.
6. Michael Duff and Christine Sutton, *The Membrane at the End of the Universe*, New Scientist, June 30, 1988.
7. Charles W. Misner, Kip S. Thorne and John Archibald Wheeler, *Gravitation*, W. H. Freeman and Company, 1973.

QUANTUM CORRECTIONS TO ELEVEN-DIMENSIONAL SUPERGRAVITY

Michael B. Green
Department of Applied Mathematics and Theoretical Physics, Silver Street, Cambridge CB3 9EW, United Kingdom
E-mail: m.b.green@damtp.cam.ac.uk

This article reviews the non-perturbative structure of certain higher derivative terms in the type II string theory effective action and their connection to one-loop effects in eleven-dimensional supergravity compactified on a torus. New material is also included that was not presented in the talk.

Apart from the work for which he is best known, Professor Salam contributed many results in classical and quantum field theory that have only recently been seen to play key rôles in unravelling the underlying structure of string theory and its M-theory extension. Among these are results in supergravity in diverse dimensions, solitonic solutions to supergravity and the quantum physics of black holes – extraordinarily imaginative contributions that were often underrated at the time but which are now of widespread interest. This talk is dedicated to the memory of a superb physicist and a remarkable human being.

1 Introduction

Many features of M-theory follow directly from the strong algebraic constraints associated with the symmetries of the theory rather than specific details of any microscopic model, such as the matrix model. Certain of these properties must be inevitable consequences of eleven-dimensional supersymmetry. Others may depend on assuming the validity of the intricate web of duality interrelationships between the perturbative string theories.

This raises the question of the extent to which the higher order terms in a momentum expansion of the M-theory effective action are determined by eleven-dimensional supergravity compactified in various ways. It has been generally assumed that since perturbative supergravity is terribly divergent in eleven dimensions no interesting consequences follow by considering loop diagrams and that a microscopic theory is needed in order to regularize its divergences. However, this ignores the powerful supersymmetry constraints which are expected to reduce the number of arbitrary constants – but by how much? Furthermore, the eleven-dimensional quantum theory compactified to ten or fewer dimensions is supposed to be precisely equivalent to string theory and its compactifications. What mileage can be gained by exploiting this equivalence as an ansatz? The lowest order terms in the expansion of the M-theory action in powers of the momentum are the terms in the original classical supergravity action of Cremmer, Julia and Scherk[1]. Certain higher order terms have also been unambigously identified. One example is the $C^{(3)} \wedge X_8$ term [2] that arises as a one-loop effect in type IIA string theory [3] and can also be motivated by requiring the cancellation of chiral anomalies in p-branes [4]. Here $C^{(3)}$ is the Ramond–Ramond three-form and X_8 is an eight-form made out of the Riemann curvature.

In the next section I will review the evidence that certain 'protected' terms in the low energy effective type II string action compactified on S^1 are precisely determined by one-loop effects in eleven-dimensional supergravity compactified on T^2. For example the one-loop eleven-dimensional four-graviton amplitude leads to a R^4 term in M-theory [a] which is related by supersymmetry to the $C^{(3)} \wedge X_8$ term. This term can be expressed in the type IIB coordinates by making use of the equivalence of M-theory compactified on T^2 with type IIB string theory compactified on S^1 [5,6]. This leads to a term of the form $f(\rho, \bar{\rho})R^4$ term in the type IIB action [7] where f is a $SL(2, Z)$-invariant function of the complex IIB scalar field, ρ. The modular properties of the IIB theory are inhereted from the geometric $SL(2, Z)$ of T^2 in M-theory with ρ identified with the complex structure, Ω, of the torus. The interplay between the duality symmetries of nine-dimensional type II string theories and the one-loop eleven-dimensional supergravity amplitude are encoded in f which contains both perturbative terms and non-perturbative D-instanton contributions. Furthermore, even though the eleven-dimensional loop amplitude is cubically divergent the consistent identification of the compactified theory with the IIB theory determines a specific finite renormalized value for the coefficient of the R^4 term in eleven dimensions — this finite value would necessarily arise as the regularized value in any microscopic theory. In fact, since the ultraviolet divergence is proportional to the volume of the compactification torus, it disappears in the zero-volume limit, and the exact (non-perturbative) R^4 term of ten-dimensional type IIB [8] is reproduced by finite terms alone.

The expression for the modular function $f(\rho, \bar{\rho})$ in the IIB theory will be motivated in section 2.1 by matching the known R^4 terms that arise at tree level and one loop in string perturbation theory together with the expected structure of multiply-charged D-instanton contributions. A single D-instanton of charge $N = \hat{m}n$ can be identified, following T-duality, with configurations in

[a] The notation R^4 represents a particular contraction of four Riemann curvature tensors that will be reviewed below.

the IIA theory in which the world-line of a charge-n D-particle winds $\hat{m}$ times around a compact dimension. Precisely the same expression will be obtained in section 2.2 from the low momentum limit of the one-loop four graviton scattering amplitude in eleven-dimensional supergravity compactified on T^2. I will also make some comments concerning the relation of the higher-order terms in the momentum expansion of the one-loop supergravity amplitude with string loop diagrams (related comments are made in [9]).

The R^4 term is related by supersymmetry to a large number of other terms. In the language of the type IIB theory these terms include a sixteen-fermion term, $f_{16}(\rho,\bar{\rho})\lambda^{16}$, where λ is the complex spin-$\frac{1}{2}$ chiral fermion [10]. This is the analogue of the 't Hooft fermion vertex in a conventional Yang–Mills instanton background. These interaction terms can be expressed as integrals over half the ten-dimensional on-shell superspace (there is no ten-dimensional off-shell superspace formalism). Such protected terms again only receive perturbative contributions at tree-level and at one loop in string theory and appear to be determined entirely by one-loop diagrams in eleven-dimensional supergravity. A brief overview of such terms is given in section 2.3.

The expressions for the instanton contributions in the expansions of these interaction terms defines a measure on the space of charge-N D-instantons. It has been argued [11] that this should equal the partition function of the zero-dimensional Yang–Mills model, which is an integral over bosonic and fermionic matrices in the Lie algebra of $SU(N)$ [12]. This, in turn, is related to the bulk term in the Witten index for charge-N D-particles [13,14]. This circle of arguments which is presented in detail in [11] will be briefly reviewed at the end of section 2.3.

It would be very interesting to determine how much of the structure of the string perturbation expansion can be determined from eleven-dimensional supergravity beyond these protected terms. For example, the R^4 term is the lowest-order term in a momentum expansion of the exact four-graviton scattering amplitude in type IIB string theory. There are very likely to be terms of higher order in momenta that are also protected by supersymmetry. To what extent can the momentum expansion of the effective action be determined by perturbation theory in eleven dimensions? A more limited question is to what extent the momentum expansion of the string tree-level amplitude can be reproduced by eleven-dimensional supergravity perturbation theory? This depends on the systematics of the multi-loop diagrams in eleven-dimensional supergravity as will be described in section 3. These diagrams have finite pieces that depend on the moduli of the compact dimensions. New primitive divergences arise corresponding to terms which have derivatives acting on R^4, denoted symbolically as $\partial^{2n}R^4$. It would be interesting to see whether these might be determined by the same kind of arguments that led at one loop to the determination of the R^4 term. However, the arguments that I present beyond one loop are based on dimensional analysis rather than explicit evaluation of the rather complicated Feynman diagrams. In this way it is easy to see how specific multi-loop diagrams could give specific finite terms in the M-theory action that correspond to terms that arise from string perturbation theory. However, it seems likely that the values of the counterterms are fixed unambiguously only at low orders in the momentum expansion where they are protected by supersymmetry.

2 Higher-order terms in M-theory

2.1 $SL(2,Z)$-invariant R^4 terms in type IIB

The leading perturbative contributions to the four graviton scattering amplitude in type IIB superstring theory are of the form (in string frame)

$$S^{pert}_{R^4} = \int \sqrt{g^B}\left(\zeta(3)\rho_2^2 + \frac{\pi^2}{3}\right) t^8 t^8 R^4, \tag{1}$$

where $g^B_{\mu\nu}$ is the IIB string-frame metric and

$$t_8 t_8 R^4 \equiv t^{\mu_1 \dots \mu_8} t_{\nu_1 \dots \nu_8} R^{\nu_1\nu_2}_{\mu_1\mu_2} \cdots R^{\nu_7\nu_8}_{\mu_7\mu_8}, \tag{2}$$

where the rank-eight tensor, t^8, is defined in [15]. The first term in (1) is the tree-level contribution [16,17] that is of order α'^3 relative to the leading Einstein–Hilbert term. The second term is the one-loop contribution [18] and has no dependence on the dilaton in the string frame. Non-perturbative contributions to the $t_8 t_8 R^4$ term also arise from single D-instantons with charge N [8] which give an infinite series of non-perturbative contributions that have the form

$$S^{nonpert}_{R^4} = \sum_{N>0} c_N(\rho_2)(e^{2\pi i N\rho} + e^{-2\pi i N\bar\rho}) t_8 t_8 R^4. \tag{3}$$

One way of counting these D-instantons makes use of T-duality between the IIB and the IIA theories in nine dimensions. From the IIA point of view a D-instanton is associated with the world-line of a D-particle of charge n and mass $e^{-\phi^A} n$ (where ϕ^A is the IIA dilaton) winding around the compact ninth dimension of radius r_A. The euclidean action of this configuration is $S = 2\pi \hat m n(C^{(1)} + i r_A e^{\phi^A})$ where $C^{(1)}$ is the IIA one-form and $\hat m$ is the winding number of the world-line. After T-duality this leads to the expression $S = 2\pi N\rho$ for the D-instanton action where $N = \hat m n$. Although the coefficients c_N

are probably very hard to determine directly they are fixed by the requirement that the total action should be invariant under $SL(2,Z)$ transformations. This means that it must have the form,

$$S_{R^4} \equiv S^{pert}_{R^4} + S^{nonpert}_{R^4} = \int \sqrt{g^B}\,\rho_2^{1/2}\, f(\rho,\bar\rho) t_8 t_8 R^4\, d^{10}x, \tag{4}$$

where $f(\rho,\bar\rho)$ is a modular function since R is invariant under $SL(2,Z)$ transformations in the Einstein frame.

The precise coefficients of the known perturbative contributions, S^{pert}, together with the general form of the instanton corrections motivates the suggestion [8] that f is given by

$$f(\rho,\bar\rho) = \pi^{-2} \sum_{(\hat l_1,\hat l_2)\neq(0,0)} \frac{\rho_2^{3/2}}{|\hat l_1 + \rho \hat l_2|^3} = \pi^{-2}\zeta(3) E_{\frac{3}{2}}(\rho,\bar\rho), \tag{5}$$

where the nonholomorphic Eisenstein series E_s is defined by [19]

$$\begin{aligned}\zeta(2s)E_s(\rho,\bar\rho) &= \sum_{(\hat l_1,\hat l_2)\neq(0,0)} \frac{\rho_2^s}{|\hat l_1 + \rho \hat l_2|^{2s}} \\ &= \zeta(2s)\rho_2^s + \frac{\sqrt{\pi}\Gamma(s-\frac{1}{2})\zeta(2s-1)}{\Gamma(s)}\rho_2^{1-s} + \mathcal{R}_s .\end{aligned} \tag{6}$$

The last equality is a large-ρ_2 expansion and $\mathcal{R}_s$ indicates a specific sum of exponentially decreasing terms. In the special case $s = 3/2$ this sum is given by

$$\mathcal{R}_{\frac{3}{2}} = \sum_{N>0}\left(\sum_{N|\hat m}\frac{1}{\hat m^2}\right) N^{\frac{1}{2}}\left(e^{2\pi i N\rho} + e^{-2\pi i N\bar\rho}\right) \times \sum_{k=0}^{\infty}(2\pi N\rho_2)^{-k}\frac{\Gamma(\frac{1}{2}-k)}{\Gamma(-\frac{1}{2}-k)}, \tag{7}$$

where $\sum_{N|\hat m}$ indicates a sum over the divisors of N. There are only two power-behaved terms in the expansion (6) and they correspond precisely to the known tree-level and one-loop terms in the R^4 effective action of the IIB theory, while the series of exponentials in (7) correspond to D-instanton corrections with the expected instanton number $N = \hat m n$. This lends weight to the suggestion that (5) is the exact result, in which case there should be a perturbative non-renormalization theorem [8] that forbids corrections to the R^4 term beyond one loop. Recently there has been a certain amount of evidence for the validity of such a theorem [20,21] (although an apparent contradiction in the literature [22] deserves closer analysis). The measure factor $\sum_{N|n}\frac{1}{n^2}$ in the instanton sum (7) can be related [11] to the expression for the Witten index of relevance in the analysis of D-particle threshold bound states [13,14] as will be described at the end of section 2.3.

2.2 One loop in eleven dimensions

The leading low energy behaviour of the one-loop four-graviton amplitude of eleven-dimensional supergravity compactified on a torus with radii R_{10} and R_{11} was considered in [7]. Here we will consider the complete momentum dependence of the same amplitude (a similar argument was also given in [9]) and expand the expression in a power series in the Mandelstam invariants, S, T and U,

$$S = -(k_1 + k_2)^2, \qquad T = -(k_1 + k_4)^2, \qquad U = -(k_1 + k_3)^2, \tag{8}$$

so that S+T+U =0.

The terms of interest arise from the sum of all one-loop diagrams with four external gravitons and with the graviton, gravitino or antisymmetric three-form potential circulating around the loop. This sum is most succinctly calculated in a first-quantized light-cone gauge formalism in which the amplitude is described as a trace over the states of an eleven-dimensional super-particle circulating around the loop and coupled to the four external gravitons by vertex operators (as in [7]). The result is given by

$$A_4 = \frac{1}{\kappa_{11}^2}\tilde{K}[I(S,T) + I(S,U) + I(U,T)], \tag{9}$$

where $\tilde{K}$ is the linearized approximation to R^4 (which is eighth order in momenta and symmetric under the interchange of any pair of gravitons) and the function $I(S,T)$ has the form of a Feynman integral for a massless scalar field theory. It is given by

$$I(S,T) = \frac{1}{\pi^{5/2}\mathcal{V}_2}\int\prod_{r=1}^{4} d\tau_r \int d^9 q \sum_{\{l_1,l_2\}} e^{-G^{IJ}l_I l_J \tau - \sum_{r=1}^{4} p_r^2 \tau_r}, \tag{10}$$

where $\tau = \sum_{r=1}^{4}\tau_r$ and q_i $(i = 1, \cdots, 9)$ is the nine-dimensional loop momentum transverse to T^2. The parameters τ_i label the relative positions of the four vertices. The sum is over the Kaluza–Klein momenta (l_1 and l_2) in the two compact dimensions ($I, J = 1, 2$) and the momenta in the legs of the loop are given by

$$p_r = q + \sum_{s=1}^{r} k_s, \tag{11}$$

where the external momenta, k_r^μ satisfy $k_r^2 = 0$ and $\sum_{r=1}^4 k_r = 0$. The particular kinematic configuration has been chosen in which the external momenta have zero components in the directions of the torus, i.e. $k_r^I = 0$. The eleven-dimensional coupling constant, κ_{11}, has dimension, $(length)^{9/2}$ and will be set equal to 1 in most of the following. The inverse metric on T^2 is defined by

$$G^{IJ} l_I l_J = \frac{1}{\mathcal{V}_2 \Omega_2} |l_1 + \Omega l_2|^2, \tag{12}$$

where $\mathcal{V}_2$ is the volume of the torus with complex structure, $\Omega = \Omega_1 + i\Omega_2$ (where Ω_1 is an angular parameter and $\Omega_2 = R_{10}/R_{11}$). Setting all the $k_r = 0$ in (10) gives the lowest-order result of [7]. The full amplitude (9) gives rise to terms in the M-theory effective action compactified on T^2 of the form

$$S_4 = \frac{1}{\kappa_{11}^2} \int \sqrt{G^{(9)}} \mathcal{V}_2 h(\mathcal{V}_2, \Omega; \partial^2) R^4 d^9 x. \tag{13}$$

The function h is a modular function of Ω and its argument ∂^2 symbolically indicates derivatives acting on the fields in R^4, corresponding to the dependence of A_4 on the momenta k_r.

After completing a square in the exponent of (10) and then performing the shifted loop integral the expression becomes

$$I(S,T) = \frac{\pi^2}{\mathcal{V}_2} \int \prod_{r=1}^{4} d\tau_r \tau^{-9/2} \sum_{\{l_1,l_2\}} e^{-G^{IJ} l_I l_J \tau + \frac{1}{\tau}(S\tau_1\tau_3 + T\tau_2\tau_4)}. \tag{14}$$

This integral is to be evaluated in the region $S, T < 0$ where it converges and then analytically continued to the physical region.

The momentum-independent terms in (14) can be isolated by writing

$$I(S,T) = I_0 + I'(S,T), \tag{15}$$

where

$$I_0 \equiv I(0,0) = \frac{\pi^{3/2}}{\mathcal{V}_2} \int_0^\infty d\tau \tau^{-3/2} \sum_{\{l_1,l_2\}} e^{-\pi G^{IJ} l_I l_J \tau}, \tag{16}$$

which is the expression considered in [7]. It has a divergence in the limit $\tau = 0$. A double Poisson resummation reexpresses I_0 as a sum over $\hat{l}_1$ and $\hat{l}_2$ that may be identified with the winding numbers of the euclidean world-line of the super-graviton around the directions R_{11} and R_9 of the torus, respectively.

The result is

$$
\begin{aligned}
I_0 &= \pi^{3/2} \int_0^\infty d\tau \tau^{-5/2} \sum_{\{\hat{l}_1, \hat{l}_2\}} e^{-\pi G_{IJ} \hat{l}_I \hat{l}_J \frac{1}{\tau}} \\
&= \pi^{3/2} \int_0^\infty d\hat{\tau} \hat{\tau}^{\frac{1}{2}} \sum_{\{\hat{l}_1, \hat{l}_2\}} e^{-\pi G_{IJ} \hat{l}_I \hat{l}_J \hat{\tau}},
\end{aligned} \tag{17}
$$

where $\hat{\tau} = 1/\tau$. This isolates the divergence in the zero winding term ($\hat{l}_1 = \hat{l}_2 = 0$). This is presumably regularized by a microscopic theory, such as the matrix model [23], but its regularized value is also determined uniquely by requiring that I_0 reproduce the IIA and IIB string theory R^4 terms in nine dimensions. Since the loop diverges as Λ^3, where Λ is a momentum cut-off, the regularized value of this term has the form $c\kappa_{11}^{-2/9}$ where c is a dimensionless constant. The remaining terms in (17) depend on the volume and complex structure of T^2 and are finite. The $\hat{\tau}$ integral is trivial for these terms and (17) can be written as

$$
I_0 = c\kappa_{11}^{-2/9} + \mathcal{V}_2^{-3/2} \zeta(3) E_{\frac{3}{2}}(\Omega, \bar{\Omega}). \tag{18}
$$

Substituting into (15) and (9) leads to a R^4 contribution to the M-theory effective action that can be expressed in terms of the IIB theory compactified on a circle.[b]

In the limit $\mathcal{V}_2 \to 0$ the radius, r_B, of the IIB circle becomes infinite and the second term in (18) dominates, leading to an expression for the action for the decompactified IIB theory which coincides with (4) (with f defined by (5)). The IIB string tree-level contributions arise from terms in (17) with $\hat{l}_1 \neq 0$ and $\hat{l}_2 = 0$ so the loop has non-zero winding only in the eleventh dimension. The string one-loop and D-instanton terms in (4) can be extracted from the $\hat{l}_2 \equiv \hat{m} \neq 0$ terms in (17) by performing a Poisson summation that takes the winding numbers $\hat{l}_1$ into Kaluza–Klein charges n and identifying $N = \hat{m}n$. The $n = 0$ term gives rise to the term in (4) that is independent of the dilaton (the one-loop contribution) while the charge-N D-instanton contributions in (7) come from the $N \neq 0$ contributions.

Although the zero winding number term with coefficient c does not contribute in the $\mathcal{V}_2 \to 0$ limit it does contribute to the finite-$\mathcal{V}_2$ amplitude. By

[b] This makes use of the usual relations between the parameters of M-theory on T^2 and the type II string theories on S^1: $r^A = (r^B)^{-1} = R_{10}(R_{11})^{\frac{1}{2}}$, $e^{\phi^B} = (r^A)^{-1} e^{\phi^A} = R_{11}/R_{10}$, where r^A is the radius of the IIA circle in string units and ϕ^A and ϕ^B are the dilatons of the two type II theories that are related by T-duality.

transforming to the type IIA coordinates it is easy to see that c is the coefficient of the one-loop contribution to the R^4 term in the IIA theory in ten dimensions and takes precisely the same value as the one-loop term in S^{pert}, thereby ensuring the T-duality of the type IIA and IIB theories [7].

There is an intriguing analogy between this calculation and the calculation of the exact prepotential of four-dimensional $N = 2$ supersymmetric Yang–Mills theory from a one-loop amplitude in five dimensions [24].

The momentum-dependent terms in $I(S,T)$ in (15) are contained in $I'(S,T)$. This is evaluated by separating the $l_1 = l_2 = 0$ term from the rest,

$$I'(S,T) = I^0(S,T) + \sum_{n=2}^{\infty} I_n(S,T). \tag{19}$$

The term with zero Kaluza–Klein momenta, I^0, gives a contribution to the amplitude that is non-analytic in the Mandelstam invariants while I_n is a homogeneous polynomial in S and T of degree n. After translating to string coordinates the term I^0 can be identified with a corresponding one-loop term in string perturbation theory. The polynomials $I_n(S,T)$, $I_n(T,U)$ and $I_n(U,S)$ can be identified with terms that should arise from multi-loop amplitudes in the supergravity theory [9].

Compactification of the supergravity loop amplitude to eight dimensions on T^3 raises new issues associated with the presence of new instantons that are identified with wrappings of the M-theory membrane world-volume around T^3. The complete expression for the R^4 term is now obtained by combining the one-loop calculation with the constraints of U-duality [25,7,26]. Compactification to lower dimensions raises yet more issues analogous to those that arise in the compactification of the matrix model.

2.3 Sixteen-fermion and related terms

The $f(\rho,\bar\rho)t_8t_8R^4$ term is one of a large number of terms of the same dimension that are related to each other by supersymmetry. Prominent among these is the analogue of the 't Hooft multi-fermion vertex of the IIB theory,

$$S_{\lambda^{16}} = \int \sqrt{g^B}\rho_2^{1/2} f_{16}(\rho,\bar\rho)\lambda^{16} d^{10}x, \tag{20}$$

where λ is the complex chiral spin-$\frac{1}{2}$ fermion. This field transforms with weight 3/2 under the $U(1)$ denominator in the coset $SL(2,Z)/U(1)$ [27,28,29] so that a general $SL(2,Z)$ transformation acts as

$$\lambda \to \left(\frac{c\bar\rho + d}{c\rho + d}\right)^{\frac{3}{4}} \lambda. \tag{21}$$

As a consequence of $SL(2,Z)$ invariance of the action it follows that the coefficient f_{16} in (20) is a non-holomorphic modular form of weight $(12,-12)$ (where the notation indicates the holomorphic and anti-holomorphic weights, respectively).

The expression for f_{16} can also be obtained by a one-loop calculation in eleven-dimensional supergravity compactified on T^2. This time the process has sixteen external gravitini in polarization states that correspond to λ in the IIB description. With this choice of external states the loop diagram vanishes in the limit of eleven non-compact dimensions and is finite for generic T^2. It was evaluated in a recent paper [10] and the result is

$$f_{16} = \frac{1}{\pi^2}\rho_2^{3/2}\Gamma(27/2)\sum_{(\hat{l}_1,\hat{l}_2)\neq(0,0)}\frac{(\hat{l}_1+\hat{l}_2\bar{\rho})^{24}}{|\hat{l}_1+\hat{l}_2\rho|^{27}}, \tag{22}$$

where $\hat{l}_1$ and $\hat{l}_2$ are the winding numbers of the world-line around the cycles of the torus. This expression has the large ρ_2 expansion,

$$\begin{aligned}\rho_2^{1/2}f_{16} &= \pi^{-2}\Gamma(27/2)\zeta(3)(\rho_2)^2+\pi^{-2}\Gamma(23/2)\zeta(2)\\ &+2^{24}\pi^{23/2}\sum_{N>0}\left(\sum_{N|\hat{m}}\frac{1}{\hat{m}^2}\right)(N\rho_2)^{25/2}e^{2\pi iN\rho}(1+O(\rho_2^{-1})),\end{aligned} \tag{23}$$

which again indicates the presence of string perturbation theory contributions at tree-level and one-loop together with an infinite number of D-instanton contributions (the anti D-instantons enter with coefficients that are of higher order in ρ_2^{-1}). There are presently no sixteen-fermion perturbative string calculations in the literature with which to compare the two leading terms.

In [10] it was also shown that the functions f_{16} and f are related by

$$f_{16}(\rho,\bar{\rho}) = \rho_2^{12}\mathcal{D}^{12}f(\rho,\bar{\rho}). \tag{24}$$

The covariant derivative $\mathcal{D}$ is defined by

$$F_{d+2,\bar{d}} = \mathcal{D}_d F_{d,\bar{d}} = i\left(\frac{\partial}{\partial\rho}+\frac{d}{(\rho-\bar{\rho})}\right)F_{d,\bar{d}}, \tag{25}$$

where $F_{d,d'}$ is a modular form of holomorphic weight d and anti-holomorphic weight d' and

$$\mathcal{D}^k F_{d,\bar{d}} \equiv \mathcal{D}_{d+2(k-1)}\mathcal{D}_{d+2(k-2)}\cdots\mathcal{D}_d F_{d,\bar{d}}. \tag{26}$$

The relationship (24) between f and f_{16} is consistent with the consequences of linearized supersymmetry. The physical on-shell fields of linearized

IIB supergravity are contained in a superfield $\Phi(x,\theta)$ (where θ is a complex chiral Grassmann spinor of $SO(9,1)$) satisfying the holomorphic constraint $D^*\Phi = 0$ and the on-shell condition $D^4\Phi = D^{*4}\Phi^*$ where D and D^* are supercovariant derivatives [29]. This field has an expansion in powers of θ,

$$\Phi = \rho_0 + a - \frac{2i}{g}\bar{\theta}\lambda - \cdots - \frac{i}{48g}\bar{\theta}\Gamma^{\mu\nu\eta}\theta\bar{\theta}\Gamma_{\eta}{}^{\sigma\tau}\theta R_{\mu\nu\sigma\tau} + \cdots, \tag{27}$$

where $\rho_0 = \rho - a$ is a constant background that defines the coupling constant in the linearized theory, R is the linearized approximation to the curvature and $\cdots$ indicates the other terms in the expansion. The matrices Γ^μ are $SO(9,1)$ gamma matrices and $\bar{\theta} = \Gamma^0\theta$ (with no complex conjugation). The linearized supersymmetric on-shell action has the form

$$S' = \mathrm{Re}\int d^{10}x d^{16}\theta\, g^4 F[\Phi], \tag{28}$$

and the various component interactions are contained in the θ^{16} term in the expansion of $F[\Phi]$. In addition to the R^4 and λ^{16} terms there are many others – some of these are singlets under the $U(1)$ (and arose at one string loop in [30]) while others carry a net even number of units of $U(1)$ charge [10] (some of these terms have also been discussed in [31]).

Although the fully nonlinear theory cannot easily be expressed in terms of Φ the leading behaviour of the charge-N instanton term is obtained by substituting $F = N^{-7/2}e^{2\pi i N\Phi}\sum_{N|n}\frac{1}{\hat{m}^2}$ into (28). In this manner the leading instanton contributions to the R^4, λ^{16} and related terms are all contained in a simple superspace expression.

The expression for the measure on single charge-N D-instantons, $\sum_{N|n}\frac{1}{\hat{m}^2}$, can be read off from the expressions for the instanton contributions to the various processes described above. This can then be compared with the expression that should follow from the matrix description of D-instantons. According to [12] this is determined by the partition function for $U(N)$ ten-dimensional supersymmetric Yang–Mills theory reduced to zero dimensions, in which case the configuration space consists of the ten $N\times N$ bosonic matrices, A_μ, and sixteen $N\times N$ fermionic matrices, ψ^a. The partition function is given by

$$Z_{U(N)} = \int d^{10}y d^{16}\epsilon Z_{SU(N)}, \tag{29}$$

where the center of mass super-translation zero modes have been separated from the measure factor,

$$Z_{SU(N)} = \int_{SU(N)} DAD\psi \exp(-S(A,\psi)). \tag{30}$$

This integral, which enters into the matrix model of[34], has not been explicitly evaluated apart from the $N = 2$ case considered in[13,14] (which gave $Z_{SU(2)} = 5/4$). However, equating it with the measure for N coincident D-instantons obtained from the $t_8 t_8 R^4$ and related terms strongly suggests that

$$Z_{SU(N)} = \sum_{N|\hat{m}} \frac{1}{m^2}, \tag{31}$$

for all N[11]. It would obviously be of interest to check this by explicit evaluation of the finite-dimensional integral (30).

3 Tree-level string theory and multi-loop supergravity.

We have seen that the lowest-dimensional terms in the string theory effective action correspond to terms that arise at one loop in eleven-dimensional supergravity. We would now like to see whether higher-order string-theory terms can be obtained from multi-loop amplitudes in eleven dimensions. To be explicit we will compare the expansion of the ten-dimensional IIA string tree amplitude with the compactification of multi-loop diagrams of eleven-dimensional supergravity on a circle of radius R_{11}. The calculation of multi-loop diagrams is obviously very difficult but we will be able to infer certain systematic properties from dimensional considerations.

First, consider the tree amplitude for four-graviton scattering in either of the type II superstring theories which is given by

$$\begin{aligned} A_4^{tree} &= e^{-2\phi} \frac{\tilde{K}}{stu} \frac{\Gamma(1-\alpha's)\Gamma(1-\alpha't)\Gamma(1-\alpha'u)}{\Gamma(1+\alpha's)\Gamma(1+\alpha't)\Gamma(1+\alpha'u)} \\ &= e^{-2\phi} \frac{\tilde{K}}{stu} \exp\left(\sum_{n=1}^{\infty} \frac{2\zeta(2n+1)}{2n+1} (s^{2n+1} + t^{2n+1} + u^{2n+1}) \right), \end{aligned} \tag{32}$$

where the string-frame Mandelstam invariants are related to those in the M-theory coordinates by

$$s = \frac{S}{R_{11}}, \qquad t = \frac{T}{R_{11}}, \qquad u = \frac{U}{R_{11}}. \tag{33}$$

It is important that every term in the exponent can be expressed as a polynomial in s and t multiplied by stu, as can be seen from the identity,

$$s^{2n+1} + t^{2n+1} + u^{2n+1} = stu \left[\sum_{r=1}^{n} \sum_{q=0}^{2n-2r} \frac{(2n+1)!}{r!(2n+1-r)!} (-1)^q s^{2n-1-r-q} t^{r+q-1} \right], \tag{34}$$

where $n \geq 1$. This means that the massless poles only contribute to the first term in the expansion of the exponential.

When expressed in terms of the Mandelstam invariants in the M-theory metric the expression (32) has the low-energy expansion,

$$A_4^{tree} \sim \tilde{K}\left(\frac{1}{STU} + \frac{2\zeta(3)}{R_{11}^3} + \frac{2\zeta(5)}{R_{11}^5}(S^2+ST+T^2) + \frac{2\zeta(3)^2}{R_{11}^6}STU \right.$$
$$+\frac{2\zeta(7)}{R_{11}^7}(S^4+2S^3T+3S^2T^2+2ST^3+T^4) + \frac{2\zeta(3)\zeta(5)}{R_{11}^8}STU(S^2+ST+T^2)$$
$$\left. +\frac{2\zeta(9)}{R_{11}^9}(S^6+4S^5T+\cdots+T^6) + \frac{4}{3R_{11}^9}\zeta(3)^3S^2T^2U^2+\cdots\right). \quad (35)$$

The first term in the expansion combines with the kinematic factor, $\tilde{K}$, to give the tree-level amplitude that is described by the Einstein–Hilbert action of eleven-dimensional supergravity compactified on a circle of radius $R_{11} = e^{2\phi/3}$. The subsequent term, with coefficient $\zeta(3)$ is the term considered earlier that comes from the R^4 term in the effective action. We saw that this term is reproduced by the one-loop diagram of eleven-dimensional supergravity compactified on a circle. The higher-order terms in (35) come from terms in the effective action with derivatives acting on R^4. The question is whether the whole series might be reproduced by summing loop diagrams of eleven-dimensional supergravity compactified on a circle.

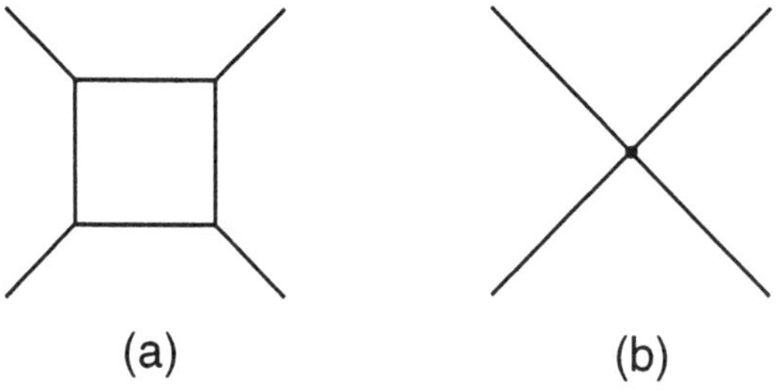

Figure 1: (a) The finite part of the one-loop diagram contributing to the R^4 term of compactified eleven-dimensional supergravity comes from the non-zero windings of the circulating particles. (b) The primitively divergent R^4 term that arises from the regularized zero winding sector is regularized to a specific finite value consistent with supersymmetry and various duality symmetries.

We will restrict ourselves here to considering some simple multi-loop contributions that should reproduce the first few terms in the low-energy expansion of the string tree-level amplitude, (35). We have already seen that the constant term proportional to $\zeta(3)(R_{11})^{-3}$ comes from the finite part of the one-loop diagram of figure 1(a). Figure 1(b) represents the counterterm for the primitive divergence of the one-loop diagram that is regularized to a finite value by the considerations outlined earlier.

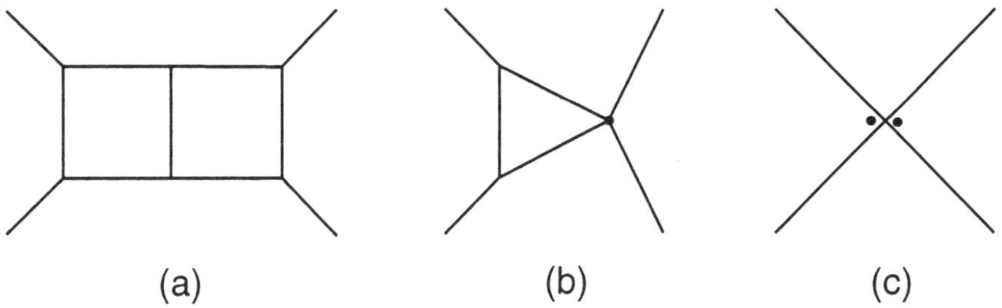

Figure 2: (a) A representation of the non-zero windings of the particles circulating in a two-loop contribution to four-graviton scattering. The dimension, $(momentum)^{20}$, of this finite contribution could give terms of the form s^3 in the ten-dimensional string tree-level amplitude. (b) A contribution in which one loop is replaced by the R^4 counterterm. has dimension $(momentum)^{17}$ and could contribute terms of the form s^2 to the string tree-level amplitude. (c) A counterterm for the two-loop primitive divergence.

Just as with the one-loop diagrams, there are finite contributions from multi-loop diagrams which come from non-zero windings of the internal loops around the compact dimension(s). At each order there is also a new primitive divergence. For example, there are many two-loop diagrams that must be summed to give the complete amplitude. These have superficial degrees of divergence Λ^{20}, where Λ is a momentum space cut-off and is $O(\kappa_{11})^4$. However, we are discussing terms in the amplitude proportional to derivatives acting on R^4 so there are at least eight powers of the external momenta, reducing the naive divergence to Λ^{12}, or less (depending on the number of derivatives). This leads to new (two-loop) primitive divergences indicated by figure 2(c), where the two dots indicate the fact that this is a combination of two-loop counterterms. More generally, at n loops there are primitive divergences of the form $\Lambda^{9n-6}R^4$, where the powers of the cut-off Λ may be substituted by powers of the external momenta. The ultra-violet divergences coming from the zero winding-number sector give rise to counterterms that are independent of R_{11}.

However, just as with the one-loop diagrams there will be finite terms arising from the effects of the internal propagators winding around the compact direction(s). These are collectively represented, in the case of the two-loop diagrams, by the ladder diagram in figure 2(a) which has dimension

$(momentum)^{20}$. After accounting for the eight powers of the external momenta in the overall R^4, twelve powers of momenta remain that must be replaced by appropriate powers of the dimensional parameters, R_{11} and S, T, U. This diagram may give a contribution to the tree-level string amplitude which is proportional to $(R_{11})^{-3}$. In such a contribution there are nine powers of momentum left to be soaked up by a combination involving Mandelstam invariants. In the analogous expressions derived from string theory, such as (35), each Mandelstam invariant comes along with one power of R_{11}^{-1}. Thus, the nine powers of momenta should be associated with a linear combination of S^3/R_{11}^3, T^3/R_{11}^3, S^2T/R_{11}^3 and T^2S/R_{11}^3. We would like to identify this with the term in (35) that is cubic in the invariants. At n loops the Feynman diagrams analogous to figure 2(a) can give finite contributions of the form s^{3n-3} in the expansion of the string tree amplitude (35).

In addition to these finite contributions there are also many other loop diagrams in which the counterterms are inserted as vertices. In fact, the diagram with the smallest primitive degree of divergence (other than the one-loop diagrams of figure 1) is the one shown in figure 2(b), in which the vertex indicated by a dot is the R^4 counterterm that was required to make sense of the one-loop diagram. Since all the particles of the supermultiplet circulate around the loop the supersymmetric partners of the R^4 vertices are also involved. These couple the two external gravitons to two $C^{(3)}$'s, or to two gravitini. Figure 2(b) is meant to represent all diagrams of this order containing a single one-loop counterterm. These have dimension $(momentum)^{17}$ so that after allowing for the eight powers of external momenta in R^4 there are nine powers of momentum left to account for. These diagrams have not been explicitly evaluated but if it is assumed that they contribute to the string tree-level process there must be an overall factor of $e^{-2\phi^A} = R_{11}^{-3}$, absorbing three of the nine momentum powers. Then dimensional counting implies that the the only possible finite function is *quadratic* in the invariants, $s, t, u \sim S/R_{11}, T/R_{11}, U/R_{11}$. This quadratic term ought to be precisely the first momentum-dependent correction to the R^4 term deduced by expanding the tree diagram. It should therefore turn out that an explicit calculation of figure 2(b) gives the finite contribution,

$$\frac{2\zeta(5)}{R_{11}^3}\tilde{K}\left(\frac{S^2+ST+T^2}{R_{11}^2}\right), \tag{36}$$

where $\tilde{K}$ is the linearized approximation to $t_8t_8R^4$ and is eighth-order in the external momenta. Although the diagram in figure 2(b) has not been explicitly evaluated it is tempting to imagine that it has the form $\tilde{K}I(S,T,U)$, where I is the expression for the loop amplitude of the same geometry in scalar field theory. This has a coefficient that includes a factor of $\zeta(5)$ just as there was a factor of $\zeta(3)$ in the finite part of the box diagram of figure 1(a). Verifying that the precise coefficient is the one in (36) would lend support to the suggestion that quantum corrections to compactified eleven-dimensional supergravity coincide with the string tree-level expressions at this order. In a similar manner the term with coefficient $\zeta(7)$ in (35) could arise from the diagam with two R^4 vertices and two propagators. However, this exhausts the contributions from one-loop diagrams with R^4 vertices and it is not at all clear how the terms of higher order in momenta in (35) would emerge in any systematic manner.

Of course, such diagrams also contribute to string loop effects. For example, there are the generalizations of the non-analytic one-loop massless threshold terms which can be written (for either of the type II theories) symbolically as

$$\int \sqrt{g_E^{(10)}} s_E^n f_n(t_E/s_E) R^4 d^{10}x = \int \sqrt{g^{(10)}} e^{\frac{n-1}{2}\phi^A} s^n f_n(t/s) R^4 d^{10}x, \qquad (37)$$

where g_E is the type II string metric in the Einstein frame. The function $f_n(t/s)$ involves up to n powers of t/s as well as factors of $\log(t/(s+t))$ and $\log(s/(s+t))$. The one-loop ($n = 1$) case is well known. Since these terms arise in string perturbation theory, which is an expansion in $e^{2\phi}$, it must be that $n = 4m + 1$, where m is an integer. In the Einstein frame these terms are independent of the dilaton so they are inert under the action of $SL(2, Z)$ in the IIB theory. They are therefore not constrained by S-duality. In the IIA theory such terms are directly related to analogous thresholds in genus-$(m+1)$ multi-loop eleven-dimensional supergravity amplitudes. These Feynman diagrams are dimensionally of the form $(\text{momentum})^{9m+3} R^4$. After compactification their contribution to the genus-$(m+1)$ string amplitude has a factor of $e^{2m\phi^A} = R_{11}^{3m}$ from powers of the coupling. The total momentum dimension of this contribution is then written as $R_{11}^{3m}(\text{momentum})^{12m+3} R^4$, which must therefore be made up of terms of the form $e^{2m\phi^A} s^{4m+1} f_n(t/s) R^4$ in the string coordinates, using the fact that $s = S/R_{11}$. This agrees in structure with the terms in (37) if the function f contains the appropriate logarithmic terms.

There are many complications in understanding in detail the systematics of the correspondence between the higher-loop supergravity diagrams and string diagrams. Whereas the R^4 and related terms of the same dimension are integrals over half the superspace, terms with more derivatives are integrals over a higher fraction of the superspace. Each power of momentum is equivalent to two powers of θ so that terms with less than eight powers of momentum acting on R^4 should be protected and may be determined in this manner. This would include the terms up to the $\zeta(3)^2$ term in (35) and possibly also the $\zeta(7)$ term.

Whether it is possible to go beyond this and relate terms in string perturbation theory at higher order in the momentum expansion to eleven-dimensional supergravity is much less obvious.

1. E. Cremmer, B. Julia and J. Scherk, *Supergravity theory in eleven dimensions*, Phys. Lett. **76B** (1978) 409.
2. M.J. Duff, J.T. Liu and R. Minasian, *Eleven-dimensional origin of string-string duality: a one loop test*, hep-th/9506126, Nucl. Phys. **452B** (1995)261.
3. C. Vafa and E. Witten, *A one loop test of string duality*, hep-th/9505053, Nucl. Phys. **B447** (1995)261.
4. Blum and J. Harvey, *Anomaly inflow for gauge defects*, hep-th/9310035; Nucl. Phys. **B416** (1994) 119.
5. P. Aspinwall, *Some Relationships Between Dualities in String Theory*, in Proceedings of 'S-duality and mirror symmetry', Trieste 1995, hep-th/9508154, Nucl. Phys. Proc. **46** (1996) 30.
6. J.H. Schwarz, *Lectures on Superstring and M-theory dualities*, hep-th/9607201;J.H. Schwarz, *An* $Sl(2,Z)$ *multiplet of type IIb superstrings*, hep-th/9508143 Phys. Lett. **360B** (1995) 13.
7. M.B. Green, M. Gutperle and P. Vanhove, *One loop in eleven dimensions*, hep-th/9706175, Phys. Lett. **409B** (1997)177.
8. M.B. Green and M. Gutperle, *Effects of D-instantons*, hep-th/9701093, Nucl. Phys. **B498** (1997)195.
9. J.G. Russo and A.A. Tseytlin, *One loop four graviton amplitude in eleven-dimensional supergravity* hep-th/9707134.
10. M.B. Green, M. Gutperle and H. Kwon, *Sixteen-fermion and related terms in M theory on* T^2, hep-th/9710151.
11. M.B. Green and Michael Gutperle, *D-particle bound states and the D-instanton measure* hep-th/9711107.
12. E. Witten, *Bound States of Strings and p-branes*, Nucl. Phys. **B460** (1996) 335.
13. S. Sethi and M. Stern, *D-brane bound states redux*, hep-th/9705046.
14. P. Yi, *Witten index and threshold bound states of D-branes*, hep-th/9704098.
15. M.B. Green and J.H. Schwarz, *Supersymmetric dual string theory (III). Loops and renormalization*, Nucl. Phys. **198B** (1982) 441.
16. M.T. Grisaru, A.E.M Van de Ven and D. Zanon, *Two-dimensional supersymmetric sigma models on Ricci flat Kähler manifolds are not finite*, Nucl. Phys. **B277** (1986) 388; *Four loop divergences for the N=1 supersymmetric nonlinear sigma model in two-dimensions*, Nucl. Phys. **B277** (1986) 409.

17. D.J. Gross and E. Witten, *Superstring modifications of Einstein's equations*, Nucl. Phys. **B277** (1986) 1.
18. M.B. Green and J.H. Schwarz, *Supersymmetrical string theories*, Phys. Lett. **109B** (1982) 444.
19. A. Terras, *Harmonic Analysis on Symmetric Spaces and Applications*, vol. I, Springer–Verlag (1985).
20. I. Antoniadis, B. Pioline and T.R. Taylor, *Calculable* $e^{1/\lambda}$ *effects*, hep-th/9707222.
21. N. Berkovits, *Construction of* R^4 *terms in N=2 D=8 superspace*, hep-th/9709116.
22. R. Jengo and C.-J. Zhu, *Two loop computation of the four particle amplitude in heterotic string theory*, Phys. Lett. **212B** (1988) 313.
23. T. Banks, W. Fischler, S.H. Shenker and L. Susskind, *M-Theory as a matrix model: a conjecture*, hep-th/9610043, Phys. Rev. **D55** (1997) 5112.
24. N. Nekrasov and A. Lawrence, *Instanton sums and five dimensional gauge theories*, hep-th/9706025.
25. M.B. Green and P. Vanhove, *D-instantons, strings and M-theory*, hep-th/9704145, Phys. Lett. **408B** (1997)122.
26. E. Kiritsis and B. Pioline, *On* R^4 *threshold corrections in IIB string theory and (p,q) string instantons*, hep-th/9707018
27. J.H. Schwarz and P.C. West, *Symmetries and transformations of chiral N=2 D=10 supergravity*, Phys. Lett. **126B** (1983) 301.
28. J.H. Schwarz, *Covariant field equations of chiral N=2 D=10 supergravity*, Nucl. Phys. **B226** (1983) 269.
29. P.S. Howe and P.C. West, *The complete N=2 D=10 supergravity*, Nucl. Phys. **B238** (1984) 181.
30. D.J. Gross and J.H. Sloan, *The quartic effective action for the heterotic string*, Nucl. Phys. **291B** (1987) 41.
31. A. Kehagias and H. Partouche, *The exact quartic effective action for the type IIB superstring*, hep-th/9710023.
32. E. Kiritsis and B. Pioline, *On* R^4 *threshold corrections in IIB string theory and (p,q) string instantons*, hep-th/9707018;
A. Strominger, *Loop corrections to the universal hypermultiplet*, hep-th/9706195.
I. Antoniadis, S. Ferrara, R. Minasian and K.S. Narain, R^4 *couplings in M and type II theories*, hep-th/9707013.
33. C. Bachas, C. Fabre, E. Kiritsis, N.A. Obers and P. Vanhove, *Heterotic/type I duality and D-brane instantons*, hep-th/9707126.
E. Kiritsis and N.A. Obers, *Heterotic/type-I duality in* $D < 10$ *dimen-*

sions, threshold corrections and D-instantons, hep-th/9709058.

34. N. Ishibashi, H. Kawai and Y. Kitazawa and A. Tsuchiya, *A large N reduced model as superstring*, hep-th/9612115; Nucl.Phys.B498:467-491,1997.
35. C. Bachas, *Heterotic versus type I*, Talk at STRINGS'97 (Amsterdam, June 16-21 1997) and HEP-97 (Jerusalem, August 19-26 1997), hep-th/9710102.

BPS STATES AND SUPERSYMMETRY

Sergio FERRARA
Theoretical Physics Division, CERN
CH-1211 Geneva 23

We describe duality invariant constraints, which allow a classification of BPS states preserving different fractions of supersymmetry. We then relate this analysis to the orbits of the exceptional groups $E_{6(6)}, E_{7(7)}$, relevant for black holes in five and four dimensions. The orbits of $O(P,Q)$ groups enter in the classification of BPS string in six dimensions.

1 Introduction

A variety of results have been obtained in recent past for the study of general properties of BPS branes in supersymmetric theories of gravity. The latter are described by string theory and M-theory [1] whose symmetry properties are encoded in extended supergravity effective field theories.

Of particular interest are extremal black holes in four and five dimensions which correspond to BPS saturated states [4] and whose ADM mass depends, beyond the quantized values of electric and magnetic charges, on the asymptotic value of scalars at infinity. The latter describe the moduli space of the theory Another physical relevant quantity, which depends only on quantized electric and magnetic charges, is the black hole entropy, which can be defined macroscopically, through the Bekenstein-Hawking area-entropy relation or microscopically, through D-branes techniques [5] by counting of microstates [6]. It has been further realized that the scalar fields, independently of their values at infinity, flow towards the black hole horizon to a fixed value of pure topological nature given by a certain ratio of electric and magnetic charges [7]. These "fixed scalars" correspond to the extrema of the ADM mass in moduli space while the black-hole entropy is the value of the squared ADM mass at this point in $D = 4$ [8,9] and the power 3/2 of the ADM mass in $D = 5$. In four dimensional theories with $N > 2$, extremal black-holes preserving one supersymmetry have the further property that all central charge eigenvalues other than the one equal to the BPS mass flow to zero for "fixed scalars".

A similar analysis exists in six dimensional theories for BPS strings. Scalar fields flow towards the BPS string horizon to a fixed value obtained by extremizing the string tension in the moduli space [10,11]. The extremum of the string tension is a duality invariant quantity which is the $O(P,Q)$ Lorentzian norm of the integral charge vector.

The entropy formula for black holes turns out to be in all cases a U-duality invariant expression (homogeneous of degree two in $D = 4$ and of degree $3/2$ in $D = 5$) built out of electric and magnetic charges and as such can be in fact also computed through certain (moduli-independent) topological quantities which only depend on the nature of the U-duality groups and the appropriate representations of electric and magnetic charges [2,3]. More specifically, in the $N = 8$, $D = 4$ and $D = 5$ theories the entropy was shown to correspond to the unique quartic E_7 and cubic E_6 invariants built with the 56 and 27 dimensional representations respectively [12,8]. In this report we respectively describe the invariant classification of BPS states preserving different numbers of supersymmetries in $D = 4$ 5 and 6 and then relate this analysis to the theory of BPS orbits of the exceptional groups $E_{7(7)}$, $E_{6(6)}$ and of the pseudo-orthogonal $O(P, Q)$ groups.

2 BPS Conditions for Enhanced Supersymmetry

In this section we will describe U-duality invariant constraints on the multiplets of quantized charges in the case of BPS black holes whose background preserves more than one supersymmetry [13].

We will still restrict our analysis to four and five dimensional cases for which three possible cases exist *i.e.* solutions preserving 1/8, 1/4 and 1/2 of the original supersymmetry (32 charges).

The invariants may only be non zero on solutions preserving 1/8 supersymmetry. In dimensions $6 \leq D \leq 9$ black holes may only preserve 1/4 or 1/2 supersymmetry, and no associated invariants exist in these cases.

The description which follows also make contact with the D-brane microscopic calculation, as it will appear obvious from the formulae given below. We will first consider the five dimensional case.

In this case, BPS states preserving 1/4 of supersymmetry correspond to the invariant constraint $I_3(27) = 0$ where I_3 is the E_6 cubic invariant [8]. This corresponds to the E_6 invariant statement that the **27** is a null vector with respect to the cubic norm. As we will show in a moment, when this condition is fulfilled it may be shown that two of the central charge eigenvalues are equal in modulus. The generic configuration has 26 independent charges.

Black holes corresponding to 1/2 BPS states correspond to null vectors which are critical, namely

$$\partial I(27) = 0 \tag{1}$$

In this case the three central charge eigenvalues are equal in modulus and a generic charge vector has 17 independent components.

To prove the above statements, it is useful to compute the cubic invariant in the normal frame, given by:

$$\begin{aligned} I_3(27) &= Tr(Z\mathbf{C})^3 \\ &= 6(e_1 + e_2)(e_1 + e_3)(e_2 + e_3) \\ &= 6s_1 s_2 s_3 \end{aligned} \tag{2}$$

where:

$$\begin{aligned} e_1 &= \frac{1}{2}(s_1 + s_2 - s_3) \\ e_2 &= \frac{1}{2}(s_1 - s_2 + s_3) \\ e_3 &= \frac{1}{2}(-s_1 + s_2 + s_3) \end{aligned} \tag{3}$$

are the eigenvalues of the traceless antisymmetric 8×8 matrix. We then see that if $s_1 = 0$ then $|e_1| = |e_2|$, and if $s_1 = s_2 = 0$ then $|e_1| = |e_2| = |e_3|$. To count the independent charges we must add to the eigenvalues the angles given by $USp(8)$ rotations. The subgroup of $USp(8)$ leaving two eigenvalues invariant is $USp(2)^4$, which is twelve dimensional. The subgroup of $USp(8)$ leaving invariant one eigenvalue is $USp(4) \times USp(4)$, which is twenty dimensional. The angles are therefore $36 - 12 = 24$ in the first case, and $36 - 20 = 16$ in the second case. This gives rise to configurations with 26 and 17 charges respectively, as promised.

Taking the case of Type II on T^5 we can choose s_1 to correspond to a solitonic five-brane charge, s_2 to a fundamental string winding charge along some direction and s_3 to Kaluza-Klein momentum along the same direction.

The basis chosen in the above example is S-dual to the D-brane basis usually chosen for describing black holes in Type IIB on T_5. All other bases are related by U-duality to this particular choice. We also observe that the above analysis relates the cubic invariant to the picture of intersecting branes since a three-charge 1/8 BPS configuration with non vanishing entropy can be thought as obtained by intersecting three single charge 1/2 BPS configurations[22,23,24].

By using the S–T-duality decomposition we see that the cubic invariant reduces to $I_3(27) = 10_{-2}10_{-2}1_4 + 16_1 16_1 10_{-2}$. The 16 correspond to D-brane charges, the 10 correspond to the 5 KK directions and winding of wrapped fundamental strings, the 1 correspond to the N-S five-brane charge.

We see that to have a non-vanishing area we need a configuration with three non-vanishing N-S charges or two D-brane charges and one N-S charge.

Unlike the 4-D case, it is impossible to have a non-vanishing entropy for a configuration only carrying D-brane charges.

We now turn to the four dimensional case.

In this case the situation is more subtle because the condition for the 56 to be a null vector (with respect to the quartic norm) is not sufficient to enhance the supersymmetry. This can be easily seen by going to the normal frame where it can be shown that for a null vector there are not, in general, coinciding eigenvalues. The condition for 1/4 supersymmetry is that the gradient of the quartic invariant vanish.

The invariant condition for 1/2 supersymmetry is that the second derivative projected into the adjoint representation of E_7 vanish. This means that, in the symmetric quadratic polynomials of second derivatives, only terms in the 1463 of E_7 are non-zero. Indeed, it can be shown, going to the normal frame for the 56 written as a skew 8×8 matrix, that the above conditions imply two and four eigenvalues being equal respectively.

The independent charges of 1/4 and 1/2 preserving supersymmetry are 45 and 28 respectively.

To prove the latter assertion, it is sufficient to see that the two charges normal-form matrix is left invariant by $USp(4) \times USp(4)$, while the one charge matrix is left invariant by $USp(8)$ so the $SU(8)$ angles are $63 - 20 = 43$ and $63 - 36 = 27$ respectively.

The generic 1/8 supersymmetry preserving configuration of the 56 of E_7 with non vanishing entropy has five independent parameters in the normal frame and $51 = 63 - 12$ $SU(8)$ angles. This is because the compact little group of the normal frame is $SU(2)^4$. The five parameters describe the four eigenvalues and an overall phase of an 8×8 skew diagonal matrix.

If we allow the phase to vanish, the 56 quartic norm just simplifies as in the five dimensional case:

$$\begin{aligned} I_4(56) &= s_1 s_2 s_3 s_4 = (e_1 + e_2 + e_3 + e_4)(e_1 + e_2 - e_3 - e_4) \\ &\times (e_1 - e_2 - e_3 + e_4)(e_1 - e_2 + e_3 - e_4) \end{aligned} \tag{4}$$

where e_i $(i = 1, \cdots, 4)$ are the four eigenvalues.

1/4 BPS states correspond to $s_3 = s_4 = 0$ while 1/2 BPS states correspond to $s_2 = s_3 = s_4 = 0$.

An example of this would be a set of four D-branes oriented along 456, 678, 894, 579 (where the order of the three numbers indicates the orientation of the brane). Note that in choosing the basis the sign of the D-3-brane charges is important; here they are chosen such that taken together with positive coefficients they form a BPS object. The first two possibilities ($I_4 \neq 0$ and $I_4 = 0$, $\frac{\partial I_4}{\partial q^i \neq 0}$) preserve 1/8 of the supersymmetries, the third ($\frac{\partial I_4}{\partial q^i} = 0$, $\frac{\partial^2 I_4}{\partial q^i \partial q^j}|_{Adj\ E_7} \neq 0$) 1/4 and the last ($\frac{\partial^2 I_4}{\partial q^i \partial q^j}|_{Adj\ E_7} = 0$) 1/2.

It is interesting that there are two types of 1/8 BPS solutions. In the supergravity description, the difference between them is that the first case has non-zero horizon area. If $I_4 < 0$ the solution is not BPS. This case corresponds, for example, to changing the sign of one of the three-brane charges discussed above. By U-duality transformations we can relate this to configurations of branes at angles such as in [14].

Going from four to five dimensions it is natural to decompose the $E_7 \to E_6 \times O(1,1)$ where E_6 is the duality group in five dimensions and $O(1,1)$ is the extra T-duality that appears when we compactify from five to four dimensions. According to this decomposition, the representation breaks as: $\mathbf{56} \to \mathbf{27}_1 + \mathbf{1}_{-3} + \mathbf{27'}_{-1} + \mathbf{1}_3$ and the quartic invariant becomes:

$$\begin{aligned} \mathbf{56}^4 &= (\mathbf{27}_1)^3\mathbf{1}_{-3} + (\mathbf{27'}_{-1})^3\mathbf{1}_3 + \mathbf{1}_3\mathbf{1}_3\mathbf{1}_{-3}\mathbf{1}_{-3} + \mathbf{27}_1\mathbf{27}_1\mathbf{27'}_{-1}\mathbf{27'}_{-1} \\ &+ \mathbf{27}_1\mathbf{27'}_{-1}\mathbf{1}_3\mathbf{1}_{-3} \end{aligned} \tag{5}$$

The **27** comes from point-like charges in five dimensions and the $\mathbf{27'}$ comes from string-like charges.

Decomposing the U-duality group into T- and S-duality groups, $E_7 \to SL(2,\mathbf{R}) \times O(6,6)$, we find $\mathbf{56} \to (\mathbf{2},\mathbf{12}) + (\mathbf{1},\mathbf{32})$ where the first term corresponds to N-S charges and the second term to D-brane charges. Under this decomposition the quartic invariant (4) becomes $\mathbf{56}^4 \to \mathbf{32}^4 + (\mathbf{12}.\mathbf{12'})^2 + \mathbf{32}^2.\mathbf{12}.\mathbf{12'}$. This means that we can have configurations with a non-zero area that carry only D-brane charges, or only N-S charges or both D-brane and N-S charges.

It is remarkable that $E_{7(7)}$-duality gives additional restrictions on the BPS states other than the ones merely implied by the supersymmetry algebra. The analysis of double extremal black holes implies that I_4 be semi-definite positive for BPS states. ¿From this fact it follows that configurations preserving 1/4 of supersymmetry must have eigenvalues equal in pairs, while configuratons with three coinciding eigenvalues are not BPS.

To see this, it is sufficient to write the quartic invariant in the normal frame basis. A generic skew diagonal 8×8 matrix depends on four complex eigenvalues z_i. These eight real parameters can be understood using the decomposition [25]:

$$\mathbf{56} \to (\mathbf{8_v},\mathbf{2},\mathbf{1},\mathbf{1}) + (\mathbf{8_s},\mathbf{1},\mathbf{2},\mathbf{1}) + (\mathbf{8_c},\mathbf{1},\mathbf{1},\mathbf{2}) + (\mathbf{1},\mathbf{2},\mathbf{2},\mathbf{2}) \tag{6}$$

under

$$E_{7(7)} \to O(4,4) \times SL(2,\mathbf{R})^3 \tag{7}$$

Here $O(4,4)$ is the little group of the normal form and the $(\mathbf{2},\mathbf{2},\mathbf{2})$ are the four complex skew-diagonal elements. We can further use $U(1)^3 \subset SL(2,\mathbf{R})^3$ to further remove three relative phases so we get the five parameters $z_i = \rho_i e^{i\phi/4}$ $(i = 1, \cdots, 4)$.

The quartic invariant, which is also the unique $SL(2,\mathbf{R})^3$ invariant built with the $(\mathbf{2},\mathbf{2},\mathbf{2})$, becomes [13]:

$$\begin{aligned} I_4 &= \sum_i |z_i|^4 - 2\sum_{i<j} |z_i|^2|z_j|^2 + 4(z_1z_2z_3z_4 + \bar{z}_1\bar{z}_2\bar{z}_3\bar{z}_4) \\ &= (\rho_1+\rho_2+\rho_3+\rho_4)(\rho_1+\rho_2-\rho_3-\rho_4) \\ &\times (\rho_1-\rho_2+\rho_3+\rho_4)(\rho_1-\rho_2-\rho_3+\rho_4) \\ &+ 8\rho_1\rho_2\rho_3\rho_4(cos\phi - 1) \end{aligned} \tag{8}$$

The last term is semi-definite negative. The first term, for $\rho_1 = \rho_2 = \rho$ becomes:

$$-[4\rho^2 - (\rho_3+\rho_4)^2](\rho_3-\rho_4)^2 \tag{9}$$

which is negative unless $\rho_3 = \rho_4$. So 1/4 BPS states must have

$$\rho_1 = \rho_2 > \rho_3 = \rho_4\ , \quad cos\phi = 1 \tag{10}$$

For $\rho_1 = \rho_2 = \rho_3 = \rho$, the first term in I_4 becomes:

$$-(3\rho+\rho_4)(\rho-\rho_4)^3 \tag{11}$$

so we must also have $\rho_4 = \rho$, $cos\phi = 1$ which is the 1/2 BPS condition.

An interesting case, where I_4 is negative, corresponds to a configuration carrying electric and magnetic charges under the same gauge group, for example a 0-brane plus 6-brane configuration which is dual to a K–K-monopole plus K–K-momentum [15,16]. This case corresponds to $z_i = \rho e^{i\phi/4}$ and the phase is $tan\phi/4 = e/g$ where e is the electric charge and g is the magnetic charge. Using (4) we find that $I_4 < 0$ unless the solution is purely electric or purely magnetic. In [17] it was suggested that $0+6$ does not form a supersymmetric state. Actually, it was shown in [18] that a $0+6$ configuration can be T-dualized into a non-BPS configuration of four intersecting D-3-branes. Of course, I_4 is negative for both configurations. Notice that even though these two charges are Dirac dual (and U-dual) they are not S-dual in the sense of filling out an $SL(2,\mathbf{Z})$ multiplet. In fact, the K–K-monopole forms an $SL(2,\mathbf{Z})$ multiplet with a fundamental string winding charge under S-duality [19].

Let us now consider BPS strings in $D = 6$. In this case the duality group is $O(P,Q)$ with $P = 1,5$ for the (1,0) and (2,0) chiral theories respectively and Q denotes the number of matter tensor multiplets; $P = Q = 1$ and $P = Q = 5$ for the non chiral (1,1) and (2,2) theories. For non-chiral theories, charge vectors with non-vanishing Lorentzian norm correspond to 1/4 BPS states, while null vectors correspond to 1/2 BPS states [13].

For chiral theories, fixed scalars which extremize the BPS tension correspond to a time–like charge vector, while moduli values giving tensionless strings correspond to charge vectors to be space–like [11].

3 Duality Orbits for BPS States Preserving Different Numbers of Supersymmetries

In this last section we give an invariant classification of BPS black holes preserving different numbers of supersymmetries in terms of orbits of the **27** and the **56** fundamental representations of the duality groups $E_{6(6)}$ and $E_{7(7)}$ resperctively [20,21].

In five dimensions the generic orbits preserving 1/8 supersymmetry correspond to the 26 dimensional orbits $E_{6(6)}/F_{4(4)}$ so we may think the generic 27 vector of E_6 parametrized by a point in this orbit and its cubic norm (which actually equals the square of the black-hole entropy).

The light-like orbit, preserving 1/4 supersymmetry, is the 26 dimensional coset $E_{6(6)}/O(5,4) \odot T_{16}$ where $\odot$ denotes the semidirect product.

The critical orbit, preserving maximal 1/2 supersymmetry (this correspond to $\frac{\partial I_4}{\partial q^i} \neq 0$) correspond to the 17 dimensional space

$$\frac{E_{6(6)}}{O(5,5) \odot T_{16}} \tag{12}$$

In the four dimensional case, we have two inequivalent 55 dimensional orbits corresponding to the cosets $E_{7(7)}/E_{6(2)}$ and $E_{7(7)}/E_{6(6)}$ depending on whether $I_4(56) > 0$ or $I_4(56) < 0$. The first case corresponds to 1/8 BPS states whith non-vanishing entropy, while the latter corresponds to non BPS states.

There is an additional 55 dimensional light-like orbit ($I_4 = 0$) preserving 1/8 supersymmetry given by $\frac{E_{7(7)}}{F_{4(4)} \odot T_{26}}$.

The critical light-like orbit, preserving 1/4 supersymmetry, is the 45 dimesnsional coset $E_{7(7)}/O(6,5) \odot (T_{32} \oplus T_1)$

The critical orbit corresponding to maximal 1/2 supersymmetry is described by the 28 dimensional quotient space

$$\frac{E_{7(7)}}{E_{6(6)} \odot T_{27}} \tag{13}$$

Let us now consider BPS strings in six dimensions [11]. Let's first discuss the chiral theories. In both theories the central charge has only one eigenvalue.

This is trivial for the (1,0) theory since there is only one central charge, while in the (2,0) theory is due to the fact that the central charge matrix is the **5** of $USp(4)$. Therefore chiral theories can only have one half BPS strings. A null charge vector does not correspond to supersymmetry enhancement in this case. Time–like and space–like orbits correspond to the quotient space:

$$\frac{O(P,Q)}{O(P-1,Q)} \quad \text{and} \quad \frac{O(P,Q)}{O(P,Q-1)} \tag{14}$$

with $P = 1$ and 5 respectively.

In the case of the non chiral (1,1) and (2,2) theories the central charge matrix has two eigenvalues in both cases. In the (2,2) theory this is due to the fact that the central charge matrix is in **5** + **5'** of $USp(4) \times USp(4)'$. Null $O(P,Q)$ vectors correspond in this case to one half BPS states while vectors with non-vanishing norm are one quarter BPS. Null vectors correspond to the orbits $O(P,Q)/IO(P-1,Q-1)$. For a generic one quarter BPS orbit in the (2,2) case, the 10 charges can be thought as given by the two eigenvalues plus 8 angles which parametrize the coset:

$$\frac{O(5)}{O(4)} \times \frac{O(5)'}{O(4)'} \tag{15}$$

We actually see that the counting of parameters in terms of invariant orbits reproduces the counting previously made in terms of normal frame parameters and angles. The above analysis makes a close parallel between BPS states preserving different numbers of supersymmetries with time-like, space-like and light-like vectors in Minkowski space [20].

References

1. For a review, see for instance: M. J. Duff, R. R. Khuri and J. X. Lu, *String solitons*, Phys. Rep. **259** (1995) 213;
   M. J. Duff, Kaluza-Klein theory in perspective, in *Proceedings of the Nobel Symposium Oskar Klein Centenary, Stockholm, September 1994* (World Scientific, 1995), E. Lindstrom editor, hep-th/9410046;
   G. Horowitz, UCSBTH-96-07, gr-qc/9604051;
   J. M. Maldacena, Ph.D. thesis, hep-th/9607235;
   M. Cvetic, UPR-714-T, hep-th/9701152.
2. L. Andrianopoli, R. D'Auria and S. Ferrara, Phys. Lett. **B 403** (1997) 12.
3. L. Andrianopoli, R. D'Auria and S. Ferrara, Phys. Lett. **B 411** (1997) 39.

4. G. Gibbons, in *Unified theories of Elementary Particles. Critical Assessment and Prospects*, Proceedings of the Heisemberg Symposium, München, West Germany, 1981, ed. by P. Breitenlohner and H. P. Dürr, Lecture Notes in Physics Vol. 160 (Springer-Verlag, Berlin, 1982);
G. W. Gibbons and C. M. Hull, Phys. lett. **109B** (1982) 190;
G. W. Gibbons, in *Supersymmetry, Supergravity and Related Topics*, Proceedings of the XVth GIFT International Physics, Girona, Spain, 1984, ed. by F. del Aguila, J. de Azcárraga and L. Ibáñez, (World Scientific, 1995), pag. 147;
R. Kallosh, A. Linde, T. Ortin, A. Peet and A. Van Proeyen, Phys. Rev. **D46** (1992) 5278;
R. Kallosh, T. Ortin and A. Peet, Phys. Rev. **D47** (1993) 5400;
R. Kallosh, Phys. Lett. **B282** (1992) 80;
R. Kallosh and A. Peet, Phys. Rev. **D46** (1992) 5223;
A. Sen, Nucl. Phys. **B440** (1995) 421; Phys. Lett. **B303** (1993) 221; Mod. Phys. Lett. **A10** (1995) 2081;
J. Schwarz and A. Sen, Phys. Lett. **B312** (1993) 105;
M. Cvetic and D. Youm, Phys. Rev. **D53** (1996) 584;
M. Cvetic and A. A. Tseytlin, Phys. Rev. **D53** (1996) 5619;
M. Cvetic and C. M. Hull, Nucl. Phys. **B480** (1996) 296.
5. A. Strominger and C. Vafa, Phys. Lett. **B379** (1996) 99;
C. G. Callan and J. M. Maldacena, Nucl. Phys. **B472** (1996) 591;
G. Horowitz and A. Strominger, Phys. Rev. Lett. **B383** (1996) 2368, hep-th/9602051;
R. Dijkgraaf, E. Verlinde, H. Verlinde, Nucl.Phys. **B486** (1997) 77;
P. M. Kaplan, D. A. Lowe, J. M. Maldacena and A. Strominger, hep-th/9609204;
J. M. Maldacena, hep-th/9611163.
6. L. Susskind, hep-th/9309145;
L. Susskind and J. Uglum, Phys. Rev. **D50** (1994) 2700;
F. Larsen and F. Wilczek, Phys. Lett. B375 (1996) 37.
7. S. Ferrara, R. Kallosh and A. Strominger, Phys. Rev. **D52** (1995) 5412;
A. Strominger, Phys. Lett. **B383** (1996) 39.
8. S. Ferrara and R. Kallosh, Phys. Rev. **D54** (1996) 1514.
9. S. Ferrara and R. Kallosh, Phys. Rev. **D54** (1996) 1525.
10. L. Andrianopoli, R. D'Auria and S. Ferrara, Int. Jour. Mod. Phys. **A** Vol. 12 N. 21 (1997) b3759.
11. L. Andrianopoli, R. D'Auria, S. Ferrara and M.A. Lledo, CERN-TH/98-48, UCLA/98/TEP/4, hep-th/9802147.
12. R. Kallosh and B. Kol, Phys. Rev. **D53** (1996) 5344.

13. S. Ferrara and J.M. Maldacena, hep-th/9706097.
14. V. Balasubramanian, F. Larsen and R. Leigh, hep-th/9704143.
15. R. Khuri and T. Ortin, Phys. Lett. **B 373** (1996) 56.
16. H. Sheinblatt, hep-th/9705054.
17. J. Polchinski, hep-th/9611050.
18. W. Taylor, Nucl. Phys. **B508** (1997) 122.
19. A. Sen, Phys. Rev. Lett. **79** (1997) 1619.
20. S. Ferrara and M. Günaydin, hep-th/9708025.
21. H. Lü, C.N. Pope and K.S. Stelle, hep-th/9708109.
22. M. Berkooz, M.R. Douglas and R.G. Leigh, *Nucl. Phys.* **B 480** (1996) 265.
23. V. Balasubramanian, F. Larsen and R.G. Leigh, hep-th/9704143.
24. H. Lü, C.N. Pope, T.A. Tran and K.W. Xu, hep-th/9708055.
25. L. Andrianopoli, R. D'Auria, S. Ferrara, P. Fré, M. Trigiante, hep-th/9707087, Nucl. Phys. **B 509** (1998) 463.

MICRO STATES OF THE 5-DIMENSIONAL BLACK HOLE

Spenta R. Wadia[a b c]
Theoretical Physics Division, CERN, CH - 1211 Geneva 23, Switzerland

In this contribution I record the contribution of Abdus Salam to the development of some aspects of theoretical high energy physics in India. I also briefly review the hyper-multiplet moduli of the $D_1 - D_5$ brane system which give rise to the effective SCFT describing the micro states of the 5-dim. black hole.

1 The Legacy of Abdus Salam

The name of Abdus Salam is associated with the the Standard Model of elementary particles which is undoubtedly a major scientific achievement of the 20th century. This fact is a source of inspiration and encouragement to scientists from developing countries because Salam grew up in one such country and was able to overcome certain disadvantages of their history. In his writings and speeches Salam has emphasized that the material contrast between the developed and developing countries is a passing phase of history. On many occasions he has reminded us of the periods in hist ory when tables were turned and developing countries like India, Iran, China and others have contributed profoundly to science and technology during earlier periods of history.

Besides his scientific achievement another historic contribution of Salam is the establishment of the ICTP. I will not detail the general contribution of the ICTP to the development of science in the develop ing world except the important supportive role it has played in the development of some aspects of modern theoretical high energy physics (String Theory) in India.

From the beginning of the 2nd String Revolution in 1984, Salam was enthusiastic about this subject especially because it included the possibility of explaining the vexing riddles of quantum gravity. Whenever the subject seemed to stray into more mathematical explorations he would ask, "But, what about gravity ?". We have learnt at this meeting that he asked the same question towards the very end of his life. Salam understood the importance of the string theory developments and he supported this activity with enthusiasm. He was very happy to know that a string theory group was taking shape in India. Once in a while he felt overwhelmed by the developments and worried whether

[a]e-mail: wadia@nxth04.cern.ch, wadia@theory.tifr.res.in
[b]On leave from the Tata Institute of Fundamental Research, Homi Bhabha Road, Mumbai-400005, India
[c]Also at Jawaharlal Nehru Centre for Advanced Scientific Research, Jakkur P.O., Bangalore 560064, India

we could cope with that. Once while I sat next to him in an ICTP seminar he whispered in my ear,"do you know that even Steve Weinberg finds it hard to cope with the subject". There was mischief in his eyes and what he was actually saying (my interpretation) is that if Steve Weinberg cannot cope with String Theory how can you guys!

Many members of our group became regular participants in the ICTP summer school activity and all the members of our group at the TIFR are or have been Associate Members of the ICTP. Not only were they listeners to the work done in the western world but they were also among the lecturers...telling their peers and seniors what they have been doing back home. Our students were able to interface with the best people that came to the ICTP. This helped them find excellent post-doctoral positions in the west. It is hard to quantify the impact of all this on our early efforts.

Salam was very easy to meet and talk to and in spite of his very busy schedule he found time to discuss physics. His interest was keen and genuine. It was always a great pleasure for me to summarize for him the recent developments during my yearly visits to the ICTP. Besides the physics sessions we had he was also interested in discussing politics of the Indian sub-continent with me. Around 1985 some of us were part of the Group for Nuclear Disarmament (GROUND) which was based in Bombay and which was actively involved in the issue of nuclear weapons in the sub-continent. Salam was very enthusiastic about this movement. He too was very concerned about the India-Pakistan nuclear weapons issue and was disturbed by the trends in both countries.

In the passing away of Salam we have lost a great man whose achievement and example will be a beacon to all of us in the years to come.

2 Black Holes

The scientific part of my talk is going to be on some recent developments in black hole physics and string theory. It is unfortunate that Salam did not live to see the derivation of the Bekenstein-Hawking formula from string theory and the related work on black hole thermodynamics. I mention that many years ago Salam had speculated that highly excited reggeons may behave like black holes.

Black holes are a subject of much interest in various areas of astro-physics and physics. There is little doubt among astronomers about black hole candidates in binary systems and there is also evidence that the matter of galaxies is belched in 'active galactic nuclei' which are black holes with a million times the solar mass. Why do black holes occur? The standard answer is gravitational collapse and the existence of black hole solutions in general relativity. The basic characteristic of a black hole within the framework of general relativity is that it has a curvature singularity where all the matter is supposed to have fallen in and that this singularity is 'surrounded' by an event horizon. This is a null surface that divides the space time into two distinct regions and plays a fundamental role in the description of the properties of a black hole.

The general theory of relativity has been very successful as a classical theory, but a consistent quantization is still lacking. Our present understanding indicates that this problem does not have a solution within the framework of local quantum field theory. One may argue that the consistency problem of quantum gravity arises only at the Planck length $l_p^2 = \frac{\hbar G}{c^3}$ (G is Newton's constant and c is the speed of light) and hence one may expect general relativity to be a consistent description for low energy phenomena involving macroscopic objects, like black holes. However this turns out not to be true. When one turns on quantum effects black holes behave like black bodies which absorb and emit radiation with a characteristic temperature determined by the macroscopic parameters of the black hole. Hence it is inevitable that they can radiate and evaporate. Complete black hole evaporation means that the final state of whatever went into a black hole is thermal radiation. Hawking's calculation [1] implies that this radiation is exactly thermal and this is in conflict with unitary evolution of pure states in quantum mechanics. This fact is called the 'information paradox'. Besides the temperature the thermodynamic state of the black hole is also characterized by its 'entropy', given by the Bekenstein-Hawking formula

$$S = \frac{A}{4l_p^2} \tag{1}$$

A refers to the area of the horizon of the black hole. In cgs units $l_p = 10^{-33}$ cms. This is one of the most remarkable formulas of physics because it involves the Planck length at low energies. The formula is of geometrical origin and if one equates it to Boltzmann's formula for the entropy of a macroscopic system then it is natural to ask the question about its microscopic origin.

While graviton loops, propagating in a background Minkowski space time, render quantized gravity inconsistent at Planck distances, the existence of black holes leads to the information paradox on a macroscopic scale, and raises fundamental questions about the true degrees of freedom of quantum gravity. It is well known that string theory resolves the problem of ultra violet divergences. The basic physical reason is that the strings cannot come closer than the string length l_s, which is related to the Planck length (in 4-dims.) as $l_p = gl_s$, where g is the string coupling. In this article we will present recent mounting evidence that string theory has the ingredient to show that the Bekenstein-Hawking formula is a consequence of Boltzman's law [2,3,4,5,6]

$$S = \ln \Omega \tag{2}$$

The number of black hole states Ω at given values of macroscopic parameters are derived from the 'underlying' microscopic degrees of freedom of string theory. The thermodynamics of near extremal black holes can also be derived from the known properties of the highly degenerate black hole wave function using the standard principles of quantum statistical mechanics [7,8,9,10,11,12,13,15]. The issue of directly accessing the black hole micro states by 'external exp eriments' can also be addressed with success. It seems that unitarity in quantum mechanics is not violated at least in the examples amenable to rigorous study. What seems to undergo a revision is our ideas of space time at the microscopic level. The success of string theory in explaining and accounting for the various features of black holes should be considered as providing evidence for the "String Paradigm".

3 Black Hole Thermodynamics

Before we discuss the recent progress on these questions it is worth recalling the main points leading to black hole radiance within the framework of General Relativity. For simplicity of presentation we discuss the Schwarzschild black hole characterized by its mass M. One considers the propagation of matter in the background of a black hole. For large M , the metric in the vicinity of the horizon, in the Kruskal coordinates is (up to a scaling) that of flat Minkowski space time. These coordinates are related to the Schwarzschild coordinates by the well known Rindler transform. Now we know how to do quantum field theory in Minkowski coordinates. The Fock vacuum here is what is usually called the Minkowski vacuum and would be the vacuum of an observer who is falling freely into a black hole. Under the Rindler transform this description gets radically transformed: the Minkowski vacuum appears as a density matrix to the observer who does quantum field theory in Schwarzschild (Rindler) coordinates. This density matrix has the associated Hawking temperature $T = \frac{\hbar c^3}{8\pi MG}$ of the black hole and the spectrum is that of a black body. The appearance of the density matrix stems from the fact that the Hilbert space of the Rindler observer is a direct sum of Hilbert spaces corresponding to observables 'inside' and 'outside' the horizon. Since the observables inside the horizon are not accessible to the Rindler observer, one has to perform a trace over them. As yet one cannot say that there is information loss because the thermal spectrum arises from an average that the Rindler observer performs. The problem arises when the black hole reaches the end point of evaporation, i.e. when most of its mass has been radiated away and the end point is a purely thermal state. In such a situation one cannot even in principle recover the correlations present in the imploding pure state that created the black hole. This is the information paradox: a pure quantum mechanical initial state ends in a thermal (mixed) state.

We note that in the case of the Schwarzschild black hole the area of the horizon in (1) is given by $A = 16\pi M^2$. This plus the formula for the Hawking temperature leads to the 1st law of black hole thermodynamics.

Hawking asserts that the black hole thermodynamics and information loss is intrinsic to general relativity because of the focusing properties of the gravitational field. It is this contention that has been questioned most vehemently by 't Hooft [16] and Susskind [2] and many other physicists who believe that black hole thermodynamics must have a statistical basis in a unitary quantum theory, just like the standard thermodynamics of a macroscopic system.

As we have already mentioned in this article we will review some evidence in favor of the latter view point.

4 The 5-dim. Black Hole of IIb String Theory

We will base our discussion on a black hole which can be analyzed using present techniques. This black hole occurs as a 5-dim. solution of the supergravity theory which describes the low energy limit of type IIb string theory. We shall denote by (x_1, x_2, x_3, x_4) the 4 non-compact dimensions of space and by $(x_5, x_6, x_7, x_8, x_9)$ the compact dimensions which form a 5-torus. The 5th circle has a radius R and the 4-torus has a volume V. Here we choose $R \gg V^{1/4} \sim l_s$. The theory has various massless fields, but for our purpose we consider only the metric, the dilaton, and the anti-symmetric 3-form which gives rise to 'electric' and 'magnetic' charges Q_1 and Q_5 respectively in 4 space dimensions. The (t, x_5) component of the metric gives rise to a Kaluza-Klein charge N. The dilaton field is a finite constant at the horizon.

The black hole metric in 5-dims. is given by

$$ds_5^2 = -f(r)^{-2/3}\left(1 - \frac{r_0^2}{r^2}\right) dt^2 + f(r)^{1/3}\left[\left(1 - \frac{r_0^2}{r^2}\right)^{-1} dr^2 + r^2 d\Omega_3^2\right] \tag{3}$$

where

$$f = \left(1 + \frac{r_0^2 \sinh^2\alpha}{r^2}\right)\left(1 + \frac{r_0^2 \sinh^2\gamma}{r^2}\right)\left(1 + \frac{r_0^2 \sinh^2\sigma}{r^2}\right) \tag{4}$$

Note that in 5 space time dims., the above represents a black hole with the inner horizon at $r = 0$, and the outer horizon at $r = r_0$. Note that the location of the inner and outer horizons is a coordinate dependent statement and it is possible to choose coordinates so that the outer horizon is located at $r = A^{1/3}$.

The charges Q_1, Q_5, N are given by the formulae

$$Q_1 = \frac{Vr_0^2}{2g} \sinh 2\alpha \tag{5}$$

$$Q_5 = \frac{r_0^2}{2g} \sinh 2\gamma \tag{6}$$

$$N = \frac{R^2 V r_0^2}{2g^2} \sinh 2\sigma \tag{7}$$

In order to have a meaningful semi-classical limit where the string couplig $\rightarrow 0$, the charges Q_1, Q_5 and N have to be chosen large holding gQ_1, gQ_5 and $g^2 N$ fixed. This also implies that the area of the horizon is large compared to the string scale l_s as it should be for a black hole. Let us now present some relevant formulae of black hole thermodynamics for the above black hole. The energy, entropy and the temperature are given by

$$E = \frac{RVr_0^2}{2g^2} (\cosh 2\alpha + \cosh 2\gamma + \cosh 2\sigma) \tag{8}$$

$$S = \frac{2\pi RVr_0^3}{g^2} \cosh\alpha \cosh\gamma \cosh\sigma \tag{9}$$

$$T^{-1} = 2\pi r_0 \cosh\alpha \cosh\gamma \cosh\sigma \tag{10}$$

We note that in the limit $\sigma \rightarrow \infty$ and $r_0 \rightarrow 0$, holding $r_0 \exp\sigma$ fixed, we have a BPS solution which corresponds to an extremal black hole with $T = 0$ but a large non-zero entropy $S = \sqrt{Q_1 Q_5 N}$. We are interested in the case with a slight deviation from extremality, when the temperature is tiny but non-zero and the specific heat at fixed values of the charges is positive. In this limit we can present a sound understanding of black hole thermodynamics in terms of string theory. This circumstance is in contrast to the situation of the Schwarzschild black hole which has negative specific heat so that the black hole becomes hotter as it radiates. However in spite of the difference the near extremal black hole also presents the information paradox as it radiates towards its ground state.

Our strategy is to give a microscopic model of the above black hole. The situation is very much like giving a microscopic basis to the soliton of the chiral model of pions. We of-course know that the soliton solution of the weakly coupled chiral model is the baryon of $SU(N)$ QCD and it is composed of N quarks. In the chiral model the baryon number appears as a topological invariant in terms of the chiral field while in QCD the baryon number is a counting problem. One simply assigns a baryon charge $\frac{1}{N}$ to the quarks. If the black hole entropy is indeed given by Boltzmann's law then it is natural to ask about the string theory analouges of the quarks of QCD.

In the next section we will introduce D-branes which are special soliton solutions of string theory. The main hypothesis is that these solitons are the constituents of the black holes. Just as the quarks of QCD interact via gluons, the D-brane interactions are mediated via open strings which end on them. It is worth emphasizing that these degrees of freedom are not evident in the standard formulation of general relativity in much the same way that in the chiral model of pions it is hard to see any trace of the quark building blocks.

5 String Theory Solitons: D-branes

D-branes are like the heavy quarks of string theory [17]. Except that unlike the quarks of QCD they appear as soliton solutions of string theory. The mass of a D-brane is $M \sim \frac{\alpha\prime}{g}$, where g is the string coupling. For small g they are indeed very heavy objects. Their gravitational field is given by $MG \sim g$ because Newton's constant scales as $G \sim g^2$. D-branes also carry one unit of Ramond charge. It should be noted that perturbative excitations of string theory have zero Ramond charge. In that sense Ramond charge is a soliton charge. In theories with more than one supersymmetry these solitons are stable because they saturate the so called BPS bound, which essentially says that their mass, in appropriate units is equal to their Ramond charge.

Another feature of D-branes is that these solitons are not necessarily point like. Their world volume can be anywhere between zero to nine dimensional. D-branes are generalized domain walls and their precise characterization is in terms of Dirichlet boundary conditions imposed on the directions of the open strings which end on them. This also explains why they are called D-branes. The other important feature is that a brane configuration is invariant under only half of the supersymmetries of the bulk theory.

6 Effective Gauge Theory

For a review of the initial material in this section see [18]. The type IIb string theory in which (3) is a solution has D_1 branes and D_5 branes as solitons. The D_1 branes are wrapped around the 5th circle and the D_5 branes are wrapped around the 5-torus in the directions 5,6,7,8 and 9. Note that the 5th direction is common to both the branes. There are Q_1 D_1 branes and Q_5 D_5 branes. There are interactions between these branes which are mediated by open strings that end on the branes. It is one of the most remarkable facts, that properties of branes which are near each other (or are sitting on top of each other) are described by the massless modes of the open string theory, namely gauge theories. Let us enumerate the relevant massless fields for the above configuration of branes. Clearly we have 3 types of open massless strings depending upon which brane they end on: 1) (1,1) strings with both ends on a D_1 brane. There are $4Q_1^2$ massless components which arrange themselves as 4 $Q_1 \times Q_5$ matrices. The 4 refers to the directions (6,7,8,9) which lie in the 5 branes. We have set the strings along the non-compact space directions to zero. Since these branes are indistinguishable we can rotate the matrix by an element of the unitary group $U(Q_1)$. 2) (5,5) strings with both ends on a D_5 brane. These massless strings arrange themselves once again into 4 $Q_5 \times Q_5$ matrices with $U(Q_5)$ as the symmetry group. 3) The third set of massless strings (1,5) and (5,1) have one end on a D_1 brane and another on a D_5 brane. They form $4Q_1Q_5$ massless modes with a $U(Q_1) \times U(Q_5)$ symmetry group. It turns out that the interacting system of branes is a $U(Q_1) \times U(Q_5)$ non-abelian gauge theory in the 2-dims. (x_5,t) with N=(4,4) supersymmetry. The restriction to 2-dims. roughly speaking comes about because x_5 is the only direction common to the 2 sets of branes. Why N=(4,4) supersymmetry? This is because half the 32 supersymmetries of the bulk IIb theory are broken by the set of 5-branes and the other half by the set of 1-branes. This leaves us with 8 supersymmetries whose charges are 4-left movers and 4-right movers in the 2-dim. field theory.

The analysis of the low lying modes of the above gauge theory is quite involved and it was only recently completed [15]. We shall not discuss it here in any detail, except summarize the result. The modes of the gauge theory which have long wavelengths from the view point of the bulk theory are described by a N=(4,4) super-conformal field theory (SCFT) with central charge c=6. The bosonic co-ordinates of the SCFT Y^m form a vector representation of $SO(4)_I$, which is the tangent space group of the internal directions (6,7,8,9). Their 4-fermionic partners are also spinors of $SO(4)_E$, which is the rotational symmetry of the 4 external space dimensions. The bosonic coordinates are scalars under $SO(4)_E$ and the fermionic coordinates are scalars under $SO(4)_I$. An important feature of this SCFT is that its fields are fractionally moded w.r.t. the original 5th circle of radius R and the min. momentum is given by $\frac{1}{RQ_1Q_5}$ which is vanishingly small in the limit when the charges are very large. This fact is consistent with the fact that the absorption cross section at zero wave number is non-zero and equal the area of the horizon of the blac k hole. One of the most important aspects of this story is that as long as we are dealing with low energy phenomena the above SCFT is independent of the strength of the coupling constant of the gauge and hence the open string theory. This is one of the miracles of supersymmetric field theory with N=(4,4) supersymmetry.

Let us now discuss the wave function of the extremal black hole. The states can be easily constructed within the sub-space of the total fock space of the SCFT. This sub-space is defined by the requirement that the black hole state is either extremal (BPS) or atmost near extremal. As we had noted before the extremal black hole which is also a BPS state is characterized by the Kaluza-Klein momentum $P = \frac{N}{R}$. If we identify this momentum with $L_0 - \overline{L_0}$ which generates rotations along x_5 in the SCFT then we get

$$P = \frac{NQ_1Q_5}{RQ_1Q_5} \tag{11}$$

The BPS condition implies that $\overline{L_0} = H - P = 0$, and hence we get the level of the SCFT to be $L_0 = NQ_1Q_5$. This means that the ground state of the black hole is characterized by all the degenerate states of the SCFT which are at this level. There is a standard formula for the number of states at a certain level n in a SCFT of central charge c, provided n is very large.

$$\Omega \sim \exp 2\pi\sqrt{cn/6} \tag{12}$$

Substituting $n = NQ_1Q_5$ and $c = 6$ in (12), reproduces the Bekenstein-Hawking formula, viz. $S = 2\pi\sqrt{Q_1Q_5N}$

In order to describe a near extremal black hole we relax the BPS condition $\overline{L_0} = 0$, and choose the conditions

$$L_0 = Q_1Q_5(N + n_l) \tag{13}$$

$$\overline{L_0} = Q_1Q_5n_l \tag{14}$$

$n_l \ll N$ and can be fractional. The above condition ensures that the deviation from the BPS state is small and the condition $P = \frac{N}{R}$ is maintained. The entropy of the above configuration turns out to be

$$S = 2\pi\sqrt{Q_1Q_5}(\sqrt{N + n_l} + \sqrt{n_l}) \tag{15}$$

This formula also exactly matches the entropy deduced from the near extremal black hole solution (3).

7 Black Hole Thermodynamics

Now that we have a precise understanding of the micro states of a near extremal black hole, we can address the important question of Hawking radiation. In order to discuss that we need to know 2 things. Firstly we have to know whether the states of the SC FT can be treated as a thermal ensemble. In our discussion we will simply assume this. This is not really a serious assumption because we are dealing with a system with very large degrees of freedom. Secondly we need to know how the modes of the closed string couple to the SCFT. Since the latter has fractional moding on a circle of radius R, at low energies, only the the zero modes of the closed field w.r.t the 5th dim. have a non-zero coupling. Further this coupling is located at the origin of the 4-dim. space, $x_i = 0$, $i = 1, 2, 3, 4$. The next question is about the derivation of the effective coupling. Here we appeal to symmetry principles, (like in the discussion of the effective pion n ucleon coupling in QCD) which in our problem are the global $SO(4)_I$ and $SO(4)_E$ symmetries. As an example one can write down the coupling of the closed string states which correspond to deformations of the internal torus. These are scalar particles in 5-dims. denoted by the field h_{mn},

$$T_{eff} \int dt \int_0^{2\pi R\tilde{Q}_1\tilde{Q}_5} d\tilde{\sigma}(\delta_{mn} + h_{mn}(t))\partial_\alpha \tilde{Y}^m \partial^\alpha \tilde{Y}^n . \tag{16}$$

$T_{eff} = \frac{1}{\alpha'^2 g}\sqrt{\frac{V_4}{Q_1 Q_5}}$, is the effective string tension. Note that it is very small compared to the tension of the original open strings that mediated the interactions between the branes. A very important point that is implicit in the above interaction is the fact that ultra high energy processes in the brane effective theory correspond to low energy processes in the transverse space, which is the space time of the black hole.

Using the above interaction one can calculate the S-matrix element in first order perturbation theory between initial and final states of the black hole constructed out of the states of the SCFT. After this straightforward calculation of the S-matrix, between black hole micro-states we make the statistical assumption that that the black hole is described by a density matrix

$$\rho = \frac{1}{\Omega}\sum_{\{i\}} |i\rangle\langle i| \tag{17}$$

where the sum $\{i\}$ is over all possible distributions keeping N_L and N_R fixed. It is this formula that leads to the entropy 15, $S = -Tr\rho\ln\rho = -\sum_{\{i\}}\frac{1}{\Omega}\ln\frac{1}{\Omega} = \ln\Omega$.

Density matrices like the one described above are not unfamiliar in particle physics. They arise, e.g. in calculating the decay rate of an unpolarized particle into unpolarized products. As there, in the present case also, the total "unpolarized" transition probability is given by

$$P_{\text{decay}}(i \rightarrow f) = \frac{1}{\Omega} \sum_{\{i\},\{f\}} |\langle f|S|i\rangle|^2 \tag{18}$$

The division by Ω represents averaging over initial states, while the final states are simply summed over. The passage to the decay rate $d\Gamma$ is usual and one can be easily derived. The relevant formulae are

$$\Gamma_H = \sigma_{abs}\rho(\frac{w}{T_H})^{-1}\frac{d^4k}{(2\pi)^4} \tag{19}$$

where σ_{abs} is the absorption cross section given by

$$\sigma_{abs} = \frac{g^2 Q_1 Q_5 \pi^3}{V} w\rho(\frac{w}{T_H})\rho(\frac{w}{T_L})^{-1}\rho(\frac{w}{T_R})^{-1} \tag{20}$$

and $\rho(x) = (\exp x - 1)^{-1}$ is the Bose factor. $T_L = \frac{r_0 V \exp\frac{\sigma}{2}}{\pi g^2 Q_1 Q_5}$ and $T_R = \frac{r_0 V \exp -\frac{\sigma}{2}}{\pi g^2 Q_1 Q_5}$ are the temperatures of the left and right moving excitations on the effective string and T_H is the Hawking temperature given by $2T_H^{-1} = T_R^{-1} + T_L^{-1}$. This precisely agrees with the same quantities derived using general relativity!

8 Concluding Remarks

It seems clear from the above sketch that the case of the 5-dim. near extremal black hole of IIb string theory is amenable to a rigorous treatment in terms of the constituent brane model. The low energy processes calculated using his model coincide with the calculations of general relativity. The constituent model produced the answers by calculating the square of the S-matrix between highly degenerate states of the SCFT. After this we assumed the representation of the black hole in terms of a density matrix and calculated the absorption cross section and the corresponding decay rate. This is exactly as in the thermodynamic treatment of a macroscopic body and hence we conclude that within the parameters of the above calculations not only do we have a microscopic model of the black hole degrees of freedom that explain the Bekenstein-Hawking formula, but a resolution of the information paradox itself.

One of the most important issues for future work is the bearing or the imprint of the black hole micro states on the geometry of the space transverse to the brane system. In order to understand this issue we have to probe the effective 'geometry' felt by external probes. The other issue is that of being able to observe the micro states (hair) by performing suitable experiments from outside the black hole. The answer to both these questions will point to a new understanding of space time which would be a d erived concept rather than an ab initio arena to describe physics.

Acknowledgments

I would like to thank Fawad Hassan and Gautam Mandal for many useful discussions.

Refernences

1. S. Hawking and R. Penrose, *The Nature of Space and Time*, Princeton Univ. Press, (1996); Sci. Am. **275** (1996) 44.
2. L. Susskind, *Some Speculations about Black Hole Entropy in String Theory*, hep-th/9309145.
3. A. Sen, *Extremal Black Holes and Elementary String States*, hep-th/9504147.
4. A. Strominger and C. Vafa, Phys. Lett. **B379** (1996) 99, hep-th/9601029.
5. C. V. Johnson, R. R. Khuri and R. C. Myers, Phys. Lett. **B378** (1996) 78, hep-th/9603061.
6. R. Breckenridge, D. Lowe, R. Meyers, A. Peet, A. Strominger and C. Vafa, Phys. Lett. **B381** (1996) 423, hep-th/9603078;
   R. Breckenridge, R. Meyers, A. Peet and C. Vafa, Phys. Lett. **B391** (1997) 93, hep-th/9602065.
7. C. Callan and J. Maldacena, Nucl. Phys. **B475** (1996) 645, hep-th/9602043.
8. J. Maldacena and L. Susskind, Nucl. Phys. **B475** (1996) 679, hep-th/9604042;
9. A. Dhar, G. Mandal and S. R. Wadia, Phys. Lett. **B388** (1996) 51, hep-th/9605234.
10. S. R. Das and S. D. Mathur, Nucl. Phys. **B482** (1996)153, hep-th/9607149.
11. J. Maldacena and A. Strominger, *Black Hole Grey Body Factors and D-brane Spectroscopy*, hep-th/9609026; *Universal Low-Energy Dynamics*

for Rotating Black Holes, hep-th/9702015.

12. C. Callan, S. Gubser, I. Klebanov and A. Tseytlin, *Absorption of Fixed Scalars and the D-brane Approach to Black Holes*, hep-th/9610172.
13. I. Klebanov and M. Krasnitz, *Fixed Scalar Grey Body Factors in Five and Four Dimensions* hep-th/9612051; *Testing Effective String Models of Black Holes with Fixed Scalars* htp-th/9703216.
14. S. F. Hassan and S. R. Wadia, Phys. Lett. **B402** (1997) 43, hep-th/9703163.
15. S. F. Hassan and S. R. Wadia, *Gauge Theory Description of D-brane Black Holes: Emergence of the Effective SCFT and Hawking Radiation*, hep-th/9712213 (to appear in Nucl. Phys. B).
16. G. 'tHooft, *The Scattering matrix Approach for the Quantum Black Hole*, gr-qc/9607022.
17. J. Polchinski, *TASI Lectures on D-branes*, hep-th/9611050.
18. J. Maldacena, *Black Holes in String Theory*, Ph.D. Thesis, hep-th/9607235.

BOUND STATES OF BRANES

K. S. NARAIN
The Abdus Salam International Centre for Theoretical Physics,
Trieste, Italy

We discuss the dualities between various superstring theories and show that the bound states of extended solitonic objects (branes) play a crucial role in providing non-trivial checks on the idea of duality. In particular we discuss the formation of bound states of p- and $(p+2)$- Dirichlet branes via condensation of tachyonic modes.

I would like to thank the organizers for inviting me to speak at this very special occasion. Professor Salam's contributions to physics and in establishing the ICTP are well known and many other speakers have already talked about it. I have had the fortune to work at ICTP, first as a post-doc and later as a permanent staff member during Salam's period so I would like to say a few words about this experience. Although I never collaborated with him, his enthusiasm and encouragement for any new ideas in physics was a great inspiration to all of us around him. The problem of gravity and unification of all forces had occupied him for a long time and indeed he himself had contributed seminally in this area. With the eighties revolution of string theory, he felt strongly that string theory was the right direction towards a solution of this problem and he encouraged all of us to work in this field. Unfortunately during the last few years of his life, his health had started deteriorating and it became increasingly difficult for him to keep up with the developments. But it was not uncommon for him to stop one of us in the corridors and ask about the latest developments in the field and when he felt excited about some new result he would ask the person to go to his office and explain the work in greater detail. I can only imagine his contagious excitement and enthusiasm, had he been active, at these new times when we are acquiring some understanding of non-perturbative string theory. In this talk I would like to describe one of the aspects of these new developments. The talk will be very qualitative. For quantitative details the reader is recommended to look at the references given at the end.

In the eighties, we knew of 5 different superstring theories in 10-dimensions. These were type IIA and IIB having N=2 supersymmetry (non-chiral and chiral respectively) and 3 theories with N=1 supersymmetry, type I with $SO(32)$ gauge group, and heterotic theories with gauge group $SO(32)$ and $E_8 \times E_8$ respectively. Some of these theories were related to each other upon compactification. For example compactifying on a circle IIA and IIB were realized to be T-dual (radius of the circle going to inverse radius) to each other and similarly the two heterotic theories were T-dual to each other. However the type II theories, type I theory and the heterotic theories had no relation to each other. Indeed the perturbative spectrum of these theories is completely different.

Take for instance, type I and heterotic $SO(32)$ theories. Their massless spectrum in 10-dimensions is identical and so is the two derivative effective action of these massless fields upto field redefinition. This was to be expected since the N=1 supergravity theory in 10-dim with $SO(32)$ gauge group, which is the effective low energy theory of massless modes is unique. To be explicit, the bosonic massless modes are the graviton $G_{\mu\nu}$, the antisymmetric tensor field $B_{\mu\nu}$, the dilaton ϕ and the $SO(32)$ gauge fields A_μ. The field redefinitions which map the two effective actions are:

$$G_h = e^{-\phi_I} G_I, \qquad B_h = B_I, \qquad A_h = A_I, \qquad \phi_h = -\phi_I. \tag{1}$$

where we have suppressed the μ, ν indices and the subscripts h and I refer to fields appearing in heterotic and type I respectively. Recall that dilaton plays a special role in string theory; the expectation value of the dilaton determines the string coupling constant via $g_{st} = e^{<\phi>}$ in terms of which string perturbation theory is organized. Therefore the above field redefinition maps strong coupling of heterotic theory to weak coupling of type I and vice versa.

However, the similarity ends here. The perturbative massive string spectrum in the two theories is completely different. This, however, is not a serious problem, because the field redefinitions above relate strongly coupled heterotic theory to weakly coupled type I and vice versa. One could argue that the massive modes are not stable and as the coupling constant increases they decay so fast they could no longer be thought of as particle states. The difficulty with this, however, is that there are some states that are protected by supersymmetry. These are the so-called BPS states which form short multiplets of supersymmetry algebra and therefore are protected even at the non-perturbative level. These states carry central charge appearing in the supersymmetry algebra and their masses are given by the central charge.

In particular, if one compactifies each of these theories on a circle, one can break the gauge group to its maximal abelian subgroup by giving expectation value to Wilson lines. The total number of $U(1)$'s appearing are 18 where 16 come from the Cartan subalgebra of $SO(32)$ and 2 come from the Kaluza-Klein reduction of G and B fields. The perturbative string states include also the charged states. On the heterotic side, these 18 charges span an 18-dimensional lattice $\Gamma_{1,17}$ which is even and self dual (w.r.t. metric with 1 positive and 17 negative signatures). For each of these charge vectors $Q \in \Gamma_{1,17}$ subject to the condition $Q^2 \geq -2$ there exist BPS states with a definite Q^2 dependent multiplicity.

On the other hand in type I these charges are completely different. In particular there is no state which is charged with respect to $U(1)$ coming from the reduction of B field. The reason is easy to understand. The B field on the heterotic side couples naturally to the closed string and the states that are charged with respect to the corresponding $U(1)$ are the winding modes of heterotic string when it winds around the circle. On the other hand in type I, the B field comes from Ramond-Ramond sector which does not couple to the fundamental string. Indeed the open string and the unoriented closed string that enter in type I do not have any winding number associated to them. In the heterotic side these winding modes have a mass (in string units) proportional to the radius R_h of the circle. By using the relations above it is easy to see that this mass in type I string units is proportional to R_I/g^I_{st}. Since their masses are non analytic in type I string coupling constant they cannot be visible in perturbation theory. Thus if these states appear in type I theory they are necessarily non- perturbative. In order to establish duality, one must therefore prove that such non-perturbative (solitonic) states exist on the type I side with right multiplicities.

In fact one can construct classical BPS solutions that carry charges with respect to Ramond-Ramond fields (in the present case B field) and their tensions go as $1/g_{st}$. Such objects carrying Ramond-Ramond charges admit an exact conformal field theory description in terms of open strings obeying certain Dirichlet boundary conditions [1] and hence are known as D-branes. To compute the multiplicities of states carrying higher units of these charges, one needs to quantize the zero modes of these solutions and count the number of normalizable wave functions giving rise to bound states of D-branes. Counting the number of such objects therefore constitutes a very stringent test of strong-weak duality. In the present context one needs to compute the number of bound states of D-strings wound around the circle and carrying arbitrary charges w.r.t. other 17 $U(1)$ gauge fields. This has been done recently by studying the 2-dim. effective theory of zero-modes of a system of n D-strings which turns out to be an (8,0) supersymmetric $O(n)$ gauge theory. With a very plausible assumption, that in the infrared limit, this theory flows to an orbifold conformal field theory, one can show that the resulting number of bound states agrees with the multiplicities on the heterotic side.[2,3]

A similar situation arises in the conjectured duality between heterotic theory compactified on T^4 and IIA theory compactified on $K3$ surface. $K3$ breaks half the supersymmetry so that in 6-dim. both sides have the same number of supersymmetries namely (1,1). Furthermore the number of vector multiplets in both sides turns out to be 20. Again the uniqueness of the resulting low energy supergravity theory in 6-dim. implies that equations of motion on the two sides must agree upto field redefinitions. Once again this redefinition involves inversion of the string coupling constant. By comparing the BPS charged states on the two sides one finds that while the heterotic charges are labelled by vectors of an even self dual lattice $\Gamma_{4,20}$, on the type IIA side since the gauge fields come from Ramond-Ramond sector the charged states are bound states of D-branes wrapped on various non-trivial cycles of $K3$. In a very interesting work,[4] Bershadsky et al counted the number of such states and showed that it agrees with that on heterotic side. These checks have been particularly important in establishing the relation between type II and heterotic theory.

In this talk, I will consider a very simple example of the D-brane bound state problem which enters in establishing another strong-weak coupling duality, namely the $SL(2,Z)$ duality of type IIB in 10-dim. The bosonic fields of IIB are

$$\begin{aligned} \mathrm{NS-NS}: &\quad G_{\mu\nu}, B_{\mu\nu}, \phi \\ \mathrm{R-R}: &\quad \chi, B'_{\mu\nu}, D^{(4)}_{+} \end{aligned}$$

where χ, B' and $D^{(4)}_{+}$ are 0-, 2-, and self-dual 4-forms respectively. Combining the 2 scalar fields into one complex field $\tau = \chi + ie^{-\phi}$ one finds that the equations of motion at the classical level have an $SL(2,R)$ symmetry under which the two 2-forms B and B' transform in fundamental representation and $\tau \to \frac{a\tau+b}{c\tau+d}$, where a, b, c, d are the entries of the $SL(2,R)$ matrices. This continuous classical symmetry is expected to break to, at most, a discrete subgroup at the full quantum level due to charge quantizations. The conjecture is that $SL(2,Z)$ subgroup remains the symmetry of the quantum theory. One of the elements of $SL(2,Z)$ is the inversion $\tau \to -1/\tau$ and simultaneous exchange of B and B'. This is therefore again a strong-weak coupling duality which exchanges the fundamental IIB string which couples to B with the D-string which couples to R-R B' field. More generally an arbitrary $SL(2,Z)$ element transforms the fundamental string to a (m,n) string (with m and n co-prime) which has m units of B charge and n units of B' charge. Furthermore the tension of the (m,n) string goes as $\sqrt{m^2 + \frac{n^2}{g_{st}^2}}$. Triangular inequality implies that this system of m fundamental strings and n D-strings should form a true bound state with non-zero binding energy.

The above question was answered by Witten,[5] who studied the 2-dim effective theory of zero modes of a system of n D-strings which turns out to be a (8,8) supersymmetric $U(n)$ gauge theory. The m units of the fundamental string charge (for $m < n$) are excited by turning on an electric flux corresponding to a source in m-th antisymmetric tensor representation of $U(n)$. He showed that the resulting system had a unique supersymmetric ground state in which $U(n)$ is broken down to $U(1)$ and that its tension is indeed given by the formula expected from $SL(2,Z)$ invariance.

In the following we describe a variant of the above problem which is related to it by various T-duality and $SL(2,Z)$ duality. Consider compactifying the above system on a transverse T^2. Now by doing 2 T-dualities on the two directions of the T^2 and noting that the B' field which is transversal to T^2 under these T-dualities is mapped to $D_+^{(4)}$ with two of its indices along T^2 while the transverse B field remains unchanged, we conclude that the above system of m fundamental strings and n D-strings is mapped to a system of m fundamental strings and n D-3-branes which are wrapped on the T^2 (note that D-3-branes carry the charge w.r.t. to the 4-form $D_+^{(4)}$). Now if one transforms by the inversion element of $SL(2,Z)$ (called S-duality) the m fundamental strings are mapped to m D-strings while the D-3-branes remain unaffected as $D_+^{(4)}$ is invariant under $SL(2,Z)$. Thus by this sequence of operations we have transformed the original system to that of m D-strings and n D-3-branes. More generally by compactifying further on various tori and doing appropriate T-dualities this system is equivalent to a system of m D-p-branes and n D-(p+2)-branes. The advantage of this description is that one knows the effective field theory that describes this system and which has explicitly the degrees of freedom corresponding to positions of all the branes and therefore one can try to understand more dynamically the mechanism responsible for forming the bound state. The tension of this bound state is proportional to $\frac{1}{g_{st}}\sqrt{n^2V^2+m^2}$ where V is the volume of T^2 (after performing the T-dualities).

In ref 1, Polchinski proposed a CFT description of D-branes that described its interactions with other D-branes as well as with the bulk fields that arise from the perturbative states of the type II theories (such as gravitons, R-R forms etc.) in the weak coupling limit. In this limit the D-branes are infinitely massive and one can take them to be at rest. The interaction of D-branes then is described by open strings that end on the D-branes. The low energy dynamics of branes is governed by the lowest modes of such open strings. Now the lowest modes of open strings describe N=1 Yang-Mills theory in 10-dimensions. Since the ends of the open string in the present case are constrained to lie on the brane world volume, the resulting low energy theory is obtained by dimensionally reducing the 10-dimensional N=1 Yang-Mills theory to the world volume of the brane. If we have m number of p-branes, then considering the oriented open strings ending on different p-branes one gets a $U(m)$ gauge theory with 16 supercharges. Similarly for n $(p+2)$-branes one has a $U(n)$ gauge theory with 16 supercharges.

The instability of this system is seen by considering the open string whose one end is on one of the p-branes and the other on one of the $(p+2)$-branes. This open string has mixed boundary conditions for two spatial directions which are part of the $(p+2)$-brane world volume but not that of the p-brane. In these two directions one end of the open string obeys Neumann boundary condition while the other end obeys Dirichlet boundary condition. By writing down the mode expansion for these two directions, subject to the above boundary conditions, one finds that the oscillator modes are half-integer moded. The resulting zero point shift can be calculated easily and one finds that the supersymmetry is broken and the lowest mode of such a string (when the p-brane is lying inside the $(p+2)$-brane world volume) is tachyonic with a mass square proportional to $-\frac{\alpha'}{4}$, where α' is the string tension.

The presence of a tachyon basically signals the instability of the vacuum. If we denote by Φ the tachyon field (this transforms as the (m,n) representation of $U(m)\times U(n)$ gauge group in the $(p+1)$ dimensional world volume of the p-brane), then its potential energy starts off as mass square times $|\Phi|^2$, which is negative. To minimize the energy, Φ must start acquiring a non-zero expectation value. The actual minimum however will depend on the higher order terms in the potential. The fact that there is a BPS bound on the tension of the system under consideration, shows that there should be indeed a minimum of the potential where the tachyon has a finite non-zero expectation value. Moreover, since the tachyon transforms as the fundamental representation of $U(m)\times U(n)$, this expectation value breaks the gauge group all the way down to the center of mass $U(1)$ under which the tachyon is neutral.

This is the basic mechanism for forming this bound state. However if one wants to do a quantitative check, which means finding the minimum of the potential, one needs to know the details of the higher order terms in the potential. For the system described above, the tachyon mass square is of order $1/\alpha'$ and by a simple dimensional analysis one can show that all higher powers of Φ in the potential contribute at the same order in $1/\alpha'$. However one can consider a modified system where one starts with a bound state carrying 1 unit of p-brane charge and 1 unit of $(p+2)$-brane charge and bring another $(p+2)$-brane close to it. The former bound state can be obtained by turning on a minimal quantized magnetic flux in the $(p+2)$-brane along the T^2. The lowest mode of the open string stretched between this bound state and another $(p+2)$-brane is again tachyonic but now the tachyon mass square is of order $1/V$ where V is the volume of T^2 and therefore the expectation value of Φ at the minimum is of order $1/\sqrt{V}$. One can then set up a reliable perturbation scheme in $1/V$. In ref 6 these computations have been done for various situations where there is a perturbation parameter and the minimum of the potential gives the BPS bound.

The calculation described above is a check on the $SL(2, Z)$ duality of type IIB theory, which moreover identifies the dynamical mechanism responsible for the formation of the relevant bound states. One can however turn this check to conclude some interesting facts about boundary conformal field theories. As mentioned above the CFT description of D-branes is given in terms of open strings that end on the D-branes. The tree level physics of open strings is given by a conformal field theory on the string world sheet which is a disc. The tachyon operator being an open string state is a boundary conformal operator. Furthermore since it is tachyonic it is a relevant operator. Giving expectation value to the tachyon therefore corresponds to turning on this relevant operator and the existence of a minimum of the potential implies the existence of a new non-trivial fixed point. Since the tachyon operators correspond to open strings with ends on the p and $(p+2)$ branes, they appear with definite Chan-Paton indices. This implies that the boundary relevant operators are ordered. In fact one can see easily that this ordering corresponds to the algebra of simple roots of A_n Dynkin diagrams. Thus the boundary CFT (more precisely super conformal field theory) problem at hand is a generalization of Kondo type problem. The existence of bound states implies the existence of non-trivial fixed points in this generalized class of Kondo like problem and the BPS mass of the bound states gives a precise prediction on the disc partition functions (known as G-functions) at the fixed points.

References

1. J. Polchinski, *Phys. Rev. Lett.* **75**, (1995) 4724.
2. C. Bachas, C. Fabre, E. Kiritsis, N.A. Obers and P. Vanhove, *Nucl. Phys.* **B509**, (1998) 33; and C. Bachas and P. Vanhove, "On the rules of D-brane Calculus", Ecole Polytechnique preprint, CPTH-S587.1297.
3. E. Gava, F. Morales, K.S. Narain and G.Thompson, "Bound States of Type I D-strings", hep-th/9801128.
4. M. Bershadsky, V. Sadov and C. Vafa, *Nucl. Phys.* **B463**, (1996) 420.
5. E. Witten, *Nucl. Phys.* **B460**, (1996) 335.
6. E. Gava, K.S. Narain and M.H. Sarmadi, *Nucl. Phys.* **B504**, (1997) 214.

Minimal Cycles, Black-Holes and QFT's

Cumrun Vafa
Lyman Laboratory of Physics, Harvard University
Cambridge, MA 02138, USA

We review some aspects of the geometry of minimal cycles in string compactifications and their applications in constructing new critical theories in six and lower dimensions as well as in accounting for black-hole entropy.

1 Some Reflections on Abdus Salam

Before I start with the scientific aspect of this note, I would like to mention a few words about my impressions of Abdus Salam and his contributions to physics and to the advancement of science in developing countries.

Science and scientific tradition has a long history. At different times different countries have led the cause of advancing human knowledge. Many of the countries that today are called "developing countries" or "third world countries" have at an earlier time been at the forefront of knowledge and have made important contributions to science. Indeed the present scientific tradition is based on the foundations set by these earlier advances. The very fact that science has developed in this way also clearly demonstrates that science does not belong to a restricted group of people. It belongs to all humanity. Thus all humanity and in particular scientists share a responsibility of making pursuit of science as much of a global, borderless endeaver, as they can.

In the present century if one is to name a single person who has contributed most to the development of science in the developing countries the natural choice would be Abdus Salam. With the creation of International Center for Theoretical Physics he has contributed tremendously to making modern science accessible to a much larger audience of humanity. Through this center and through numerous visits to developing countries he encouraged and challenged the citizens of these countries to take up the cause of advancing science. In fact ICTP has gone far beyond a center for developing countries, and has become a center which attracts first rate scientists from all over the world and has become a regular sight for important conferences in physics and in mathematics.

The creation of ICTP was a rather difficult task. He always had a tough task of convincing authorities for the need of having such a center and the need for improvement of science in the developing countries. The irony of this is that he had an easier time convincing authorities of developed countries in contributing in this direction than those of the developing countries. To the policy makers of the developing countries he emphasized an important fact: It may appear that advancing modern abstract science in developing countries is a futile task, when one can import any advanced technological product. However the progress of a country will ultimately come from having a strong scientific base, even though it may appear in the short run to be far from any concrete application.

Even though this aspect of his life has had such a crucial impact in the lives of so many scientists, his contributions to theoretical physics are no less important. In particular his contributions to particle physics resulted in his discovery of what we now call the Salam-Weinberg standard model. Since this aspect of his life and his scientific contributions in general are already very well known I will not write any further on this point.

2 Basic Idea

[a] Perhaps the most important aspect of the recent revolution in string theory has been the appreciatiation of the role that extended objects play in physics. These objects can come in various dimensionalities, and are typically viewed as solitonic degrees of freedom. If they have p spatial dimensions, they are called p-branes (extending the terminology from membrane where $p = 2$). The existence of such states, together with the properties of the internal geometry M where the string is compactified upon leads to many interesting and novel physical phenomena. In particular the p-brane can wrap around a q-dimensional cycle $C \subset M$ of the internal manifold leading to a $p - q$ dimensional brane in the uncompactified space. The basic idea is to investigate what kinds of C's there are and what physical consequences they lead to. There are basically two classes of cycles that we will consider; the cycles C that can shrink to zero size, and are thus called the *vanishing cycles*, and cycles that are typically big and cannot shrink beyond a certain minimum size. We shall see that the first type will lead to interesting quantum field theory questions whereas the latter type have bearing on questions involving black holes and their entropy.

3 Vanishing Cycles and QFT interpretations

Consider a string compactification with a p-dimensional vanishing cycle. If we wrap a p-brane around such a cycle, the result state will be a 0-brane, i.e. a particle, from the viewpoing of the uncompactified spacetime. Moreover the

[a] I will not include any references in this short note. The interested readers can consult the many good review articles already available.

mass of this state is given by

$$M = T \cdot V(C)$$

where T denotes the tension of the worldvolume of the p-brane and $V(C)$ denotes the volume of C. The assumption that the cycle is vanishing means that $V(C) \to 0$, which thus implies that we end up with massless particles. Depending on various cases and the different values of p these massless particles will have different implications. For example if $p = 2$ and these vanishing 2-cycles arise in a type IIA string compactified on $K3$ down to six dimensions, this gives rise to massless vector multiplet charged under a $U(1)$, which thus naturally enhance the gauge symmetry from $U(1) \to SU(2)$. If there are more than one vanishing 2-cycle and they intersect one another according to the Dynkin diagrams of the $A-D-E$ groups, the resulting physics is an enhanced A-D-E gauge symmetry. If we further compactify down to four dimensions and we have a 2-dimensional locus with A_{k-1} singularity and another locus with A_{p-1} singularity and they meet at a point, the mixed wrapped 2-cycles will now lead to $(\mathbf{k}, \mathbf{p})$ bi-fundamental matter of $SU(k) \times SU(p)$ gauge group in four dimensions. In fact by arraging various intersecting singularities we can *engineer* and study the properties of a large class of quantum field theories in this way. Many non-perturbative questions of quantum field theories get translated to perturbative string questions.

Sometimes it may happen that a q-dimensional cycle that shrinks, but the theory in question has no q-brane. In such a case the lightest states comes from a $q + r$ dimensional brane wrapped around it, with the smallest available r, which typically is $r = 1$. In this case we would end up with a tensionaless string. An example of this is type IIB compactification on $K3$ with vanishing 2-cycle, in which case we have no 2-branes, but one can consider wrapping the available 3-brane of type IIB around the 2-cycle and this results in a tensionaless string in 6 dimension. Such cases are a source of new phenomena in quantum field theories, and are believed to be resulting in new non-trivial conformal quantum field theories. The example just mentioned would be a non-trivial six dimensional critical quantum field theory. Similar examples can be constructed also upon compactification to lower dimensions.

4 Black Hole applications

Now we consider the case where the cycle C cannot be made too small. These cases are also correlated with the fact that the cycle C can be deformed inside the compactification geometry, while preserving its minimal area. Let us consider $K3$ compactification of type IIA with a fixed volume. Consider a genus

g Riemann surface Σ in it with minimal area (which implies it is embedded holomorphically with some choice of complex structure on $K3$). In this case one can show that the volume of C is always bigger than

$$V(C) \geq \text{const.}\sqrt{g}$$

Let us now consider wrapping a 2-cycle in type IIA theory compactificat on $K3$ about a genus g Riemann surface Σ. The resulting state will be a 0-brane (i.e. a particle) in 6 dimension with mass propotional to $V(C)$. Now let us consider the case where g is large. In this case, using the above bound, we find that its mass becomes huge (of the order of $\sqrt{g}$. The resulting state can be viewed as a very massive charged particle in six dimensions. In fact it looks like a macroscopic black hole. One could then ask about comparisons between the area of the horizon and the number of such cycles C and see if there is any relation, as would be expected based on Bekenstein-Hawking entropy formula. In this particular case the corresponding black hole solution has a singular horizon and so no exact reliable area can be extracted from it. Instead if we consider type IIB compactification on $K3 \times S^1$ and wrap a 3-brane around a Riemann surface Σ of genus g in $K3$ times the S^1 and in addition consider a state with some momentum p along S^1 the corresponding black hole will have a non-singular horizon and one could ask what the comparison between the area of the horizon and the number of microscopic degrees of freedom of the 3-brane in that state.

The way to compute the microscopic degrees of freedom of this 3-brane is to first consider the effective 1+1 dimensional theory one gets after wrapping the 3-brane around the Riemann surface Σ. The number of degrees of freedom in the 1+1 dimensional theory will be related to number of ways the cycle C (together with a choice of gauge field on it) can be deformed in $K3$. This gives a theory in 1+1 dimension with an effective $c = 6g$ degrees of freedom ($4g$ coming from bosonic modes and $2g$ from fermionic modes). Then one looks at the number of states in this 1+1 dimenisonal theory on a circle with momentum p. This goes as

$$N \sim \exp[2\pi\sqrt{pc/6}] = \exp[2\pi\sqrt{pg}]$$

which agrees with the predicted entropy

$$S = \frac{1}{4}A = 2\pi\sqrt{pg}.$$

Similarly one can extend this analysis and construct other classes of black holes in 5 and 4 dimensions and compare the microscopic entropy with the predicted macroscopic entropy.

This research was supported in part by NSF grant PHY-92-18167.

ORGANISING COMMITTEE

Nicola CABIBBO

Ente Nazionale per le Nuove
Tecnologie, Energie e l'Ambiente
(ENEA)
Viale Regina Margherita 125
00198 Rome
Italy

Luciano MAIANI

Istituto Nazionale di Fisica Nucleare
(INFN)
Piazza dei Caprettari 70
00186 Rome
Italy

John ELLIS

Theory Division
CERN
CH-1211 Geneva 23
Switzerland

J. Robert SCHRIEFFER

National High Magnetic Field
Laboratory
Florida State University
1800 E. Paul Dirac Drive
P.O. Box 4005
Tallahassee, Florida 32306-2740
USA

Faheem HUSSAIN

High Energy Section
Abdus Salam International Centre for
Theoretical Physics
Strada Costiera 11
34100 Trieste
Italy

George THOMPSON

High Energy Section
Abdus Salam International Centre for
Theoretical Physics
Strada Costiera 11
34100 Trieste
Italy

Thomas W.B. KIBBLE

The Blackett Laboratory
Imperial College of Science,
Technology and Medicine
Prince Consort Road
London SW7 2BZ
UK

Miguel Angel VIRASORO

Abdus Salam International Centre for
Theoretical Physics
Strada Costiera 11
34100 Trieste
Italy

Christopher LLEWELLYN-SMITH

CERN
CH-1211 Geneva 23
Switzerland

Steven WEINBERG

Department of Physics
University of Texas at Austin
Austin, Texas 78712-1081
USA

LECTURERS

Daniele AMATI

International School for Advanced Studies (SISSA)
Via Beirut 2-4
34123 Trieste
Italy

Thomas W.B. KIBBLE

The Blackett Laboratory
Imperial College of Science, Technology and Medicine
Prince Consort Road
London SW7 2BZ
UK

Michael DUFF

Department of Physics
Texas A&M University
College Station, Texas 77843-4242
USA

Kumar Shiv NARAIN

High Energy Section
Abdus Salam International Centre for Theoretical Physics
Strada Costiera 11
34100 Trieste
Italy

John ELLIS

Theory Division
CERN
CH-1211 Geneva 23
Switzerland

Jiri NIEDERLE

Institute of Physics
Academy of Sciences of the Czech Republic
Na Slovance 2
180 40 Prague 8
Czech Republic

Sergio FERRARA

Theory Division
CERN
CH-1211 Geneva 23
Switzerland

Jogesh C. PATI

Department of Physics and Astronomy
University of Maryland
College Park, Maryland 20742-4111
USA

Michael GREEN

Department of Applied Mathematics and Theoretical Physics (DAMPT)
University of Cambridge
Silver Street
Cambridge CB3 9EW
UK

Seifallah RANDJBAR-DAEMI

High Energy Section
Abdus Salam International Centre for Theoretical Physics
Strada Costiera 11
34100 Trieste
Italy

LECTURERS

Dennis W. SCIAMA

International School for Advanced
Studies (SISSA)
Via Beirut 2-4
34123 Trieste
Italy

Nathan SEIBERG

School of Natural Sciences
Institute for Advanced Study
Olden Lane
Princeton, New Jersey 08540
USA

Ergin SEZGIN

Department of Physics
Texas A&M University
College Station, Texas 77843-4242
USA

Cumrun VAFA

Department of Physics
Harvard University
Cambridge, Massachusetts 02138
USA

Gabriele VENEZIANO

Theory Division
CERN
CH-1211 Geneva 23
Switzerland

Erik Peter VERLINDE

Institute for Theoretical Physics
University of Utrecht
Princetonplein 5
3508 TA Utrecht
Netherlands

Spenta R. WADIA

Theoretical Physics Group
Tata Institute of Fundamental Research
Homi Bhabha Road
Mumbai 400 005
India

PARTICIPANTS

Tariq ABDULLAH

Centre for Solid State Physics
University of the Punjab
Quaid-i-Azam Campus
54590 Lahore
Pakistan

M. Anis ALAM

Department of Physics
University of the Punjab
Quaid-i-Azam Campus
Lahore 54590
Pakistan

Rathin ADHIKARI

Theory Group
Physical Research Laboratory
Navrangpura
Ahmedabad 380 009
India

Ahmed ALI

DESY (Deutsches Electronen
Synchrotron)
Notkestrasse 85
22603 Hamburg
Germany

Kamaluddin AHMED

Department of Physics
Quaid-i-Azam University
Islamabad
Pakistan

Francis K.A. ALLOTEY

Ghana Atomic Energy Commission
P.O. Box 80
Legon
Accra
Ghana

Kamil Gediz AKDENIZ

Department of Physics
Istanbul University
Vezneciler
Istanbul 34459
Turkey

Paulo ALMEIDA

Departamento de Matematica
Instituto Superior Tecnico
Av. Rovisco Pais 1
1096 Lisbon
Portugal

Daniel Afedzi AKYEAMPONG

Department of Mathematics
University of Ghana
P.O. Box 62
Legon
Accra
Ghana

Ivan ANDRIC

Department of Theoretical Physics
Rudjer Boskovic Institute
Bijenicka Cesta 54
10001 Zagreb
Croatia

PARTICIPANTS

Mohamed BAGHDADI

Laboratoire de Physique Theorique
Université Mohammed V
B.P. 1014
Avenue ibn Batouta
Rabat
Morocco

Ayaz A. BAKASOV

Eidgenossische Technische Hochschule
Lab. Physikalsiche Chemie
Universitatsstrasse 22
CH-8092 Zurich
Switzerland

Miriam BALABAN

School for Scientific Communication
International Federation of Science
Editors
Mario Negri Sud
66030 Santa Maria Imbaro (CH)
Italy

Ram Prakash BAMBAH

Centre for Advanced Study in
Mathematics
Panjab University
Chandigarh 160 014
India

Fernando BARBERO

Instituto de Matematicas y Fisica
Fundamental, Consejo Superior de
Investigacions Cientificas
Serrano 113, B15
28006 Madrid
Spain

Velimir BARDEK

Department of Theoretical Physics
Rudjer Boskovic Institute
Bijenicka Cesta 54
10001 Zagreb
Croatia

Driss BEN SARI

Université Mohammed V
Faculté des Sciences
B.P. 1014
Avenue ibn Batouta
Rabat
Morocco

Luciano BERTOCCHI

Abdus Salam International Centre for
Theoretical Physics
Strada Costiera 11
34100 Trieste
Italy

Massimo BIANCHI

Dipartimento di Fisica
Università di Roma II "Tor Vergata"
Via della Ricerca Scientifica 1
00133 Rome
Italy

Wasiq M. BOKHARI

Department of Physics and Astronomy
University of Pennsylvania
209 S. 33rd Street
Philadelphia, Pennsylvania 19104-6396
USA

PARTICIPANTS

James BOYCE

Department of Physics
King's College
University of London
Strand
London WC2R 2LS
UK

Brahmananda DASGUPTA

Saha Institute of Nuclear Physics
Bidhannagar
Calcutta 700 064
India

Paolo BUDINICH

Laboratorio Interdisciplinare per le
Scienze Naturali ed Umanistiche
Strada Costiera 11
34014 Trieste
Italy

Robert DELBOURGO

Department of Physics
University of Tasmania
G.P.O. Box 252C
Hobart 7001
Tasmania
Australia

Nigel CALDER

26 Boundary Road, Northgate
Crawley
Sussex RH10 2BT
UK

Angelo DELLA SELVA

INFN - Sezione Napoli
Dipartimento di Scienze Fisiche
Mostra d'Oltremare
80125 Napoli
Italy

Julian CHELA-FLORES

Abdus Salam International Centre for
Theoretical Physics
Strada Costiera 11
34100 Trieste
Italy

Gallieno DENARDO

Abdus Salam International Centre for
Theoretical Physics
Strada Costiera 11
34100 Trieste
Italy

Bhawani CHOWDHRY

Department of Electronics
Mehran University of Engineering and
Technology
Jamshoro
Pakistan

Ibrahima DIOP MAR

Clinique Fann-Hock
Rue 70 x 55
Dakar
Senegal

PARTICIPANTS

Norman DOMBEY

Department of Physics
University of Sussex
Falmer, Sussex
Brighton BN1 9QH
UK

Saaed A. DURRANI

School of Physics and Astronomy
University of Birmingham
Edgbaston
Birmingham B15 2TT
UK

George F.R. ELLIS

Department of Applied Mathematics
University of Cape Town
7700 Rondebosch
Cape Town
South Africa

FAYYAZUDDIN

Department of Physics
Quaid-i-Azam University
Islamabad
Pakistan

Gordon FELDMAN

Department of Physics and Astronomy
Johns Hopkins University
3400 N. Charles Street
Baltimore, Maryland 21218-2695
USA

Luciano FONDA

Area di Ricerca
Sincrotrone Trieste S.P.A.
Padriciano 99
34012 Trieste
Italy

Gordon FRASER

CERN Courier
Ch-1211 Geneva 23
Switzerland

Luciene FREITAS

Instituto de Fisica Teorica
Universidade Estadual Paulista
Campus de Guratingueta
Av. Dr. Ariberto P. da Cunha, 333
Guratingueta 12500 - 000 - SP
Brazil

Sergio FUBINI

Theory Division
CERN
CH-1211 Geneva 23
Switzerland

Giuseppe FURLAN

Dipartimento di Fisica Teorica
Università degli Studi di Trieste
Strada Costiera 11
34100 Trieste
Italy

PARTICIPANTS

GianCarlo GHIRARDI

Dipartimento di Fisica Teorica
Università degli Studi di Trieste
Strada Costiera 11
34100 Trieste
Italy

Eugenia HALMAGEAN

National Research and Development
Institute of Microtechnology
Str. Erou Iancu Nicolae 32B
72996 Bucharest
Romania

André-Marie HAMENDE

Third World Academy of Sciences
Enrico Fermi Building
Via Beirut 6
34014 Trieste
Italy

Adnan HAMOUI

Department of Mathematics
Kuwait University
Faculty of Sciences
P.O. Box 5969
13060 Safat
Kuwait

Peter HARPER

National Cataloguing Unit
Archives of Contemporary Scientists
The Library, University of Bath
Claverton Down
Bath BA2 7AY
UK

Abdul Moksud HARUN AR RASHID

Department of Physics
University of Dhaka
Ramna
Dhaka 1000
Bangladesh

Mohamed H.A. HASSAN

Third World Academy of Sciences
Enrico Fermi Building
Via Beirut 6
34014 Trieste
Italy

Pervez A. HOODBHOY

Department of Physics
Quaid-i-Azam University
Islamabad
Pakistan

Hafeez Rehman HOORANI

Department of Physics
Carnegie Mellon University
5000 Forbes Avenue
Pittsburgh, Pennsylvania 15213-3890
USA

Paul S. HOWE

Department of Mathematics
King's College
University of London
Strand
London WC2R 2LS
UK

PARTICIPANTS

Roberto IENGO

International School for Advanced
Studies (SISSA)
Via Beirut 2-4
34123 Trieste
Italy

Karl KOLLER

Ludwig-Maximilians-Universitat
Munchen
Sektion Physik
Theresienstrasse 37
D-80333 Munich
Germany

Giorgio IMMIRZI

Dipartimento di Fisica
Università degli Studi di Perugia
Via Pascoli 1
06100 Perugia
Italy

Narendra KUMAR

Raman Research Institute
C.V. Raman Avenue
Sadashivanagar
Bangalore 560 080
India

Jamal Nazrul ISLAM

Research Centre for Mathematical and
Physical Sciences
University of Chittagong
Chittagong 4331
Bangladesh

Kamlesh KUMARI

Department of Physics
University of Rajasthan
Jaipur 302 004
India

Thiagarajan JAYARAMAN

Institute of Mathematical Sciences
CIT Campus
Tharamani
Chennai 600 113
India

Elsayed LASHIN

Department of Physics
Ain Shams University
Abbassia
Cairo 11566
Egypt

Sameen Ahmed KHAN

INFN
Università degli Studi di Padova
Dipartimento di Fisica
Via Marzolo, 8
35131 Padova
Italy

George LEIBBRANDT

Department of Mathematics and
Statistics
University of Guelph
Guelph N1G 2W1
Ontario
Canada

PARTICIPANTS

Miriam LEWIS

Anton-Baumgartnerstrasse 44/B5/073
A-1230 Vienna
Austria

Luis MASPERI

Instituto Balseiro
Centro Atomico Bariloche
Apartado 439
Rio Negro
San Carlos de Bariloche 8400
Argentina

Gerhard MACK

Universitat Hamburg
II Institut fur Theoretische Physik
Luruper Chaussee 149
D-22761 Hamburg
Germany

Rexhep MEIDANI

Section of Theoretical Physics
University of Tirana
Faculty of Natural Sciences
Tirana
Albania

Awele MADUEMEZIA

Department of Physics
University of Ibadan
Faculty of Science
Oyo State
Ibadan
Nigeria

Vesna MIKUTA-MARTINIS

Department of Theoretical Physics
Rudjer Boskovic Institute
Bijenicka Cesta 54
10001 Zagreb
Croatia

Nadeem Aslam MALIK

Department of Applied Mathematics
and Theoretical Physics (DAMTP)
University of Cambridge
Silver Street
Cambridge CB3 9EW
UK

Ghulam MURTAZA

Department of Physics
Quaid-i-Azam University
Islamabad
Pakistan

Mladen MARTINIS

Department of Theoretical Physics
Rudjer Boskov Institute
Bijenicka Cesta 54
10001 Zagreb
Croatia

Khavtgain NAMSRAI

Institute of Physics and Technology
Mongolian Academy of Sciences
Enkhtaivan Avenue 54B
210651 Ulaanbaatar 51
Mongolia

PARTICIPANTS

Syed M.M.R. NAQVI

Department of Physics
University of Karachi
University Road
Karachi 75270
Pakistan

NGUYEN VAN HIEU

National Centre for Science and
Technology of Vietnam
P.O. Box 607
Boho
Hanoi
Vietnam

Yuri NOVOZHILOV

Department of Theoretical Physics
St. Petersburg State University
198908 St. Petersburg
Russian Federation

Carmen Alicia NUNEZ

Instituto de Astronomia y Fisica del
Espacio
Casilla de Correo 67
Succursal 28
1428 Buenos Aires
Argentina

Jean NUYTS

Department of Theoretical Physics
State University of Mons
20 Place du Parc
B-7000 Mons
Belgium

Lochlainn O'RAIFEARTAIGH

School of Theoretical Physics
Dublin Institute for Advanced Studies
10 Burlington Road
Dublin 4
Ireland

Silvio PALLUA

Department of Theoretical Physics
University of Zagreb
P.O. Box 162
Bijenicka C. 32
10000 Zagreb
Croatia

Constantinos PANAGIOTAKOPOULOS

Physics Division
School of Technology
Aristotle University of Thessaloniki
54006 Thessaloniki
Greece

Mina K. PARIDA

Department of Physics
Northeastern Hill University
Umshing-Mawkynroh
Shillong 793 022
India

Nello PAVER

INFN-Sezione Trieste
Area di Ricerca
Padriciano 99
34012 Trieste
Italy

PARTICIPANTS

Hugo Celso PEREZ ROJAS

ICIMAF (Instituto di Cibernetica,
Matematica y Fisica)
Academia de Ciencias de Cuba
Calle 309, Esq 15, Vedado
Havana 10400
Cuba

Mohammad SALIMULLAH

Department of Physics
Jahangirnagar University
Savar
Dhaka
Bangladesh

Asghar QADIR

Department of Mathematics
Quaid-i-Azam University
Islamabad
Pakistan

Mohammed SHAMSHER ALI

Department of Physics
University of Dhaka
Ramna
Dhaka 1000
Bangladesh

Raghavan RAMACHANDRAN

Institute of Mathematical Sciences
CIT Campus
Tharamani
Chennai 600 113
India

Louis Naim SHEHATA

Department of Math and Theoretical
Physics
Atomic Energy Authority
Nuclear Research Center
Cairo 13759
Egypt

Rafiqul I.M.A. RASHID

Department of Physics
University of Dhaka
Ramna
Dhaka 1000
Bangladesh

Abul Hasan SIDDIQI

Department of Mathematics
Aligarh Muslim University
Aligarh 202 002
India

RIAZUDDIN

King Fahd University of Petroleum
and Minerals
U.P.M. Box 654
Dhahran 31261
Saudi Arabia

Yousef SOBOUTI

Institute for Advanced Studies in Basic
Science
P.O. Box 45195-159
Gavah Zang
Zanjan
Islamic Republic of Iran

PARTICIPANTS

Gheorghe STRATAN

Department of Fundamental Physics
Institute of Atomic Physics
P.O. Box MG-6
Bucharest
Romania

Naeem Ahmad TAHIR

Universitat Erlangen-Nurnberg
Physikalisches Institut
Erwin-Rommel Strasse 1
D-91052 Erlangen
Germany

Jozef WERLE

Institute of Theoretical Physics
Warsaw University
Hoza 69
Warsaw 00681
Poland

Urjit Ajitlal YAJNIK

Department of Physics
Indian Institute of Technology
Powai
Mumbai 400 076
India

Jorge Antonio ZANELLI

Centro de Estudios Cientificos de
Santiago
Av. Presidente Errazuriz 3132
Casilla 16443
Santiago 9
Chile

John ZIMAN

27 Little London Green
Oakley
Aylesbury, Bucks HP18 9QL
UK